Modern Seed Technology

Modern Seed Technology

Editor

Alan G. Taylor

MDPI • Basel • Beijing • Wuhan • Barcelona • Belgrade • Manchester • Tokyo • Cluj • Tianjin

Editor
Alan G. Taylor
Cornell AgriTech, School of
Integrative Plant Science,
Horticulture Section
Cornell University
Geneva, NY
United States

Editorial Office
MDPI
St. Alban-Anlage 66
4052 Basel, Switzerland

This is a reprint of articles from the Special Issue published online in the open access journal *Agriculture* (ISSN 2077-0472) (available at: www.mdpi.com/journal/agriculture/special_issues/Seed_Technology).

For citation purposes, cite each article independently as indicated on the article page online and as indicated below:

LastName, A.A.; LastName, B.B.; LastName, C.C. Article Title. *Journal Name* **Year**, *Volume Number*, Page Range.

ISBN 978-3-0365-1770-4 (Hbk)
ISBN 978-3-0365-1769-8 (PDF)

Contents

Editorial

Modern Seed Technology

Alan G. Taylor [1,*], Masoume Amirkhani [1] and Hank Hill [2]

[1] Cornell AgriTech, School of Integrative Plant Science, Horticulture Section, Cornell University, New York, NY 14456, USA; ma862@cornell.edu
[2] Germains Seed Technology, 8333 Swanston Lane, Gilroy, CA 95020, USA; hhill@germains.com
* Correspondence: agt1@cornell.edu

Citation: Taylor, A.G.; Amirkhani, M.; Hill, H. Modern Seed Technology. *Agriculture* **2021**, *11*, 630. https://doi.org/10.3390/agriculture11070630

Received: 28 June 2021
Accepted: 2 July 2021
Published: 6 July 2021

Modern Seed Technology (MST) includes a wide range of technologies and practices to upgrade seed quality, enhance seedling and plant growth, and assessing seed quality using imaging technology. Another key topic of MST is Seed Enhancements. First defined as post-harvest methods that improve germination and seedling growth or facilitate the delivery of seeds at the time of sowing [1]. The broader topic of MST includes pre-harvest treatments to hasten seed maturation and post-sowing methods to enhance seed viability and vigor for greenhouse and field production. This special issue of MST has a total of 12 papers with 10 research papers and 2 review articles. Papers were submitted from five countries: Brazil, China, Denmark, Pakistan, and four papers were invited from colleagues in the United States, Multi-State project W-4168. The papers in the special issue of MST were grouped into four categories: Pre- and Post-sowing Seed Enhancements, New Crop Seed Technology, Seed Treatments, and Systemic Uptake, Seed Priming and Seed Imaging. This editorial encompasses perspectives from academia (Taylor and Amirkhani) and industry (Hill) for the future vision of Modern Seed Technology.

The first category has a paper in each sub-heading: Pre- and Post-sowing Seed Enhancements and New Crop Seed Technology.

The first opportunity to manipulate seed quality is while the seeds are still on the mother plant. The use of chemical defoliants can accelerate corn (*Zea mays* L.) seed maturation and drying and thus avoid loss of quality by an early frost. Dean et al. at Iowa State University describe the effect of a selective chemical defoliant on the migration of oil bodies, a sub-cellular event that is a prerequisite for viability and vigor [2]. The major finding was the lack of differences in migration of these oil bodies between treated and nontreated controls. Thus, chemical defoliant did not harm corn seed quality, while still protecting the seed from the damage of an early frost.

The importance of the above article is that most published research concerning Modern Seed Technologies is on post-harvest seed technology because of the emphasis on seed enhancement. Therefore, the opportunity is missed to enhance quality prior to harvesting. The authors feel that future MST research should have a better balance between pre-and post-harvest technology. Moreover, a combination of pre-and post-harvest strategies in the same investigation has the greatest potential to enhance seed performance. Thus, we expand the definition of pre-harvest strategies to include plant-breeding efforts to improve seed quality and vigor as will be cited later.

The second paper by Qin and Leskovar at Texas A&M University focused on improved transplant quality of containerized vegetable crop plants by the addition of humic substances (HS), as a biostimulant, to the plug media [3]. Humic acid has been known for some time to enhance germination and seedling growth. The incorporation of 1% HS (v/v) into the growing media was demonstrated to have a biostimulant effect and enhanced several plant parameters, and modulated both root and shoot growth. The HS biostimulant effect was particularly effective in mitigating the negative effects of drought and heat stress on growth.

The above article focuses on enhancing germination and plant growth under the environmental stress of drought and heat stress because of their negative impact on stand establishment and ultimately yield. A paper from the first two authors of this article also demonstrated the positive effects of a bio-stimulant. They used a seed coating formulation composed of soy flour and vermicompost that served as a biostimulant under optimal growth conditions [4]. These biostimulant- seed treatments and coatings need to be tested under environmental stress to explore their full potential as the above authors demonstrated.

The third paper was from Mi et al., at Cornell and was on hemp (*Cannabis sativa* L.) as a new crop, or at least the reintroduction of a crop first grown in China 6000 years ago. The research was focused on the cultural practices for growing baby leaf hemp including the effect of seed size on germination and fresh and dry seedling weight [5]. Three hemp varieties were studied. The seed size distribution was determined by hand sorting with round hole sieves based on width. The distribution pattern was similar for all three varieties with a normal distribution skewed with a small percentage of small seeds. The small seed sizes had a lower percent germination and slower seedling growth than the larger-sized seeds. Thus, discarding the small percentage of small-sized seeds would upgrade the quality of the lot.

In conclusion, though the importance of seed size has been known for centuries, there is little scientific research published on hemp, and information available online may have questionable validity. Moreover, the hemp seed industry is relatively young compared to the vegetable and field crop industries, so researching the effects of seed size is important to both the seed industry and hemp growers. Continued seed technology research is needed on hemp including the development of treatments to control soil-borne pathogens responsible for damping-off. The goal is to have labeled seed treatments in the conventional and organic production of hemp.

The second category is on Seed Treatments, and Systemic Uptake (of seed treatments).

This category contains half of the papers in this special issue. The first paper by Afzal, Javed, Amirkhani, and Taylor is a joint paper from the University of Agriculture, in Pakistan and Cornell AgriTech and is a review paper on seed coating technologies [6]. For the first time, equipment and processes are described for five major seed coating technologies: dry coating, seed dressing, film coating, encrustments, and seed pelleting. Comparisons are made between each coating type with respect to weight increase after application, relative amounts of loading active ingredients, and time required performing each coating. The trend is to reduce chemical seed treatments and move to active ingredients that are organically approved. The major impetus is that organic seed treatments must be used for certified organic crop production. For organic certification, seed treatment binders and filler coating components must also be approved for organic use. This review paper presented a list of plant protectant groups, seed treatment binders, and fillers, and denotes those materials that may be approved. Seed coatings can be custom designed. Dry seed coating compositions may be required for the application of beneficial fungi that cannot withstand hydration and dehydration without loss of viability. In particular, the Entomopathogenic fungi (EPF), *Metarhizium* and *Beauveria* both require dry-coating technologies in the seed-coating process. Thus, the other four coating techniques: seed dressing, film coating, encrustments, and seed pelleting cannot be used for EPF seed treatment application as water is used in each.

The future of plant protection may well lie in the discussion above. The seed becomes the delivery system for crop protection. The controlled release of microencapsulated pesticides is just one example [7]. Already seed coating enables the additions of fungicides and insecticides to be applied in a far lower dosage on a per acre basis than with in-furrow or foliar applications [8]. Discussion of current progress will allow the seed industry to scale up and implement these new technologies in agriculture.

The second seed-treatment paper in this category is by Averitt et al. and is based on soybean lines with modified seed composition achieved through the use of mutant lines [9].

The larger context is that plant breeding may be used to improve seed quality and stand establishment when standard varieties have inherent low seed-quality potential and are also susceptible to both biotic and abiotic stress. For example, white-seeded snap bean (*Phaseolus vulgaris* L.) varieties are used in the processing vegetable industry but have lower seed quality potential than dark-seeded varieties. Dickson at Cornell summarized research using conventional plant breeding to improve white-seeded bean seed quality over 40 years ago [10]. Plant breeding may also be used to alter the composition of reserve materials in seeds for the purpose of improved taste in vegetable crops, and genetic improvements have greatly enhanced the flavor and shelf-life of fresh market sweetcorn [11].

In many cases where plant breeding alters seed composition for enhanced human and animal consumption, seed quality is compromised. This paper examines the use of soybean genotypes with low phytic acid (LPA) in comparison with normal phytic acid (NPA), and LPA lines have lower germination and low field emergence. The research presented in this paper focused on the use of chemical seed treatment fungicides and seed treatment combinations to compensate for the inherent low seed quality. Collectively, selected seed treatment combinations improved the field emergence of LPA genotypes. Further, seed priming (described later) by itself had a negative impact on stand establishment in LPA genotypes, while first priming followed with a formulation of three seed treatment fungicides improved field emergence.

The next two papers focus on seed coat- permeability and systemic uptake of seed treatments. The experimental approach in both papers used fluorescent tracers to mimic active ingredients to visualize movement within seed and seedlings and thus avoid the use of chemical pesticides. These two papers build on the characterization of the physical/chemical properties responsible for seed-coat permeability of crop seeds. Taylor and Salanenka developed a system to classify seed coat permeability based on the diffusion of ionic and nonionic compounds through the seed coat or seed covering layers [12]. Seed-coat permeability of seeds were grouped as permeable, selective permeability, and nonpermeable. Seeds with permeable seed coats allowed both ionic and nonionic compounds to diffuse through the seed coat, such as soybean and snap beans, while selective seed coat permeability only allowed nonionic compounds to pass including tomato (*Solanum lycopersicum* L.), onion (*Allium cepa* L.), and corn (*Zea mays* L.). Nonpermeable seeds blocked both ionic and nonionic compounds from entering the embryo from the environment and included cucurbits and lettuce (*Lactuca sativa* L.). A simple lab test was proposed to test the seed-coat permeability of any plant species [12].

The first paper on Systemic Uptake by Mayton et al., at Cornell AgriTech, was on tomato seed coat permeability and drilled down on a compound's lipophilicity measured as the log K_{ow} for optimal seed uptake [13]. This research was all possible with the synthesis of a series of 11 fluorescent; n-alkyl piperonyl amides ranging from log K_{ow} 0.02 to 5.66 by Stephen Donovan (co-author). The optimal log K_{ow} for tomato seed uptake was in the range of 2.9 to 3.8. However, less than 5% of the applied compound was measured in the embryo. Therefore, for control of internal seed-borne pathogens, both the log K_{ow} is important for targeting pathogens residing in the embryo and adequate dosage for efficacy.

The next paper by Wang et al., at Cornell AgriTech, investigated the uptake of 32 fluorescent tracers representing 10 chemical families on soybean seed and seedling uptake [14]. Most zanthene and coumarin compounds tested displayed both seed and seedling uptake. Though the log K_{ow} of a compound is well established to govern root uptake, the log K_{ow} alone could not predict seed uptake. Therefore, the physical/chemical properties for uptake of organic compounds by plant roots are not the same as uptake in seeds during the early stages of germination. Seedling uptake of zanthene compounds, Rhodamine B and Rhodamine 800, a NIR fluorescent tracer were further studied and detected in the true leaves of soybean.

The third category is on Seed Priming as seed enhancements.

There were two papers on Seed Priming. Seed priming is a general term that includes several techniques to hydrate seeds under controlled conditions so physiological processes of germination can occur without the completion of radicle emergence (Phase III, or visible germination) [15]. Common to all seed priming techniques is that radicle emergence is arrested due to restricted water uptake.

In the two papers in this section, seeds were allowed to imbibe in a dilute solution of potassium nitrate [16] or zinc sulfate [17], but germination was arrested prior to drying. In these studies, the concentration of KNO_3 or $ZnSO_4$ in solution was not sufficient to lower osmotic potential to arrest Phase III germination [16]. Thus the seed priming techniques described in the two papers may be considered as seed steeping [18]. There is not a review paper on seed priming in this special issue, so the reader is referred to previous reviews published from 1977 to 2010 cited in [15].

The first seed priming paper by Ali et al. used a range of potassium nitrate concentrations and 0.75% was optimal for germination, seedling growth, and other physiological attributes [16]. The objective of enhancing tomato seed germination is not new and an early paper reported the use of potassium nitrate and other salt solutions to enhance tomato seed germination almost 60 years earlier [19]. Another objective of seed priming is to improve germination under low temperatures.

The second priming paper by Imran et al., [17] investigated spinach seeds imbibed in dilute ZnSO4 solutions. The optimal concentration was found to be 6 mM resulting in enhanced germination at 8 °C. Collectively, both 'nutrient priming' techniques provided enhanced germination and seedling performance. Optimal efficacy required a precise concentration.

The last subject area was Seed Imaging using multispectral imaging (MSI) and near-infrared spectroscopy (NIRS).

The first paper in this section from Mortensen et al., at Aarhus University, Denmark was an invited review paper on both MSI and NIRS [20]. These technologies are nondestructive and noninvasive tools and have the potential in seed testing for rapid and reproducible results. Applications of MSI in seed testing include varietal identity and purity, detecting seed damage from mechanical abuse and insects, and seed health in detecting fungal infection. Both MSI and NIRS have the potential to detect seed viability on a single-seed basis, and germinating seeds validated predicted seed viability. Combining imaging with seed sorting technology could effectively upgrade seed-lot quality by detecting and removing nonviable seeds.

The second paper in this section by Rego et al. in Brazil focused on seed health using MSI for detecting seed-borne fungi in cowpea (*Vigna unguiculata* L.). MSI was able to detect seeds inoculated with *Fusarium*, *Rhizoctonia*, and *Aspergillus* [21]. A key finding was that if seeds were first imbibed and then frozen at −20 °C, pathogen detection was enhanced.

The last paper in this category is from Bello and Bradford at UC Davis. The paper was on investigating and detecting a physiological abnormality in *Brassica oleracea* called "blindness" [22]. MSI was used along with two other modern seed testing techniques: chlorophyll fluorescence and oxygen consumption. All data collection was done on a single-seed basis. In general, more immature seeds were detected by chlorophyll fluorescence; and at specific wavelengths from the MSI were associated with greater occurrence of blindness. The bigger story is that nondestructive and noninvasive imaging technologies have the potential to detect poor-quality seed lots and poor-quality seeds within a seed lot. Seed imaging integrated with seed sorting technology could upgrade seed-lot quality.

In summary, the first and third authors of this article experienced an evolution in seed technology research and development over the past 40 years. Papers in this special issue of Modern Seed Technology are an excellent illustration of current research findings in several categories from many seed research groups throughout the world. Drs. Taylor and Amirkhani are proud to contribute several papers to this special issue of Modern Seed Technology. Future research in this area will be driven by the integration of new technologies from other disciplines with seed technology. We look forward to future developments that move from evolutionary to revolutionary in exploiting seeds as the delivery systems in agriculture.

Author Contributions: Conceptualization, A.G.T. and H.H.; investigation, A.G.T. and M.A.; writing—original draft preparation, A.G.T.; writing—review and editing, M.A. and H.H.; visualization, A.G.T., M.A. and H.H. All authors have read and agreed to the published version of the manuscript.

Funding: This material is based upon work that is supported by the United States Hatch Funds under Multi-state Project W-4168 under accession number 1007938, and Multi-state Project NE-1832 under the accession number 1021019.

Acknowledgments: We would like to sincerely thank all authors who submitted papers to the special issue of *Agriculture* entitled "Modern Seed Technology", to the reviewers of these papers for their constructive comments and thoughtful suggestions, and the editorial staff of *Agriculture*.

Conflicts of Interest: The authors declare no conflict of interest.

References

1. Taylor, A.G.; Allen, P.S.; Bennett, M.A.; Bradford, K.J.; Burris, J.S.; Misra, M.K. Seed Enhancements. *Seed Sci. Res.* **1998**, *8*, 245–256. [CrossRef]
2. Dean, A.; Wigg, K.; Zambiazzi, E.; Christian, E.; Goggi, S.; Schwarte, A.; Johnson, J.; Cabrera, E. Migration of Oil Bodies in Embryo Cells during Acquisition of Desiccation Tolerance in Chemically Defoliated Corn (*Zea mays* L.) Seed Production Fields. *Agriculture* **2021**, *11*, 129. [CrossRef]
3. Qin, K.; Leskovar, D. Humic Substances Improve Vegetable Seedling Quality and Post-Transplant Yield Performance under Stress Conditions. *Agriculture* **2020**, *10*, 254. [CrossRef]
4. Amirkhani, M.; Mayton, H.S.; Netravali, A.N.; Taylor, A.G. A Seed Coating Delivery System for Bio-based Biostimulants to Enhance Plant Growth. *Sustainability* **2019**, *11*, 5304. [CrossRef]
5. Mi, R.; Taylor, A.G.; Smart, L.; Mattson, N. Developing Production Guidelines for Baby Leaf Hemp (*Cannabis sativa* L.) as an Edible Salad Green: Cultivar, Sowing Density and Seed Size. *Agriculture* **2020**, *10*, 617. [CrossRef]
6. Afzal, I.; Javed, T.; Amirkhani, M.; Taylor, A. Modern Seed Technology: Seed Coating Delivery Systems for Enhancing Seed and Crop Performance. *Agriculture* **2020**, *10*, 526. [CrossRef]
7. Taylor, A.G.; Bolotin, A.; Pollicove, S.; Taylor, R. Controlled Release of Seed and Soil Treatments Triggered by pH Change of Growing Media. Patent No. PCT/US2011/033420, 21 April 2011.
8. Taylor, A.G.; Eckenrode, C.J.; Straub, R.W. Seed treatments for onions: Challenges and progress. *HortScience* **2001**, *36*, 199–205. [CrossRef]
9. Averitt, B.; Welbaum, G.; Li, X.; Prenger, E.; Qin, J.; Zhang, B. Evaluating Genotypes and Seed Treatments to Increase Field Emergence of Low Phytic Acid Soybeans. *Agriculture* **2020**, *10*, 516. [CrossRef]
10. Dickson, M. Genetic Aspects of Seed Quality. *HortScience* **1980**, *15*, 771–774.
11. Lertrat, K.; Pulam, T. Breeding for Increased Sweetness in Sweet Corn. *Int. J. Plant Breed.* **2007**, *1*, 27–30.
12. Taylor, A.G.; Salanenka, Y. Seed Treatments: Phytotoxicity Amelioration and Tracer Uptake. *Seed Sci. Res.* **2012**, *22*, S86–S90. [CrossRef]
13. Mayton, H.; Amirkhani, M.; Yang, D.; Donovan, S.; Taylor, A. Tomato Seed Coat Permeability: Optimal Seed Treatment Chemical Properties for Targeting the Embryo with Implications for Internal Seed-Borne Pathogen Control. *Agriculture* **2021**, *11*, 199. [CrossRef]
14. Wang, Z.; Amirkhani, M.; Avelar, S.; Yang, D.; Taylor, A. Systemic Uptake of Fluorescent Tracers by Soybean (Glycine max (L.) Merr.) Seed and Seedlings. *Agriculture* **2020**, *10*, 248. [CrossRef]
15. Taylor, A.G. Seed Storage, Germination, Quality, and Enhancements. In *Physiology of Vegetable Crops*, 2nd ed.; Wien, H.C., Stuetzel, H., Eds.; CAB International: Wallingford, UK, 2020; p. 496. ISBN 978-1786393777.
16. Ali, M.; Javed, T.; Mauro, R.; Shabbir, R.; Afzal, I.; Yousef, A. Effect of Seed Priming with Potassium Nitrate on the Performance of Tomato. *Agriculture* **2020**, *10*, 498. [CrossRef]
17. Imran, M.; Mahmood, A.; Neumann, G.; Boelt, B. Zinc Seed Priming Improves Spinach Germination at Low Temperature. *Agriculture* **2021**, *11*, 271. [CrossRef]
18. Halmer, P. Commercial seed treatment technology. In *Seed Technology and Its Biological Basis*; Black, M., Bewley, J.D., Eds.; Sheffield Academic Press: Sheffield, UK, 2000; pp. 257–286.
19. Ells, J. The Influence of Treating Tomato Seeds with Nutrient Solution on Emergence Rate and Seedling Growth. *Proc. Amer. Soc. Hort. Sci.* **1963**, *83*, 684–687.
20. Mortensen, A.; Gislum, R.; Jørgensen, J.; Boelt, B. The Use of Multispectral Imaging and Single Seed and Bulk Near-Infrared Spectroscopy to Characterize Seed Covering Structures: Methods and Applications in Seed Testing and Research. *Agriculture* **2021**, *11*, 301. [CrossRef]

21. Rego, C.; França-Silva, F.; Gomes-Junior, F.; Moraes, M.; Medeiros, A.; Silva, C. Using Multispectral Imaging for Detecting Seed-Borne Fungi in Cowpea. *Agriculture* **2020**, *10*, 361. [CrossRef]
22. Bello, P.; Bradford, K. Relationships of Brassica Seed Physical Characteristics with Germination Performance and Plant Blindness. *Agriculture* **2021**, *11*, 220. [CrossRef]

Article

Migration of Oil Bodies in Embryo Cells during Acquisition of Desiccation Tolerance in Chemically Defoliated Corn (*Zea mays* L.) Seed Production Fields

Ashley N. Dean [1,2], Katharina Wigg [1,3], Everton V. Zambiazzi [1,4,5], Erik J. Christian [1], Susana A. Goggi [1,4,*], Aaron Schwarte [6], Jeremy Johnson [6] and Edgar Cabrera [6]

1 Department of Agronomy, Iowa State University, Ames, IA 50011, USA; adean@iastate.edu (A.N.D.); wigg@wisc.edu (K.W.); everton_zambiazzi@hotmail.com (E.V.Z.); chrstn@gmail.com (E.J.C.)
2 Department of Entomology, Iowa State University, Ames, IA 50011, USA
3 Department of Horticulture, University of Wisconsin-Madison, Madison, WI 53706, USA
4 Seed Science Center, Iowa State University, Ames, IA 50011, USA
5 Department de Agriculture, Federal University of Lavras, Lavras, Minas Gerais 73200-000, Brazil
6 Corteva Agriscience, Johnston, IA 50131, USA; aaron.schwarte@corteva.com (A.S.); jeremy.johnson@corteva.com (J.J.); cabreraedgarr80@gmail.com (E.C.)
* Correspondence: susana@iastate.edu

Citation: Dean, A.N.; Wigg, K.; Zambiazzi, E.V.; Christian, E.J.; Goggi, S.A.; Schwarte, A.; Johnson, J.; Cabrera, E. Migration of Oil Bodies in Embryo Cells during Acquisition of Desiccation Tolerance in Chemically Defoliated Corn (*Zea mays* L.) Seed Production Fields. *Agriculture* **2021**, *11*, 129. https://doi.org/10.3390/agriculture11020129

Academic Editors: Alan G. Taylor and Les Copeland

Received: 11 January 2021
Accepted: 26 January 2021
Published: 5 February 2021

Publisher's Note: MDPI stays neutral with regard to jurisdictional claims in published maps and institutional affiliations.

Abstract: Chemical defoliation of seed corn production fields accelerates seed maturation and desiccation and expedites seed harvest. Early seed harvest is important to minimize the risk of frost damage while in the field. This newly adopted seed production practice also allows seed companies to plan harvest and manage dryer space more efficiently. However, premature defoliation may interfere with the migration of oil bodies within embryo cells during desiccation and affect seed germination and vigor. The objective of this study was to investigate the effect of chemical defoliation on the migration patterns of oil bodies within embryo cells during desiccation. Chemically defoliated and non-defoliated plants from five commercial hybrid seed corn fields were sampled in 2014 and 2015. Whole ears with husks were harvested before and after defoliant application at 600 g H_2O kg^{-1} fresh weight (fw), and weekly thereafter until seed reached approximately 300–350 g H_2O kg^{-1} fw. Ten embryos extracted from center-row seeds were fixed to stop metabolic processes, then sliced, processed, and photographed using scanning transmission electron microscopy. The oil bodies within embryo cells followed normal migration patterns according to seed moisture content, regardless of defoliation treatment. Seed germination and vigor were verified and were not significantly affected by defoliation. Chemical defoliation is a viable production practice to accelerate seed corn desiccation and to manage harvest and seed dryer availability more efficiently without negatively affecting seed germination and vigor.

Keywords: corn; seed acquisition of desiccation tolerance; oil-bodies migration; physiological maturity; seed quality

1. Introduction

Seed corn (*Zea mays* L.) is harvested close to physiological maturity and dried artificially in specialized seed dryers before storage. Physiological maturity is the developmental stage at which seeds reach maximum dry weight [1,2]. At this developmental stage, seed moisture content ranges from 300 to 380 g H_2O kg^{-1} fresh weight (fw) depending on the genetic background of the plant and environmental conditions during seed development and maturation [3]. Seed corn is harvested early to avoid possible seed freezing injury caused by an early frost event [4]. The seed industry in the US Upper Midwest experiences significant monetary losses from early frost events every five to six years [5].

Many seed companies have adopted a new seed production practice of chemical defoliation to accelerate seed corn harvest. The defoliant is applied to the plants when seed corn is close to 600 g H_2O kg^{-1} fw or approximately 14 days before normal seed corn harvest. The seed moisture content of chemically defoliated plants decreases more rapidly than in untreated plants because of earlier senescence (personal observation). Chemical defoliation expedites harvest by two to five days, thus widening the harvest window of optimal seed moisture in different hybrid fields. This practice also facilitates harvest schedules and management of seed dryer space.

Although defoliants have been used in cotton to accelerate plant senescence and facilitate mechanical harvest in the US since 1945 [6–8], little is known about the use of defoliants in seed corn production. Drexel Defol® 5, a chemical defoliant salt solution used in the US, has not been readily adopted or widespread used. Moreover, the effect of this defoliation treatment on seed quality (seed germination and vigor) has not been fully investigated.

Orthodox seeds, such as corn, undergo a desiccation phase towards the end of seed development. These seeds survive desiccation through physiological changes called acquisition of desiccation tolerance [9]. Seed dehydration is an adaptive mechanism that allows seeds to survive unfavorable weather conditions common in temperate zones. These physiological changes are essential to the normal development of high seed quality. Seed quality in this work is defined as seed germination and vigor.

One important physiological change during the acquisition of desiccation tolerance is the migration and alignment of oil bodies along the cell membrane in corn embryo cells. These oil bodies are accumulated in the cytoplasm of the embryo cell during seed development and, as seeds dehydrate, they migrate to the cell membrane to protect cells from dehydration [10,11]. This migration of oil bodies and alignment alongside of the cell membrane is essential to seed quality.

The objective of this study was to document the migration of oil bodies in embryo cells from chemically defoliated and untreated plants.

2. Materials and Methods

2.1. Seed Production and Defoliation Treatment

A commercial hybrid seed field was sampled in 2015 near Nevada, Iowa. The field was planted in blocks with a 4:2 female-to-male ratio and managed by the seed company Corteva (Johnston, IA, USA) according to their established hybrid seed production practices.

The chemical defoliant Drexel Defol® 5 (42.3 ai $NaClO_3$) (Drexel Chemical Company, Memphis, TN, USA) was applied to the corn plants when seed moisture content was approximately 600 g H_2O kg^{-1} fw with a Hagie high-clearance sprayer (Hagie Manufacturing Co., Clarion, IA, USA) equipped with a 27.4 m boom and 68 L water tank. A strip that was two to three female blocks wide and 800 m long was not sprayed as a control. Two replications of twenty ears were hand-harvested from the treated and control areas, once prior to the application of the defoliant and at least once a week after application. Sampling continued until the field was mechanically harvested by the seed company when seed moisture content reached approximately 350–370 g H_2O kg^{-1} fw. Therefore, harvest dates are 31 August 2015; 4 September 2015; 11 September 2015; 18 September 2015; and 22 September 2015. Field replications were maintained separately throughout the experiment.

2.2. Seed Moisture Determination and Seed Drying

At each sampling date, the sampled ears were brought immediately into the Iowa State University Seed Science Center for processing. All ears were husked by hand within 1 h after sampling the field. To consistently evaluate seeds at the same developmental stage within the ear [12], seed moisture content was determined on forty seeds removed from the center portion of five ears. Seed were divided into two 5 cm diameter aluminum trays and placed inside of an 80 °C oven and weighed daily until seed reached constant weight.

Moisture content was calculated on a fresh weight basis by using the following formula: (fresh weight − dry weight) fresh weight^{-1}.

2.3. Ultrastructure Determinations

Ten embryos extracted from seed in the central portion of the ear were prepared for microscopy following the protocol described in Perdomo and Burris [10] with the following modifications. The extracted embryos were dissected in two through the point of attachment perpendicular to the embryo axis to allow the fixative solution to penetrate rapidly throughout the embryo axis. Embryo halves were immediately placed in freshly prepared fixative solution (3% glutaraldehyde (w/v) and 2% paraformaldehyde (w/v) in 0.1 M cacodylate buffer, pH 7.2). Embryos in fixative solution were stored in a refrigerator at 4 °C for 12–24 months before they were processed for microscopy. Once fixed, all metabolic processes within the embryo ceases.

For microscopy, samples were dissected and fixed with 3% glutaraldehyde (w/v) and 2% paraformaldehyde (w/v) in 0.1 M cacodylate buffer, pH 7.2 for 48 h at 4 °C. Fixed samples were rinsed three times in 0.1 M cacodylate buffer and then post-fixed in 1% osmium tetroxide in 0.1 M cacodylate buffer for 1 h at room temperature. The samples were rinsed in deionized distilled water and enbloc stained with 2% aqueous uranyl acetate for 1 h, dehydrated in a graded ethanol series, cleared with ultra-pure acetone, infiltrated and embedded using Spurr's recipe epoxy resin (Electron Microscopy Sciences, Ft. Washington, PA, USA). Resin blocks were polymerized for 48 h at 65 °C. Thick and ultrathin sections were made using a Leica UC6 ultramicrotome (North Central Instruments, Minneapolis, MN, USA). Thick sections were stained with 1% toluidine blue stain and imaged with an Olympus BX-40 light microscope (Olympus Scientific Solutions Technologies, Waltham, MA, USA). Ultrathin sections were collected onto copper grids and images were captured using a JEOL 2100 scanning and transmission electron microscope (Japan Electron Optic Laboratories, Peabody, MA, USA). Images were captured using an UltraScan 1000 camera (Gatan, Inc., Pleasanton, CA, USA).

2.4. Seed Quality Determination

Standard germination tests were conducted on seed from the last harvest according to the Association of Official Seed Analysts (AOSA) rules for testing seeds [13]. One hundred seeds per each treatment and field replication were planted on crepe cellulose paper media (Kimberly Clark Corp., Neenah, WI, USA) moistened with 800 mL of tap water on fiberglass trays (45 cm × 66 cm × 2.54 cm). Seeds were lightly pressed into the media to create good seed–media contact. After planting, the trays were placed inside germination carts, and the carts were placed inside a walk-in germination chamber at constant 25 °C with alternating 8 h of light and 16 h of darkness d^{-1}. Final seedling evaluation was performed at 7 days after planting.

Seed vigor was evaluated using the tray-method cold test [14]. One hundred seeds from each treatment and field replication were planted on top of crepe cellulose paper media watered with 1100 mL of water pre-chilled for 24 h at 10 °C on fiberglass trays (45 cm × 66 cm × 2.54 cm). After planting, trays were covered with approximately 1 cm of dry 80% sand: 20% soil mixture. The trays were placed inside enclosed germination carts, and the carts were placed inside a dark walk-in chamber at constant 10 °C for 7 days and then moved to a constant 25 °C walk-in germination chamber with alternating 8 h of light and 16 h of darkness d^{-1}. Normal seedlings [13] were evaluated and recorded at 7 days after placing in the constant 25 °C walk-in germination chamber.

2.5. Statistical Analysis for Seed Quality

The two field replications were maintained throughout the experiment, and data were analyzed as a completely randomized design (CRD). The main effects were harvest time and defoliation treatment. All main effects were fixed, and replications were random.

Data were analyzed using the MIXED procedure of SAS (SAS Institute Inc., Carey, NC, USA) [15]. The analysis of variance was estimated using the restricted maximum likelihood method after testing the data for normality and homozygous error variances. Mean comparisons were made using Fisher's protected least significant difference (LSD) test ($p < 0.05$).

3. Results

Light micrographs show the different radicle tissues (Figure 1). Transmission electron microscopy (TEM) micrographs were recorded from the epidermis and cortex cells of the radicle (Figure 1).

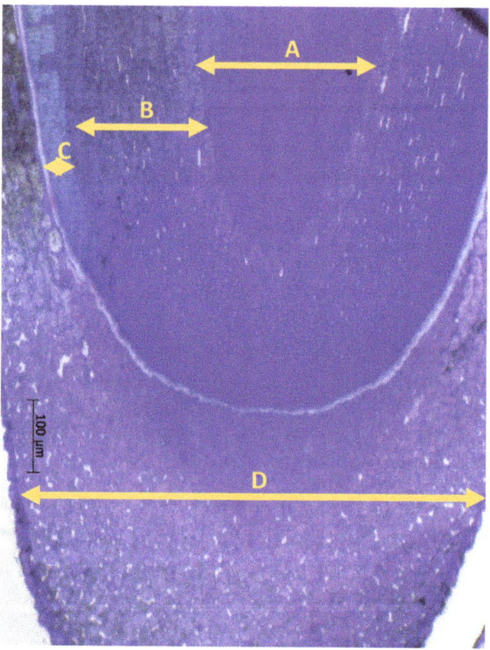

Figure 1. Light microscopy image showing the different tissues of the radicle tip: (**A**) pericycle; (**B**) cortex; (**C**) epidermis; (**D**) root cap. Magnification = 10×.

Prior to defoliation, the oil bodies in epidermis and cortex cells were located randomly throughout the cytoplasm of the cells (Figure 2). The moisture content of the seed was approximately 600 g H_2O kg^{-1} fw. Four days after defoliant application, seed moisture content decreased to 517 and 509 g H_2O kg^{-1} fw in the untreated and treated samples, respectively. The oil bodies in epidermis cells showed the initiation of migration and alignment alongside the cell membrane for both treatments, defoliated and non-defoliated plants (Figure 2). However, the oil bodies in cells from the cortex did not show oil bodies migration for the same seed moisture content.

At 11 days after defoliant application, seed moisture content decreased to 434 and 400 g H_2O kg^{-1} fw in seed samples from the untreated and treated plants, respectively. The migration and alignment of oil bodies along the cell membrane was evident in both tissues, epidermis, and cortex cells. These oil bodies remained aligned along the cell membrane, as observed 18 days after defoliant application (Figure 2). The seed moisture content at this stage was 375 and 368 g H_2O kg^{-1} fw in the untreated and treated seed samples, respectively (Figure 3). The seed field was harvested immediately after these samples were collected.

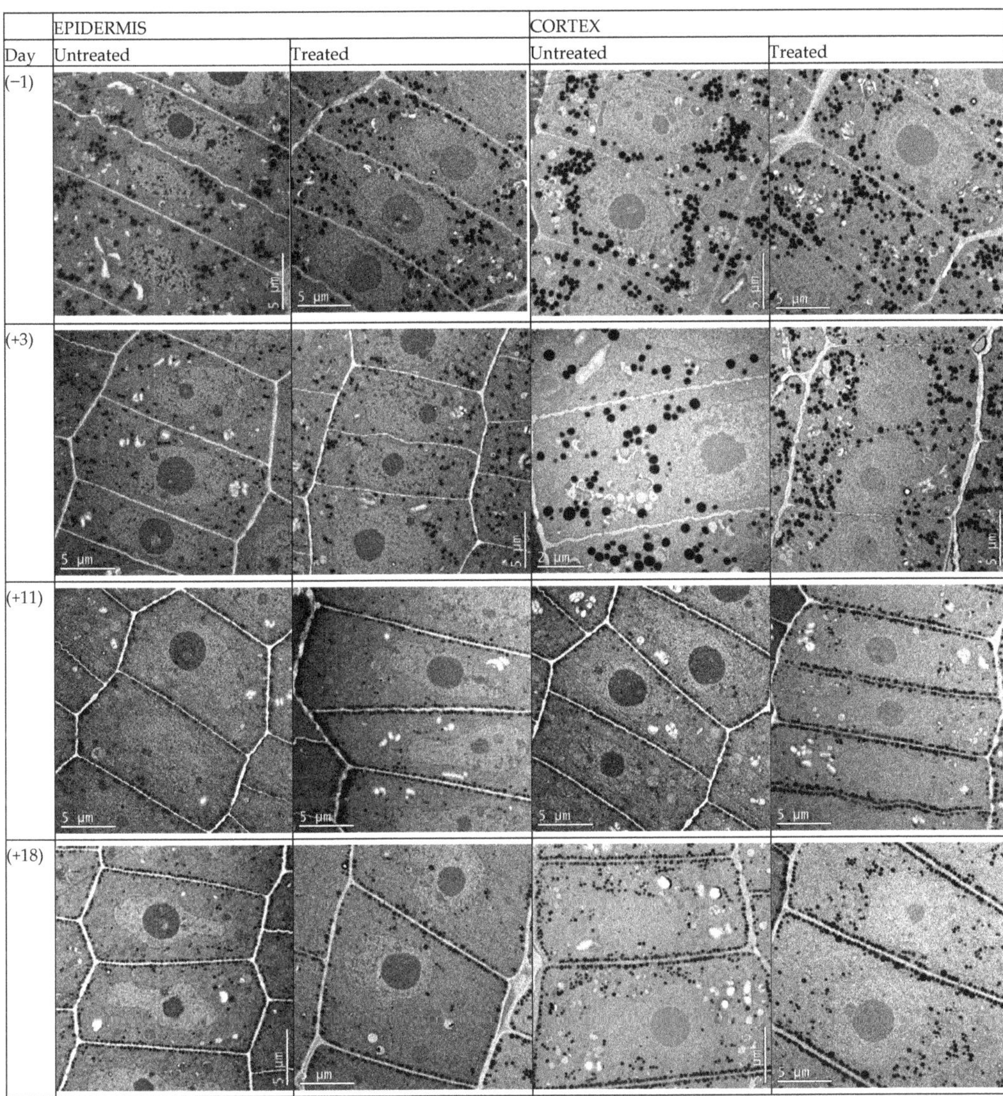

Figure 2. Transmission electron microscopy images of radicle epidermis and cortex cells. Oil body migration is recorded as seed desiccate. All images are taken at 1000×, except for cells in the cortex of untreated plants, which were photographed at 1500×. (Day) Days from defoliant application on the treated plants: day (−1) seeds were harvested and artificially dried with forced ambient air before defoliant application; days (+3), (+11), (+18) indicate seeds were harvested and artificially dried with forced ambient air at 3, 11, and 18 days after defoliant application, respectively. Seed moisture content was expressed on a fresh weight basis as gr H_2O kg seed^{-1} for all treatments. The seed moisture content of untreated plants was 605 on date (−1); 517 on date (+3); 434 on date (+11); and 375 on date (+18). The seed moisture content for seed of plants treated with a defoliant were 581 on date (−1); 509 on date (+3); 430 on date (+11); and 368 on date (+18).

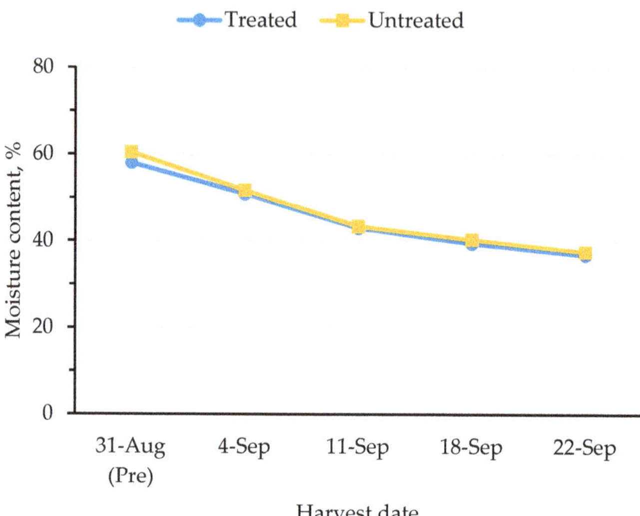

Figure 3. Mean moisture content in percentage at each harvest date for corn hybrid seeds harvested at different harvest dates in 2015. 31-Aug (Pre) refers to harvest before defoliant application; all other harvest dates are post-defoliant application. Aug and Sep indicate the months of August and September. The blue line represents seed moisture values for seed harvested from plants treated with a defoliant; the yellow line represents seed moisture values for seed harvested from the untreated control plants. Means are not significantly different ($p \leq 0.05$).

The germination (Figure 4) and cold test (Figure 5) values of seed harvested from defoliated and non-defoliated areas were not significantly different ($p \leq 0.05$).

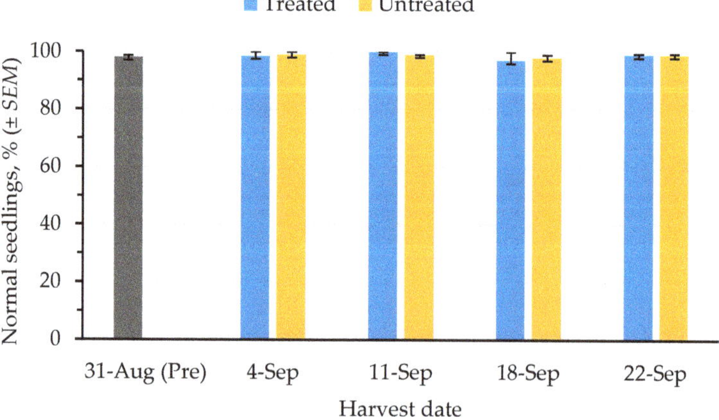

Figure 4. Mean standard germination test values in percentage for corn hybrid seeds harvested at different harvest dates in 2015. 31-Aug (Pre) refers to harvest before defoliant application; all other harvest dates are post-defoliant application. Aug and Sep indicate the months of August and September. Blue columns are the values for seed harvested from plants treated with a defoliant; yellow columns are the values for seed harvested from the untreated control plants. Bars indicate standard error of the mean (*SEM*). Means are not significantly different ($p \leq 0.05$).

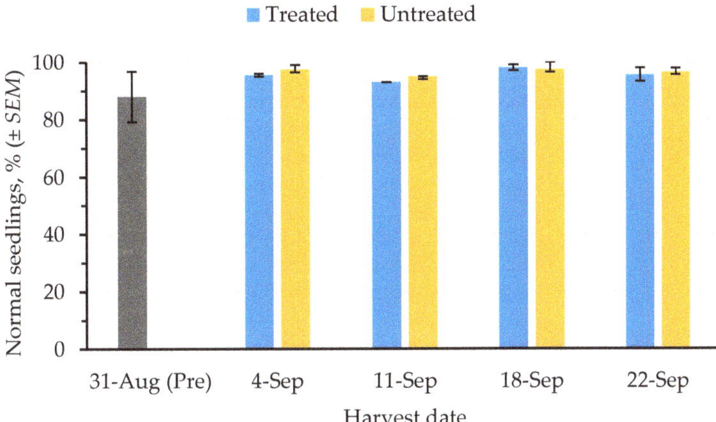

Figure 5. Mean cold test values in percentage for corn hybrid seeds harvested at different harvest dates in 2015. 31-Aug (Pre) refers to harvest before defoliant application; all other harvest dates are post-defoliant application. Aug and Sep indicate the months of August and September. Blue columns are the values for seed harvested from plants treated with a defoliant; yellow columns are the values for seed harvested from the untreated control plants. Bars indicate standard error of the mean (*SEM*). Means are not significantly different ($p \leq 0.05$).

4. Discussion

The US seed corn market is very competitive [16]. Farmers expect rapid and uniform field emergence of their crop under a wide range of environmental conditions. Cold and wet conditions at planting are common in the upper Midwest of the USA [17]. The use of seeds with high physiological potential is essential to achieve rapid and uniform emergence under these stressful environmental conditions [18]. Seed physiological potential is the maximum at physiological maturity [1,3]. Physiological maturity is defined as the developmental stage at which the seed reaches maximum dry weight [1]. Seed physiological potential for this article comprises an active seed metabolic system capable of producing a healthy seedling under a range of environmental conditions in the field (seed germination and vigor). In corn, this developmental stage coincides with black layer formation or the formation of callus tissue that marks the end of seed development and severs the connection between the seed and female parent [3].

The environmental conditions during seed development play a crucial role in seed physiological potential. Abiotic stresses such as plant defoliation during the critical stages of flowering, seed development, and seed maturation can reduce seed yield and seed physiological potential. Freezing temperatures in early fall may cause irreversible damage to cells and reduces seed physiological potential when seed moisture content is greater than 350 g H_2O kg^{-1} fw [4]. These freezing events cause intercellular and intracellular ice formation within the seed embryo, which results in irreversible damage to cells and reduces seed physiological potential [19]. Consequently, seed corn is harvested on or before physiological maturity and dried artificially. At this developmental stage, seed is also at high moisture content, approximately 300 to 400 g H_2O kg^{-1} fw [20]. Seed corn is harvested on the cob and artificially dried until seed reaches a safe moisture content for storage, approximately 120 g H_2O kg^{-1} fw [16]. Seed dryer space may become a limiting factor at the peak of seed corn harvest. In these instances, an early fall frost event can threaten the physiological potential of seed in the field.

Plant defoliation accelerates senescence and seed maturation. Defoliation early in seed development can trigger seed abortion, which lowers seed yields and seed physiological potential. The defoliation stress restricts photosynthesis and reduces the production of sugars necessary for the developing seeds. In sorghum, plants subjected to severe defoliation stress early during seed formation produced larger proportions of low specific gravity seeds with extensive hollow areas in the endosperm [21]. In corn, severe defoliation stress approximately 3 weeks after pollination accelerated seed maturation and reduced seed weight [3]. As seed approaches physiological maturity, however, defoliation accelerates seed maturation, with no negative effects on seed physiological potential.

Seed dehydration capacity is unique to orthodox seeds. These seeds are named "orthodox" because they have the capability to dehydrate to very low moisture content of 40 to 50 g H_2O kg^{-1} fw, while remaining alive. These seeds undergo a series of metabolic changes known as acquisition of desiccation tolerance. The seeds accumulate protective compounds and inactive forms of germination-promoting compounds as they lose water [22]. Also, lipid bodies from the cytoplasm of embryo cells migrate to align along the plasma membranes of the cells [10]. Cells in the root meristem exhibit a distinct migration of the lipid bodies towards the cell walls in response to desiccation. This lipid alignment is essential to seed survival and optimal seed physiological potential [11]. Seeds where lipid alignment is incomplete exhibit an increase in seed leakage during imbibition. The authors theorized that the alignment of lipid bodies along the plasma membrane leads to a more organized dehydration during seed drying [11].

5. Conclusions

In our study, plant defoliation late in seed development did not change patterns of lipid-body migration and alignment along the cell membrane. The application of a defoliant resulted in slow plant senescence and seed dehydration. The treated plant senesced a few days earlier, but the difference in moisture content between seeds from the untreated and treated plants remained within 10 to 20 g H_2O kg^{-1} fw. However, the faster seed dehydration time was enough to allow one or two days harvest-date difference between treated and untreated plants. Our study also demonstrated that chemical defoliation did not reduce seed quality, which was defined as germination and vigor in this article. The use of a defoliant allows seed companies to harvest seed earlier, thus reducing the chance of seed deterioration in the field. Farmers also benefit from this technology, as high-quality seed of multiple genetic backgrounds are available for planting.

Even though this defoliation method is not available for use in EU countries, alternative defoliation methods should be investigated to broaden seed harvest timelines and reduce the need for building additional seed dryers when dryer space is limited. These expensive buildings are an additional cost for the seed companies, which may lead to increased production costs and higher seed price.

Author Contributions: Conceptualization: E.J.C., S.A.G., A.S., J.J., and E.C.; Formal analysis: E.V.Z.; Funding acquisition: S.A.G., A.S., J.J., and E.C.; Investigation: A.N.D., K.W., E.J.C., and S.A.G.; Methodology: E.V.Z. and S.A.G.; Project administration: E.J.C., E.V.Z., and S.A.G.; Writing, review and editing: S.A.G., A.N.D., K.W., and E.V.Z. All authors have read and agreed to the published version of the manuscript.

Funding: The APC and Students were funded by Hatch Projects IOW03814, IOW04114, and IOW05594; hybrid corn seed was harvested from Corteva Agriscience Seed Production Fields.

Institutional Review Board Statement: Not applicable.

Informed Consent Statement: Not applicable.

Data Availability Statement: Data available upon request from corresponding author.

Acknowledgments: The authors thank Harry Horner and Tracey P. Stewart at the Roy J. Carver High Resolution Microscopy Facility, Iowa State University, Ames IA, for providing equipment and advice for the microscopy work reported.

Conflicts of Interest: The authors declare no conflict of interest.

References

1. Delouche, J.C. Seed maturation. In Proceedings of the Short Course for Seedsmen, Starkville, MS, USA, 5–7 April 1976; Seed Technology Laboratory, Mississippi State University: Mississippi State, MS, USA, 1976; pp. 25–33. Available online: https://ir.library.msstate.edu/handle/11668/13936 (Accessed on 30 November 2020).
2. Hunter, J.L.; Tekrony, D.M.; Miles, D.F.; Egli, D.B. Corn Seed Maturity Indicators and their Relationship to Uptake of Carbon-14 Assimilate. *Crop. Sci.* **1991**, *31*, 1309–1313. [CrossRef]
3. Tekrony, D.M.; Hunter, J.L. Effect of Seed Maturation and Genotype on Seed Vigor in Maize. *Crop. Sci.* **1995**, *35*, 857–862. [CrossRef]
4. Devries, M.; Goggi, A.S.; Moore, K.J. Determining Seed Performance of Frost-Damaged Maize Seed Lots. *Crop. Sci.* **2007**, *47*, 2089–2097. [CrossRef]
5. Burris, J.S.; Knittle, K.H. Freeze damage and seed quality in hybrid maize. Proceedings of Annual Seed Technology Conference, 7th, Ames, IA, USA, 26–27 February 1985; Burris, J.S., Ed.; Seed Science Centre, Iowa State University: Ames, IA, USA, 1985; pp. 51–74.
6. Osborne, D.J. Defoliation and Defoliants. *Nat. Cell Biol.* **1968**, *219*, 564–567. [CrossRef] [PubMed]
7. Suttle, J.C. Involvement of Ethylene in the Action of the Cotton Defoliant Thidiazuron. *Plant Physiol.* **1985**, *78*, 272–276. [CrossRef] [PubMed]
8. Chu, C.-C.; Henneberry, T.J.; Reynoso, R.Y. Effect of Cotton Defoliants on Leaf Abscission, Immature Bolls, and Lint Yields in a Short-Season Production System. *J. Prod. Agric.* **1992**, *5*, 268–272. [CrossRef]
9. Bewley, J.D.; Bradford, K.; Hilhorst, H.; Nonogaki, H. Maturation drying and "switch" to germination. In *Seeds: Physiology of Development, Germination and Dormancy*, 3rd ed.; Springer: New York, NY, USA, 2013; pp. 57–68.
10. Perdomo, A.; Burris, J.S. Histochemical, Physiological, and Ultrastructural Changes in the Maize Embryo during Artificial Drying. *Crop. Sci.* **1998**, *38*, 1236–1244. [CrossRef]
11. Córdova-Téllez, L.; Burris, J.S. Alignment of Lipid Bodies along the Plasma Membrane during the Acquisition of Desiccation Tolerance in Maize Seed. *Crop. Sci.* **2002**, *42*, 1982–1988. [CrossRef]
12. Abendroth, L.J.; Elmore, R.W.; Boyer, M.J.; Marlay, S.K. *Corn Growth and Development*; Iowa State University Extension and Outreach: Ames, IA, USA, 2011.
13. Association of Official Seed Analysts (AOSA). *Rules for Testing Seeds*; AOSA: Washington, DC, USA, 2015.
14. Association of Official Seed Analysts (AOSA). *Seed Vigor Testing Handbook*, 3rd ed.; Contribution no. 32; AOSA: Ithaca, NY, USA, 2009.
15. Littell, R.C.; Milliken, G.A.; Stroup, W.W.; Wolfinger, R.D.; Oliver, S. *SAS for Mixed Models*, 2nd ed.; SAS Institute Inc.: Cary, NC, USA, 2006.
16. Wych, R.D. Production of Hybrid Seed Corn. In *Corn and Corn Improvement*; Sprague, G.F., Dudley, J.W., Eds.; Agronomy Monographs the American Society of Agronomy, Inc.; Crop Science Society of America, Inc.; Soil Science Society of America, Inc.: Madison, WI, USA, 1988; pp. 565–606. [CrossRef]
17. Kaiser, D.E.; Coulter, J.A.; Vetsch, J.A. Corn Hybrid Response to In-Furrow Starter Fertilizer as Affected by Planting Date. *Agron. J.* **2016**, *108*, 2493–2501. [CrossRef]
18. Hegarty, T. The physiology of seed hydration and dehydration, and the relation between water stress and the control of germination: A review. *Plant Cell Environ.* **1978**, *1*, 101–119. [CrossRef]
19. Woltz, J.; TeKrony, D.M.; Egli, D.B. Corn seed germination and vigor following freezing during seed development. *Crop Sci.* **2006**, *46*, 1526–1535. [CrossRef]
20. Cordova-Tellez, L.; Burris, J.S. Embryo Drying Rates during the Acquisition of Desiccation Tolerance in Maize Seed. *Crop. Sci.* **2002**, *42*, 1989–1995. [CrossRef]
21. Goggi, A.S.; Delouche, J.C.; Gourley, L.M. Sorghum [*Sorghum bicolor* (L.) Moench] seed internal morphology related to seed specific gravity, weathering, and immaturity. *J. Seed Technol.* **1993**, *17*, 1–11.
22. Gutierrez, L.; Van Wuytswinkel, O.; Castelain, M.; Bellini, C. Combined networks regulating seed maturation. *Trends Plant Sci.* **2007**, *12*, 294–300. [CrossRef] [PubMed]

Article

Humic Substances Improve Vegetable Seedling Quality and Post-Transplant Yield Performance under Stress Conditions

Kuan Qin and Daniel I. Leskovar *

Texas A&M AgriLife Research and Extension Center, Texas A&M University, Uvalde, TX 78801, USA; qinkuan@tamu.edu
* Correspondence: d-leskovar@tamu.edu

Received: 4 June 2020; Accepted: 29 June 2020; Published: 1 July 2020

Abstract: Vegetable growers require vigorous transplants in order to reduce the period of transplant shock during early stand establishment. Organic media containing solid humic substances (HS) are amendments that have not been comprehensively explored for applications in containerized vegetable transplant production systems. In this study, HS (1% *v/v*) were applied to a peat-based growth medium to evaluate pre- and post-transplant growth modulation of four economically important vegetable species. Those were: pepper, tomato, watermelon, and lettuce. Seeding for all species was performed in two periods in order to evaluate their post-transplant yield performance under drought (water deficit vs. well-watered) and heat (hot vs. cool season) stresses. Compared with control, HS-treated plants had: (1) increased leaf and root biomass after transplanting due to faster growth rates; (2) lower root/shoot ratio before transplanting, but higher after 10 days of field establishment; and (3) increased root length and surface area. The negative effects of heat and drought stresses on crop yield were more prominent in control plants, while HS-treated transplants were able to mitigate yield decreases. The results clearly demonstrated the benefits of using solid HS as a management input to improve transplant quality in these crop species.

Keywords: containerized transplants; humic acids; relative growth rate (RGR); specific root length (SRL); heat and drought stresses; heatmaps

---

1. Introduction

In vegetable production, the use of containerized transplants is a standard practice to establish crops in open fields and protected environments. The advantages of transplants over direct seeding have been recently reviewed by Leskovar [1]: transplanting can optimize the timing and scheduling for field cultivation, shorten the cropping period, increase growth cycles, provide uniform, rapid growth and phenological synchrony (flowering, fruit set), and enhance yield and earliness. However, transplants will inevitably suffer from the mechanical damage of root tips and hairs due to the removal of seedlings from the tray, disturbing the root/shoot balance and causing transplant shock and transiently shoot growth stunting [2,3]. Poorly grown transplants will negatively affect plant performance (or tolerance) in post-field establishment environments which is often accompanied by different abiotic stresses. Therefore, a high-quality transplant should have an ability to bear transient or long-lasting field environmental changes, better survival and uniform stand establishment, and higher resource use efficiency, which will eventually achieve high and profitable yield [4]. Transplants are typically grown in multicell trays. Due to the limited volume of cells and short growing cycle (4 to 6 weeks), transplant quality is often determined by root developmental traits and root-to-shoot balance in the confined cells; high transplant quality is typically associated with vigorous root growth such as higher root length, surface area, and dry weight accumulation [4,5]. For example, lettuce seedlings grown with a proper level of N fertilization (60 mg/L) in the growing media produced better quality transplants with higher root dry weight, and subsequent yield performance as compared with seedlings grown with excessive or low N inputs [6]. It has been recognized that large root systems (represented by biomass) could benefit transplant growth with higher growth rate and improved water and nutrient capture in the soil [2].

Several management factors are known to affect transplant quality (root and shoot developmental traits), such as nitrogen fertilization rate [7], irrigation systems [8], container cell size [9], and light quality [10]. In addition, organic sources such as plant (sesame and alfalfa meal, wood fiber, coconut coir) and animal (fish meal and animal manure)-based compost, and vermicompost, are media amendments that can be potentially used in transplant production due to their potential roles for biostimulation, biofertilization, and plant pathogen suppression [11]. Organic sources can affect germination and emergence rates, and physical and chemical structures of the growth media and rhizosphere shortly after transplanting in the field, which ultimately could be translated into improved plant growth and biomass and early yield. For example, Jack et al. [12] used plant- and animal-based vermicompost (earthworm-driven) and thermogenic compost (self-heating), and found that a small level of additional sesame compost (1–2.5% *v/v*) in peat-based commercial media significantly increased tomato transplant shoot biomass. However, the use of organic substrates has to be thoroughly tested and validated since certain amended levels could negatively affect seedling growth due to their bound or unbound high salt content [11].

Humic substances (HS), resulting from the decomposition of plant and animal residues, have been widely reported to be used as organic amendments for their biostimulation (auxin-like) effects on enhancing plant root development, nutrient acquisition, and shoot growth [13]. In vegetable transplant production, HS have been used as liquid extractants (humic acids, HA) and applied as foliar sprays. Hartwigsen and Evans [14] used 2.5 and 5 g/kg HA in cucumber and squash seedlings, which resulted in significantly higher root fresh weight and lateral root length; Turkmen et al. [15] used 1 g/kg HA in tomato seedlings, which resulted in improved seedling growth and nutrient contents; Osman and Rady [16] used 0.5 g/L HA as an additive to growing media and found the dry weights, relative water contents, and NPK uptake of tomato and eggplant transplants were all increased. However, no research has been found using solid HS in seedling production. Compared with liquid HS, which can be dissolved easily and normally have quick and profound effects on plant growth [13], solid forms of HS containing humin have less intense effects, but they could increase media water holding capacity due to increased cellulose contents, and nutrient retention due to their cation exchange capacity (CEC) with much longer existence in soil solutions [17–19], which could make them suitable as supplementary amendments for growing media. Therefore, the potential use of solid HS products with the composition of both HA and humin could improve growing media properties and vegetable seedling quality traits, and the beneficial effects on transplants could last longer, even after field establishment.

In this study, we evaluated how and to what extent solid HS added to a peat-based growing media affected root and shoot developmental traits pre- and post-transplanting, as well as subsequent yield of four vegetable species: tomato, pepper, watermelon, and lettuce. We hypothesized that media amended with HS would improve root development and root-to-shoot growth modulation of containerized seedlings during the nursery period (pre-transplanting), as well as long-standing growth during field establishment (post-transplanting), which in turn will increase yield performance.

2. Materials and Methods

2.1. Plant Materials, Growing Media, and Amendment Treatments

We selected four commercial vegetable species representing high-value vegetable crops, each with two distinctive cultivar types (Figure S1): *Capsicum annuum* with cv. Hunter as bell pepper and cv. Jalafuego as jalapeño pepper; *Solanum lycopersicum* with cv. HM1823 as round tomato and cv. Sakura as cherry tomato; *Citrullus lanatus* with cv. Estrella as diploid (seeded) watermelon and cv. Fascination as triploid (seedless) watermelon; *Lactuca sativa* with cv. Sparx as romaine lettuce and cv. Buttercrunch as butterhead lettuce. Jalafuego, Sakura, Sparx, and Buttercrunch seeds were obtained from Johnny's Selected Seeds (Winslow, ME, USA); Hunter, Estrella, and Fascination from Syngenta (Minneapolis, MN, USA); and HM1823 from Clifton Seed Company (Faison, NC, USA).

Speedling (Ruskin, FL, USA) polystyrene 200-cell trays with inverted pyramid cells (Model TR200A, 2.5 × 2.5 cm^2 × 7.6 cm deep with 32 cm^3 volume per cell) were used for transplant growth in pepper, tomato, and lettuce. Watermelon seeds were sowed into 128-cell trays (Model TR128A, 3.1 × 3.1 cm^2 × 6.4 cm deep with 43 cm^3 volume per cell). Lambert Germination, Plugs and Seedlings (LM-GPS) growing media (90% sphagnum peat moss, 10% perlite and vermiculite; Lambert, Québec, Canada) were used as control (C). Lignite-derived solid humic substances (Novihum Technologies, Salinas, CA, USA), with a composition of 32% humic acid, 3% fulvic acid, and 24% humin, were mixed with the control growing media as an amendment treatment (HS) at the rate of 1% by volume (*v/v*) basis. The basic physical and chemical properties of the commercial media and humic substances were measured and are shown in Tables 1 and 2.

Table 1. Basic physical properties of the commercial media (CM) and humic substances (HS).

	Size Distribution (%)						TP [1] (%)	AS [2] (%)	CWHC [3] (%)	BD [4] (g/cm^3)
	<0.25 mm	0.25–0.50	0.50–1.00	1.00–2.00	2.00–2.80	>2.80				
CM	0.2	1.1	20.0	55.3	4.9	18.5	58.6	2.9	55.7	0.07
HS	23.4	28.4	29.0	18.0	1.1	0.1	65.8	5.3	60.5	0.61

[1] TP: total porosity; [2] AS: air space; [3] CWHC: container water holding capacity; [4] BD: bulk density.

Table 2. Basic chemical properties of the commercial media (CM), humic substances (HS), and field soil (FS).

	pH	EC [1]	OC [2]	Total N	P	K	Ca	Mg	S	Na	Fe	Zn	Mn	Cu
		(dS/m)	(%)	(%)	(mg/kg)									
CM	5.5	0.08	41.5	0.63	1.85	7.52	938	123	14.82	25.70	19.43	4.56	10.44	0.72
HS	7.4	0.44	65.7	5.48	2.90	56.22	451	334	46.52	452.18	4.18	0.04	4.59	0.04
FS	8.0	0.29	2.0	N/A [3]	47.03	801.68	11355	206	19.05	4.13	2.99	0.78	15.29	0.73

[1] EC: electrical conductivity; [2] OC: organic carbon; [3] N/A: not available.

2.2. Growth Environments and Stress Treatments

After sowing seeds, all trays received irrigation to about 60% water holding capacity and were incubated in a growth chamber (PGR15, Conviron, Winnipeg, Canada) in darkness at 25 °C for 48 h. All trays were then transferred to a greenhouse with an overhead motorized spraying boom system (total length 7.1 m with two long arms at sides and operating orbit at center; each arm has 3.2 m length with 13 sprinkler units) for delivering uniform irrigation and fertilization. Environmental conditions (temperature and humidity) inside the greenhouse were controlled by a Wadsworth control system (Arvada, CO, USA) and hourly monitored by a weather station WatchDog (Spectrum Technologies Inc., Aurora, IL, USA) (Figure 1). After six weeks of growth, seedlings were transplanted in a field with raised beds at the Texas A&M AgriLife Research and Extension Centers in Uvalde, Texas (29.21° N, 99.79° W) with a clay soil type (41% clay, 31% sand, 28% silt) (Table 2). The field was prepared using ridge tillage. Planting configuration–number of rows per bed, distance between plants and beds for peppers were double-row, 0.3 m and 1.8 m; for tomatoes were single-row, 0.46 m and 1.8 m; for watermelons were single-row, 0.6 m and 2.4 m; for lettuces were double-row, 0.25 m and 0.9 m, respectively. Drip irrigation with emitter rate at 0.87 L per hour and emitter spacing at 30 cm (Netafim, Fresno, CA, USA) was installed at 10–15 cm below the soil surface in the center bed and was used for all vegetables tested in this experiment. White plastic mulch was used for pepper and tomato, black for watermelon, and bare soil for lettuce.

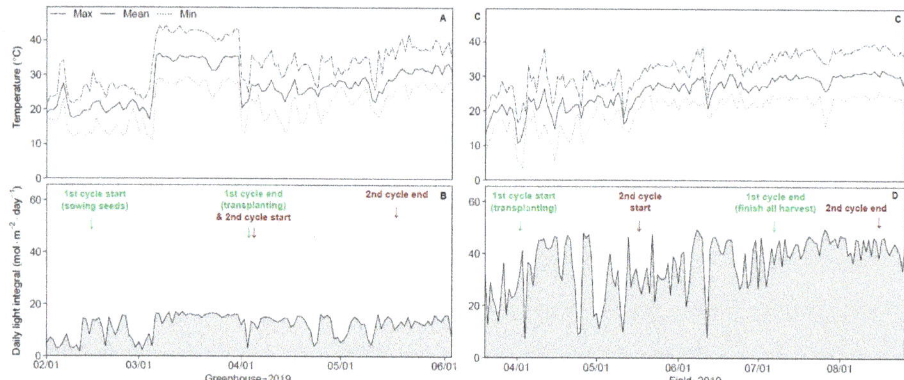

Figure 1. Temperature, daily light integral, and growing cycles (cool and hot seasons indicated by arrows) of greenhouse (**A**,**B**) and field (**C**,**D**) from 1 February 2019 to 30 August 2019.

During the field growing period, all transplants were subjected to two environmental treatment factors: heat and drought stresses. Heat stress was naturally imposed by growing seedlings during a hot season as compared with no stress with seedlings grown during a cool season. The average field maximum, mean, and minimum temperatures for the cool season were 30.4 °C, 24.1 °C, and 18.7 °C and for the hot season were 35.2 °C, 28.6 °C, and 22.9 °C, respectively (Figure 1). Drought stress was imposed by applying deficit irrigation using an evapotranspiration (ET)-based irrigation scheduling (deficit 50% ET vs. full irrigation at 100% ET). The ET crop water requirement was calculated based on the specific crop coefficients (Kc), flow rate of the drip tape, mulch covering, and precipitation [20]. The differential irrigation treatments started 10 days after transplanting, while fertilization was kept the same among treatments and other standard management practices (weeding, pest and disease control, pruning, trellis, etc.) were followed during the growing period. Within each cultivar/crop/growing season (cool vs. hot) after transplanting, the field layout was a split-plot design with four blocks–irrigation level (50% ET vs. 100% ET) as the whole-plot factor and amendment treated transplants (control vs. HS) as the split-plot factor.

2.3. *Seedling and Transplant Quality Evaluation and Yield Performance*

Seedling emergence was counted for all crops within 1 to 2 weeks after seeding. During each growing cycle, 4 plants per cultivar/crop from each treatment (C and HS) were randomly sampled from the growing trays at 4 weeks after seeding (WAS), 5 WAS, 6 WAS, and 10 days after transplanting (DAT) for seedling (plants defined as before transplanting) and transplant (after transplanting) evaluation. Plants were removed from the trays and separated by leaf, stem, and root components. The whole roots were carefully washed, scanned using an EPSON V700 scanner (Epson, Long Beach, CA, USA), and then root length (RL), root surface area (RSA), and root average diameter (RAD) were obtained by using WinRHIZO software (Regent Instruments, Québec, Canada). After taking pictures of all leaves with a 1 cm^2 square scale, ImageJ [21] was used for measuring leaf area (LA). Leaf, stem, and root dry weight (LDW, SDW, RDW) were measured after oven drying at 75 °C for 2 days. Leaf area ratio (LAR, ratio of leaf area to plant total dry weight), root/shoot ratio (R:S, ratio of root to shoot dry weight), specific root length (SRL, ratio of root length to root dry mass) were then calculated. Relative growth rate (RGR, calculated based on leaf, stem, root, and total plant) and net assimilation rate (NAR, the increases in plant dry mass per unit leaf area and time) were also calculated based on the following equations. For convenience, all abbreviations are listed in Table S1.

$$\mathrm{RGR} = ((\ln(\mathrm{DW}_{\mathrm{time1}}) - \ln(\mathrm{DW}_{\mathrm{time2}}))/(\mathrm{time1} - \mathrm{time2}) \quad (1)$$

$$NAR = ((DW_{time1} - DW_{time2}) \times ((\ln(LA_{time1}) - \ln(LA_{time2})))/((LA_{time1} - LA_{time2}) \times (time1 - time2)) \quad (2)$$

All plants were kept growing in the field under the two treatment factors (cool vs. hot season; well-watered vs. deficit irrigation) until final harvest. Pepper, tomato, and watermelon were harvested at different times during the growing season, while lettuce was once-over harvested when the majority of heads reached maturity. The total yield was calculated and the average fruit weight (AFW) for pepper, tomato, and watermelon and average head weight for lettuce were calculated based on the total number of fruits (or heads) harvested.

2.4. Statistical Analysis

Seedling and transplant evaluation parameters were analyzed considering media-amendment (Control vs. HS) as the main factor with 8 replications from both growing seasons; while yield performance was analyzed following the split-plot design. R [22] was used for performing ANOVA and means were separated by the least significant difference (LSD) test at 4 levels: $P \leq 0.1, 0.05, 0.01, 0.001$.

3. Results

Based on the two cycles of growth, there were no significant differences of seedling emergence percentage between control and humic substances (HS)-treated growing media (Table S2). Pre- and post-transplanting time-course growth data for each crop species and cultivars are presented in separate graphs.

3.1. Pepper

Compared with untreated control plants (Figure 2), bell pepper (cv. Hunter) grown in HS-added substrate had significantly higher LDW before transplanting and RDW after transplanting ($P < 0.001$). Although there were no significant differences in SDW, HS-treated seedlings had a faster stem RGR than control before transplanting ($P < 0.05$). Lower root-to-shoot ratio (R:S) was observed in HS-treated plants before transplanting compared with control, but the difference disappeared after transplanting, which may be caused by the increases in root growth (RDW). There were no significant differences in NAR, SRL, and RAD. Regarding yield responses, HS-treated transplants had higher yield compared with control under water stress (50% ET) in both cool ($P < 0.1$) and hot seasons ($P < 0.05$), but no differences were found in well-watered treatment (100% ET). HS amendments decreased bell pepper AFW under well-watered treatment in hot season ($P < 0.1$). In bell pepper, the highest RGR increase between 5 and 6 weeks of growth was mostly due to stem rather than root or leaf growth.

Similar RGR trends were observed in HS-treated jalapeño pepper (cv. Jalafuego), which in addition showed a significantly faster RGR in roots after transplanting ($P < 0.05$). Lower R:S were also observed in HS-treated plants before transplanting, but R:S significantly increased after transplanting as compared with control ($P < 0.1$), which could be explained by the significant enhancement of root growth traits (RDW, RL, RSA, $P < 0.05$). There were no significant differences in NAR, SRL, yield, and average fruit weight (AFW) due to the HS application. In field production, both bell and jalapeño peppers had lower yield and AFW in hot temperature as compared with the cool season ($P < 0.001$), and in water stress compared with no stress ($P < 0.01$) (Figure 2 and Table 3).

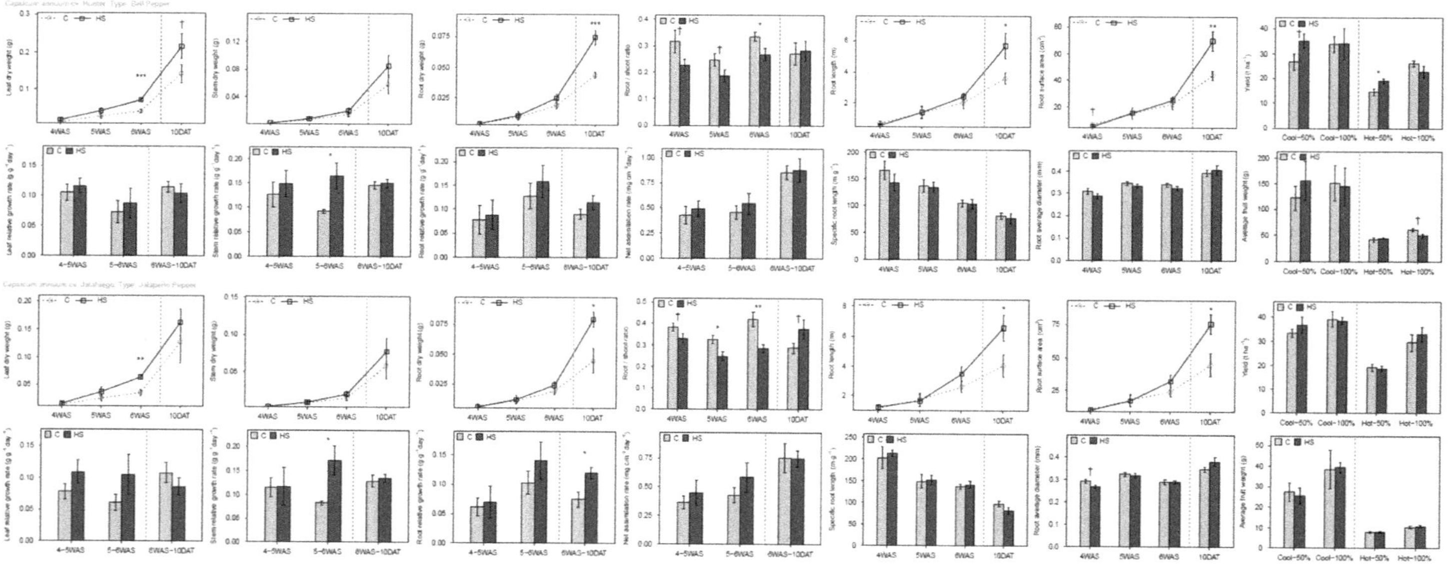

Figure 2. Pepper seedling and transplant quality traits as affected by media amendments, yield traits as affected by amendments and irrigation during the two growing seasons. †, *, **, *** show significant difference comparing HS to control (C) at $P \leq 0.1$, 0.05, 0.01, and 0.001, respectively.

Table 3. ANOVA of total yield, average fruit weight (AFW) as influenced by amendments (A) and irrigation (IR) treatments during the two growing seasons (S).

ANOVA	Pepper		Tomato		Watermelon		Lettuce	
	Bell	Jalapeño	Round	Cherry	Diploid	Triploid	Romaine	Butterhead
				Yield				
S	***	***	***	***	***	***	***	***
IR	**	***	***	***	NS	NS	NS	NS
A	NS	NS	†	*	*	*	**	NS
S × IR	NS	*	***	***	NS	NS	NS	NS
S × A	NS	NS	†	*	NS	NS	NS	*
IR × A	*	NS	*	***	NS	NS	NS	NS
S × IR × A	NS	NS	†	***	NS	NS	NS	NS
				AFW				
S	***	***	*	***	***	***	***	***
IR	NS	**	*	*	NS	NS	NS	NS
A	NS	NS	†	NS	NS	NS	**	NS
S × IR	NS	†	NS	NS	NS	NS	NS	NS
S × A	NS	NS	NS	NS	NS	NS	NS	*
IR × A	NS	NS	NS	NS	NS	NS	NS	NS
S × IR × A	NS	NS	NS	NS	NS	NS	NS	NS

†, *, **, *** show significant difference at $P \leq 0.1, 0.05, 0.01$, and 0.001, respectively; NS, not significant at $P \leq 0.1$.

3.2. Tomato

Compared with untreated control plants (Figure 3), HS-treated round tomato (cv. HM1823) had significantly higher LDW, SDW, and RDW before and after transplanting ($P < 0.05$). RGR was also higher, especially in stem; however, NAR was lower during early growth (4–5 WAS, $P < 0.05$), but these differences were reversed 10 DAT. Similarly, R:S was lower before transplanting but higher after transplanting ($P < 0.1$). In terms of root traits, RL and RSA were significantly higher, especially after transplanting ($P < 0.001$), RAD was also higher ($P < 0.1$), but SRL was lower. HS-treated transplants had higher yield compared with control under no stress conditions (100% ET and cool season) ($P < 0.1$).

In cherry tomato (cv. Sakura), HS had early beneficial effects on leaf and root growth even at 4WAS, with additional faster root RGR after transplanting and higher RL, RSA, RAD during seedling growth and transplant periods than control ($P < 0.01$). Yield was significantly higher for HS than the control under well-watered conditions ($P < 0.001$). For both round and cherry tomatoes, deficit irrigation treatment (50% ET) had significant negative effects on yield during the cool season but not during the hot season ($P < 0.001$). Under heat stress, plants exhibited a dramatic decreased in tomato yield and AFW, especially on cherry tomato ($P < 0.001$) (Figure 3 and Table 3).

3.3. Watermelon

Compared with untreated control plants (Figure 4), HS-treated diploid seeded watermelon (cv. Estrella) had lower leaf and root RGR between 4 and 5 WAS, but higher root biomass ($P < 0.1$) and RGR ($P < 0.05$) were observed 10 DAT. R:S was lower before transplanting, but these differences disappeared after transplanting. Similar trends were observed for NAR. HS-treated transplants had higher SRL but lower RAD at 6 WAS, and higher RL and RSA ($P < 0.05$) at 10 DAT. Although not significant, HS-treated plants had a numerical yield increase of diploid watermelon in the cool season regardless of irrigation treatments.

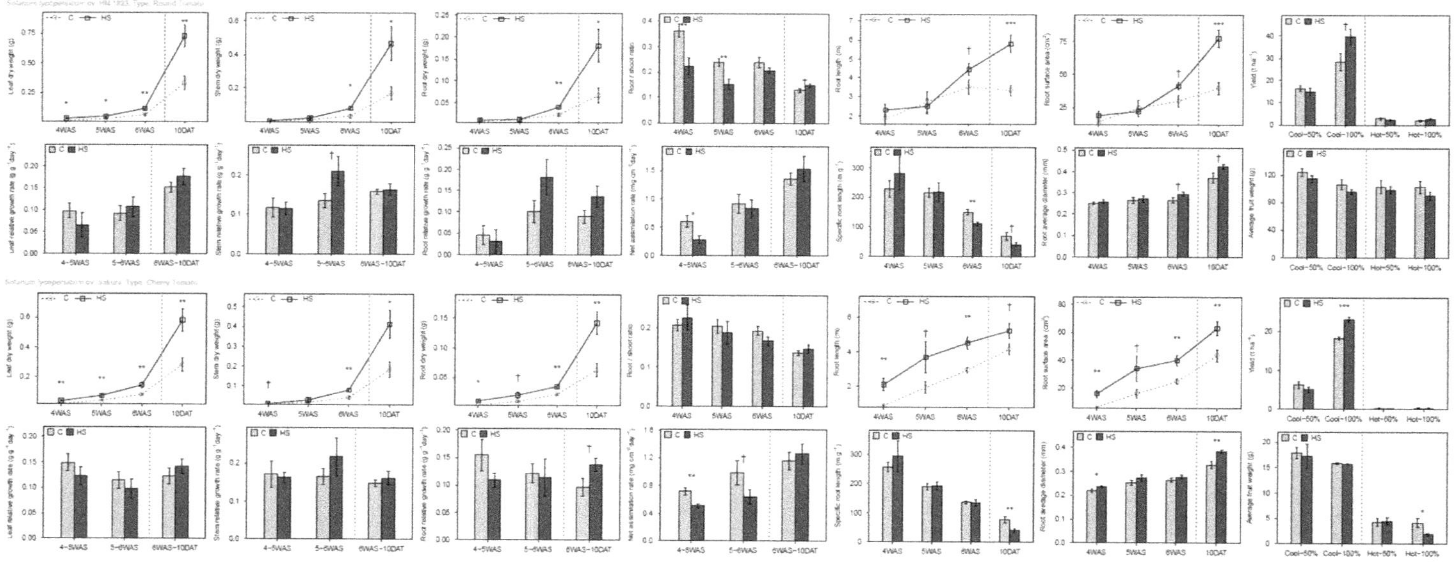

Figure 3. Tomato seedling and transplant quality traits as affected by media amendments, yield traits as affected by amendments and irrigation during the two growing seasons. †, *, **, *** show significant difference comparing HS to control (C) at $P \leq 0.1$, 0.05, 0.01, and 0.001, respectively.

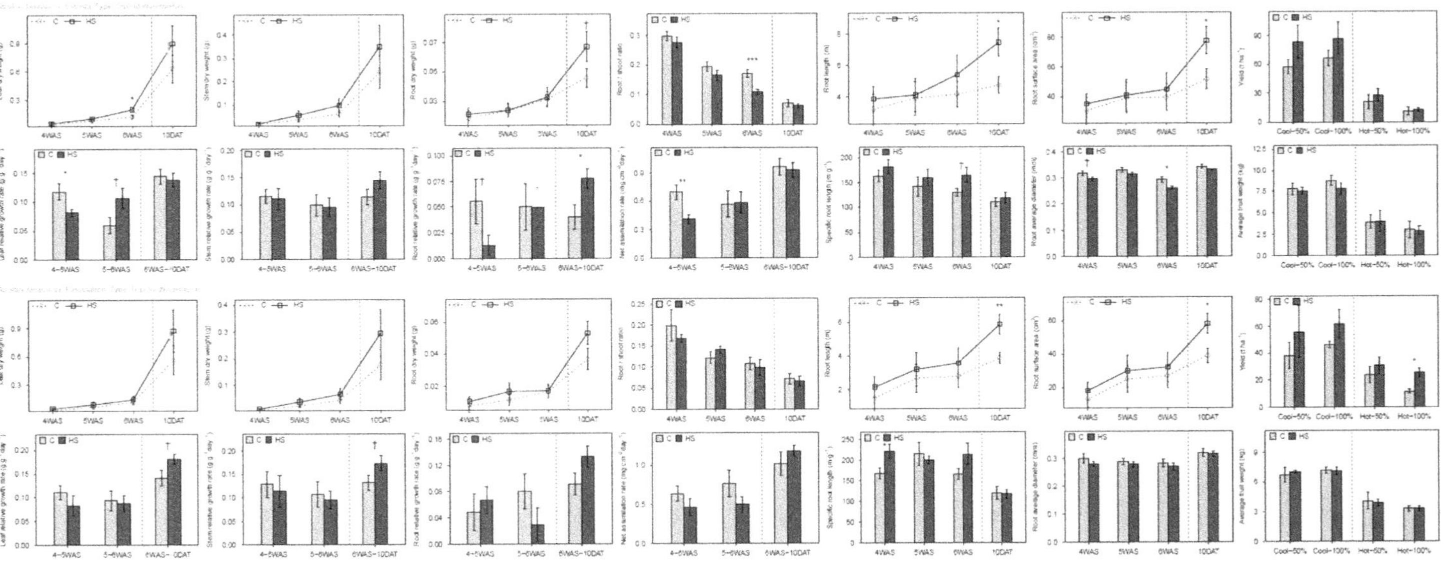

Figure 4. Watermelon seedling and transplant quality traits as affected by media amendments, yield traits as affected by amendments and irrigation during the two growing seasons. †, *, **, *** show significant difference comparing HS to control (C) at $P \leq 0.1$, 0.05, 0.01, and 0.001, respectively.

Triploid seedless watermelon (cv. Fascination) had different root and shoot growth responses as compared with Estrella. HS-treated transplants had faster leaf and stem RGR than control ($P < 0.1$) during the field establishment period (up to 10 DAT), biomass accumulation was accordingly increased although not significant. Before transplanting, RL and RSA were not affected by HS application, but they significantly increased at 10 DAT ($P < 0.05$). These root responses were consistent with those found in the diploid watermelon. SRL was higher for HS plants compared with control at 6 WAS, but similar after transplanting. During the cool season, HS-treated plants had a numerically increased yield under both irrigation rates. HS also increased yield of triploid watermelon in the hot season, particularly for the well-watered treatment ($P < 0.05$). Comparing both stresses, heat stress (high temperature) had more negative dominant effects on yield and AFW of both diploid and triploid watermelons ($P < 0.001$) as compared with water stress (Figure 4 and Table 3).

3.4. Lettuce

Compared with untreated control plants (Figure 5), HS-treated romaine lettuce (cv. Sparx) had significantly higher LDW ($P < 0.001$), faster leaf RGR ($P < 0.05$), but lower RDW before transplanting; however, RDW and root RGR were significantly higher after transplanting ($P < 0.05$). R:S was significantly lower during seedling development and after transplanting. For root traits, RL and RSA were not affected by HS, but SRL was higher before but lower after transplanting, and the reverse responses were measured for RAD. HS-treated romaine lettuce had a significant increase in yield and average head weight (AHW) in the hot season regardless of irrigation treatments ($P < 0.05$).

Butterhead lettuce (cv. Buttercrunch) had similar results as Sparx, with additional significantly lower NAR ($P < 0.05$) and no differences in final yield (though numerically lower during the cool season) comparing HS- with control-treated plants. Heat stress (hot season) had significant negative effects on yield of both romaine and butterhead lettuce types ($P < 0.001$), while the impacts from irrigation treatments were relatively low (Figure 5 and Table 3).

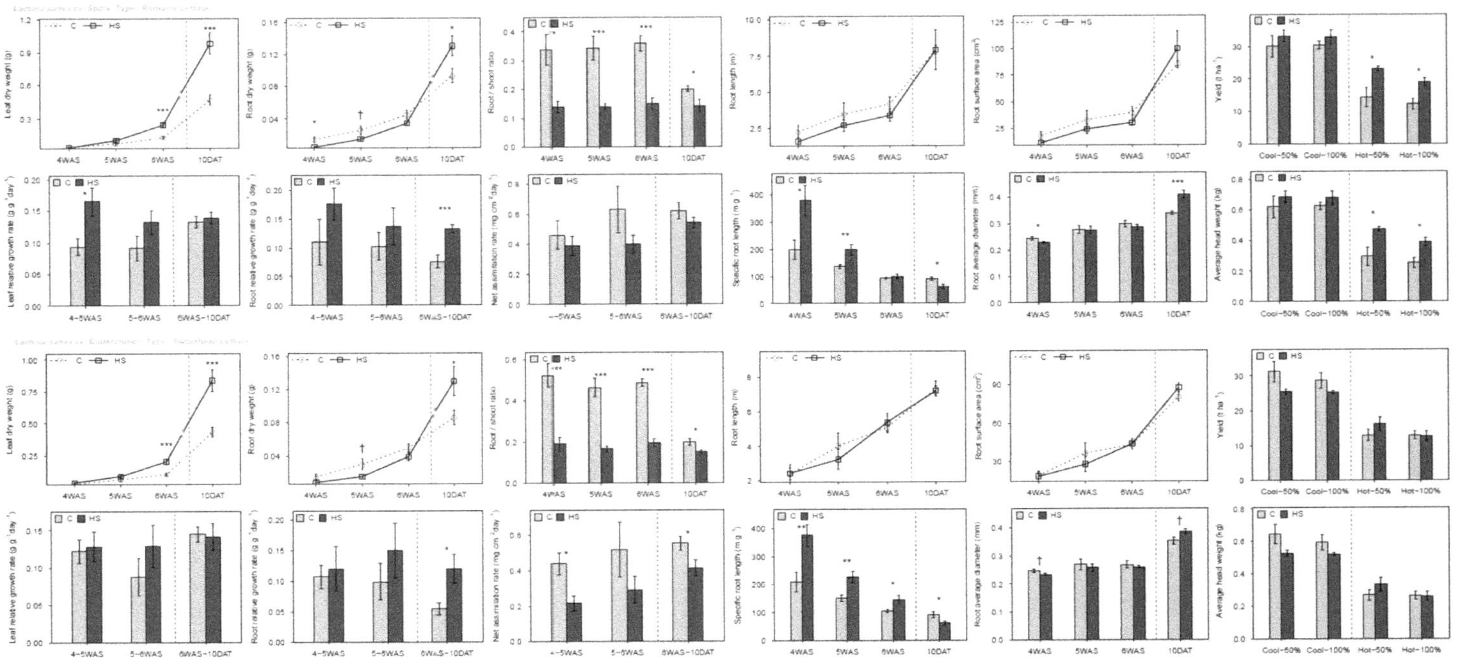

Figure 5. Lettuce seedling and transplant quality traits as affected by media amendments, yield traits as affected by amendments and irrigation during the two growing seasons. †, *, **, *** show significant difference comparing HS to control (C) at $P \leq 0.1$, 0.05, 0.01, and 0.001, respectively.

4. Discussion

Adding solid organic amendments such as compost and vermicompost (derived from organic waste) in growing media have shown benefits in transplant growth [12,23]. However, it is recognized that these amendments that contain high soluble salts could adversely affect germination by lowering the osmotic potential of the water in the media [24]. Since seed germination and seedling emergence are rapid and powerful ways to test potential substrate phytotoxicity [25], they should be fully examined before evaluating seedling or transplant quality. In our study, there were no significant differences in germination percentage and seedling emergence between control and humic substances (HS)-treated growing media, indicating that 1% (*v/v*) HS was safe and not phytotoxic on seeds tested (Table S2). The overall effects of HS amendments on leaf and root traits, RGR, NAR, yield, and average fruit weight are summarized in Table S3. We found that due to the HS application, leaf, stem, and root biomass accumulation were significantly improved, which could have resulted from higher carbon input from leaves and nutrient absorption from the root.

The HS used in this study were obtained by using the ammonoxidation procedure (lignite reacting with oxygen in aqueous ammonia) and resulted in a product with lower hydrophobicity (mainly caused by reduced aromatic compounds) and higher bioactivity than naturally slow-generated HS from lignite [26]. In addition, solid HS contain humin, which has less hydrophilic carboxyl and hydroxyl groups but higher hydrophobic alkyl groups and ash contents [18]. Raw materials also decide HS properties: lignite-derived HS are composed of highly oxidized sulfur-containing molecules and aromatic and aliphatic groups, which can give the products a higher hydrophobic protection than other raw materials (e.g., peat, compost, sludge, leonardite). This makes them more stable in terms of their existence (lifespan) in the soil solutions, having slowly beneficial effects [27,28]. This HS product contained higher N, K, Mg, and Na contents than commercial media, however, by adding HS with 1% *v/v*, the nutrient differences compared with control (solely commercial media) were minimized. The similar early growth performance (4 or 5 WAS) also indicated that there were no initial nutrient differences between control and HS-treated trays. During the seedling growth period, the fertilization amount applied for both control and HS trays were exactly the same and sufficient for seedling growth, thus the beneficial effects from HS were probably not related with nutrients. We found the increased seedling biomass in HS-treated trays mainly occurred at a later seedling growth stage (6 WAS) and during early field establishment, with prominent effects on root development. This could indicate the positive results from HS were mainly due to their biostimulation (auxin-like) effects on enhancing plant root development and nutrient acquisition [13], which occurred slowly due to the solid HS product. Since transplant quality was the main focus, below we explain in detail the effects of HS on the specific transplant growth parameters.

As a growth speed index, relative growth rate (RGR) can be affected by internal (species, seed mass, growth cycle) and external physical and environmental factors (pot volume, light, nutrients, and temperature) [29]. Based on the variability, RGR could be used as an indicator for separating functional strategies of plant growth: faster RGR indicates more competition for obtaining growing resources, slower RGR indicates more stress tolerance [30]. Variation of RGR could be predicted by NAR (representing the balance of photosynthetic and respiration rates) or LAR (representing the deployed efficiency of photosynthetic resources) [31–33]. In our study, a significantly positive correlation between RGR and NAR was found only in fruit-based vegetables (pepper, tomato, watermelon), while a significantly positive correlation between RGR and LAR was detected only in the leaf-based vegetable (lettuce) (Figure 6). This could indicate that the growth rate of fruit-based vegetables was determined by both photosynthesis and respiration, while leaf-based vegetables were mainly affected by their photosynthetic resources. In addition, HS-treated transplants had an overall higher RGR (especially root) than control transplants regardless of crop species, which showed a stronger recovery and adaptability (less transplant shock) during the field establishment period, and also indicated a higher nutrient uptake since nutrient absorption correlated with growth rate [34].

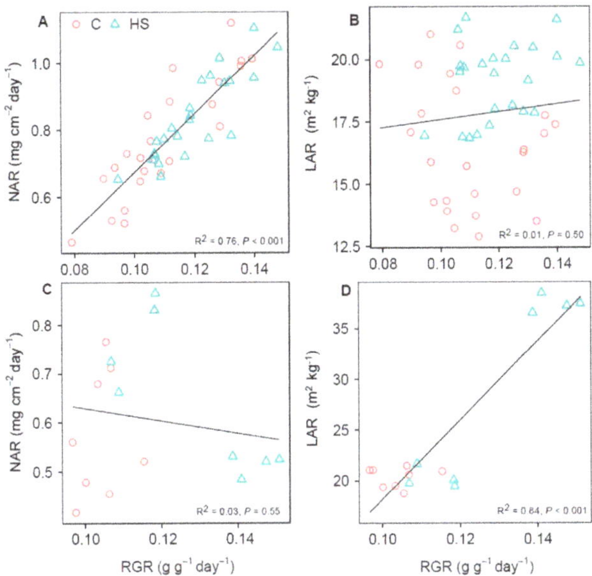

Figure 6. Linear regression plotted for net assimilation rate (NAR) and leaf area ratio (LAR) against relative growth rate (RGR) of (**A**,**B**) fruit-based vegetables (pepper, tomato, watermelon) and (**C**,**D**) leaf-based vegetable (lettuce).

Root-to-shoot ratio (R:S) is an important indicator for the allocation of plant organs against limited growing resources. In general, suitable environments rich in nutrients improve shoot (leaf and stem) growth, while poor environments with insufficient nutrients improve root relative to shoot growth. In seedling production, it is well accepted that R:S is found lower with higher substrate nutrient supply, particularly nitrogen [35]. In our study, lower R:S found in HS-treated seedlings before transplanting indicated a rich nutrient environment possibly due to the nutrient retention ability from HS. Although the boundaries of the optimum R:S are difficult to define, transplants with higher R:S are often considered to have better growth capacity and quicker establishment after transplanting [36]. HS-treated plants (except lettuce) had higher R:S than control plants after transplanting, which could explain the improvement in field establishment and yield performance.

Specific root length (SRL) is a trait that identifies the economic return (represented by root length, RL) from the cost (represented by root dry weight, RDW). The increase of SRL is often associated with nutrient limitation or dry environments [37]. However, an increase in nutrients could also lead to a higher SRL, especially when supplied in a localized nutrient patch, but the proliferation of fine root length was not accompanied by more allocation to root biomass [38]; meanwhile, this situation is species-specific [39]. SRL is strongly dependent on fine roots; with decreased RAD, SRL increased [40]. In our study, compared HS- with control-treated seedlings, RAD was lower before but higher after transplanting; in contrast, SRL changed from higher to lower (except for tomato cultivars). In seedling production before transplanting, nutrients provided in the trays are localized, thus the higher SRL was probably due to a better productive environment with HS, but after transplanting in the field, soil nutrient supply was not as localized as in trays, with lower SRL from HS-treated plants regardless of crops, indicating a less initial stress than control during the transplant shock period. In addition, a significantly increased RDW demonstrated that HS improved plant capacity for rapid root regeneration and growth for larger structural roots during field establishment.

Temperature and irrigation play important roles in vegetable production as they modulate vegetative and reproductive development. In general, flowers are the most temperature-sensitive organs, with high temperature (heat stress) decreasing pollen viability and fruit set, disturbing root functional water and nutrient uptakes, as well as causing abnormal development of shoot tip [41]. Drought stress will impair cell division and leaf area expansion, decrease leaf photosynthetic rate, and delay the conversion of vegetative to reproductive stage [42]. In our study, both heat and water stress decreased crop yield and average fruit weight (size), with heat stress having more significant effects than drought stress. Although within each crop, cultivars representing unique types had different responses, we found that stronger transplant quality due to HS application could ameliorate the adverse effects caused by the abiotic stresses, which led to a higher yield compared with control. These included: bell pepper under drought and heat stresses; round and cherry tomatoes under optimized environment (no stress); triploid watermelon under heat stress without irrigation limitation; romaine lettuce in heat stress regardless of irrigation rates.

In order to better understand the general HS effects on all crop cultivars tested and build linkages between measured seedling or transplant quality traits and subsequent yield, heatmaps (Figure 7) were created based on standardized data sets obtained before and after transplanting. Treatments were clustered based on their measured variables, and variables were clustered based on their correlations (closer meant higher positive correlations). We found that either before or after transplanting, HS treatments were clearly distinguished from control in all crops, mainly due to the higher shoot (SDW), root dry weight (RDW), root length (RL), root surface area (RSA), and yield. Yield was highly correlated with shoot growth (SDW) before transplanting, and root growth traits (RL, RDW, RSA) after transplanting, which indicated that during the seedling stage, sufficient nutrients should be kept in the growth media to improve the plant above-ground growth, while after transplanting, management practices aimed at improving root development should be considered. Besides the application of solid HS in this study, the use of other biostimulant substrates (phenols, salicylic acid, humic and fulvic acid, seaweed extracts, protein hydrolases) and microbial inoculants (plant-growth-promoting rhizobacteria and mycorrhizal fungi) have shown to boost root performance [43,44], which can be used for enhancing transplant field establishment and subsequent crop production. Overall, solid HS with shoot and root growth-promoting effects can satisfy the requirements of transplant growth and subsequent yield in both pre- and post-transplanting environments, which makes them suitable and reliable amendments for use in transplant media.

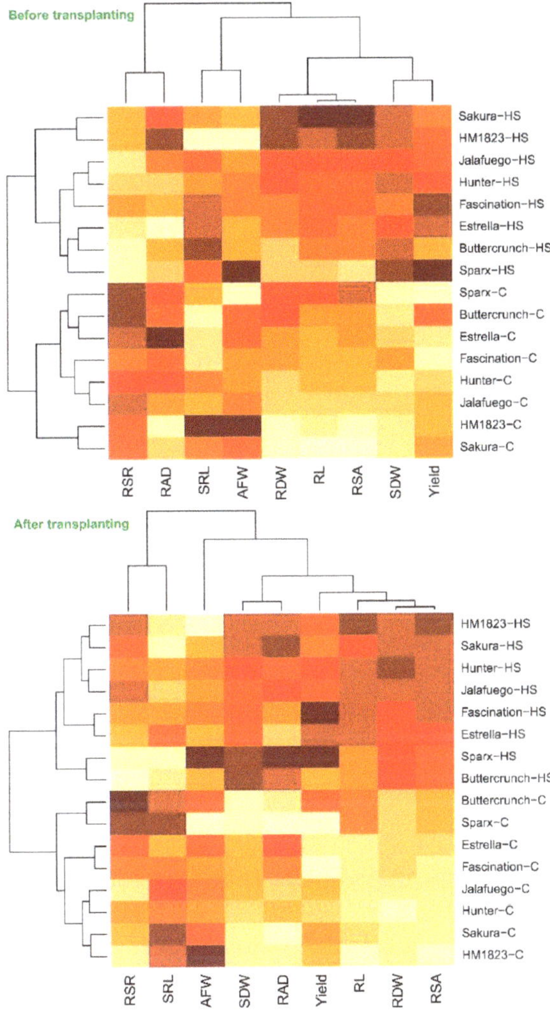

Figure 7. Heatmaps and clustering of the amendment treatments (C and HS) based on the (top) before-transplanting traits and (bottom) after-transplanting traits with the consideration of yield components. Each row represents a crop cultivar with or without HS treatment, and each column represents a measured variable, including shoot dry weight (SDW), root dry weight (RDW), root shoot ratio (RSR), root length (RL), specific root length (SRL), root average diameter (RAD), root surface area (RSA) and average fruit weight (AFW). The expression variable values of the heatmaps follow the red (high)–yellow (low) color scale. All data are standardized and measured variables are clustered based on their correlations.

5. Conclusions

In this study, humic substances (HS) added as a media amendment for growing containerized vegetable transplants were evaluated for their seedling root and shoot growth modulation effects before and after field transplanting. Compared with control, HS: (1) improved plant shoot biomass accumulation of pepper, tomato, and lettuce mostly due to faster shoot growth rates, while these effects were not prominent in watermelon; (2) enhanced pepper and watermelon root developmental traits (RDW, RL, RSA) after transplanting due to faster root growth rates, and tomato root development both before and after transplanting, while these effects were not shown in lettuce; (3) decreased net assimilation rate of tomato, watermelon, and lettuce before transplanting but improved after transplanting, while this effect was not significant for pepper; (4) improved leaf area ratio in all four crops; (5) improved specific root length of tomato, watermelon, and lettuce before transplanting but decreased it after transplanting; (6) lowered root-to-shoot ratio of all the crops before transplanting but reversed it after transplanting, except for lettuce. Based on the field performance, we found suitable R:S ranges for high-quality transplants to be as follows: 0.25–0.35 for pepper, 0.15–0.2 for tomato, 0.1 for watermelon, and 0.15–0.2 for lettuce. This study demonstrated that HS differentially modulated root and shoot growth based on crop species: root performances were outstanding in fruit-based crops (pepper, tomato, watermelon), while leaf performances were significantly improved in the leaf-based crop (lettuce). Overall, imposed heat and drought stresses had significantly negative effects on crop yield and average fruit weight, but HS-treated plants showed more improved stress tolerance than control plants by mitigating the yield loss. This study showed the potential application of solid humic substances as biostimulants for enhancing transplant quality and crop performance in four economically important vegetable species (tomato, pepper, watermelon, and lettuce).

Supplementary Materials: The following are available online at http://www.mdpi.com/2077-0472/10/7/254/s1, Figure S1: Vegetable crops and cultivars used in this study, Table S1: Abbreviations and their full names used in this study, Table S2: ANOVA and means comparison of germination percentage as affected by amendments (A), Table S3: Summary of HS effects on transplant quality traits (pre- and post-transplanting) and yield components (cool vs. hot seasons, low vs. high irrigation rates) compared to control (higher or lower at significant level $P \leq 0.1$).

Author Contributions: Conceptualization, K.Q. and D.I.L.; methodology, K.Q. and D.I.L.; software, K.Q.; validation, K.Q. and D.I.L.; formal analysis, K.Q. and D.I.L.; investigation, K.Q. and D.I.L.; resources, K.Q. and D.I.L.; data curation, K.Q.; writing—original draft preparation, K.Q.; writing—review and editing, D.I.L.; visualization, K.Q.; supervision, D.I.L.; project administration, D.I.L.; funding acquisition, D.I.L. All authors have read and agreed to the published version of the manuscript.

Funding: This research was funded by Novihum Technologies GmbH.

Acknowledgments: The material is based upon work that is supported by the National Institute of Food and Agriculture, United States Department of Agriculture, Multi-state Project W-4168. We thank Joshua T. Harvey for the proofreading and constructive comments on the paper, and Manuel Figueroa Pagan and Michael Tidwell for their assistance in field experiments.

Conflicts of Interest: The authors declare no conflict of interest.

References

1. Leskovar, D.I. Transplanting. In *The Physiology of Vegetable Crops*, 2nd ed.; Wien, H.C., Stützel, H., Eds.; CABI: Oxfordshire, UK, 2020; pp. 31–60.
2. Kerbiriou, P.J.; Stomph, T.J.; Van Bueren, E.T.L.; Struik, P.C. Influence of Transplant Size on the Above- and Below-Ground Performance of Four Contrasting Field-Grown Lettuce Cultivars. *Front. Plant Sci.* **2013**, *4*, 16. [CrossRef] [PubMed]
3. Li, X.; Zhong, Q.; Li, Y.; Li, G.-H.; Ding, Y.; Wang, S.; Liu, Z.; Tang, S.; Ding, C.; Chen, L. Triacontanol Reduces Transplanting Shock in Machine-Transplanted Rice by Improving the Growth and Antioxidant Systems. *Front. Plant Sci.* **2016**, *7*, 501. [CrossRef] [PubMed]
4. Russo, V.M. Biological Amendment, Fertilizer Rate, and Irrigation Frequency for Organic Bell Pepper Transplant Production. *HortScience* **2006**, *41*, 1402–1407. [CrossRef]
5. Leskovar, D.I.; Stoffella, P.J. Vegetable Seedling Root Systems: Morphology, Development, and Importance. *HortScience* **1995**, *30*, 1153–1159. [CrossRef]
6. Soundy, P.; Cantliffe, D.J.; Hochmuth, G.J.; Stoffella, P.J. Management of Nitrogen and Irrigation in Lettuce Transplant Production affects Transplant Root and Shoot Development and Subsequent Crop Yields. *HortScience* **2005**, *40*, 607–610. [CrossRef]
7. Leskovar, D.I.; Othman, Y.A. Low Nitrogen Fertigation Promotes Root Development and Transplant Quality in Globe Artichoke. *HortScience* **2016**, *51*, 567–572. [CrossRef]
8. Leskovar, D.I. Root and Shoot Modification by Irrigation. *HortTechnology* **1998**, *8*, 510–514. [CrossRef]

9. Leskovar, D.I.; Vavrina, C.S. Onion Growth and Yield are Influenced by Transplant Tray Cell Size and Age. *Sci. Hortic.* **1999**, *80*, 133–143. [CrossRef]
10. Hernández, R.; Eguchi, T.; Deveci, M.; Kubota, C. Tomato Seedling Physiological Responses Under Different Percentages of Blue and Red Photon Flux Ratios Using Leds and Cool White Fluorescent Lamps. *Sci. Hortic.* **2016**, *213*, 270–280. [CrossRef]
11. Pascual, J.; Ceglie, F.; Tüzel, Y.; Koller, M.; Koren, A.; Hitchings, R.; Tittarelli, F. Organic Substrate for Transplant Production in Organic Nurseries. A Review. *Agron. Sustain. Dev.* **2018**, *38*, 35. [CrossRef]
12. Jack, A.L.H.; Rangarajan, A.; Culman, S.W.; Sooksa-Nguan, T.; Thies, J.E. Choice of Organic Amendments in Tomato Transplants has Lasting Effects on Bacterial Rhizosphere Communities and Crop Performance in the Field. *Appl. Soil Ecol.* **2011**, *48*, 94–101. [CrossRef]
13. Canellas, L.P.; Olivares, F.L.; Aguiar, N.O.; Jones, D.L.; Nebbioso, A.; Mazzei, P.; Piccolo, A. Humic and Fulvic Acids as Biostimulants in Horticulture. *Sci. Hortic.* **2015**, *196*, 15–27. [CrossRef]
14. Hartwigsen, J.A.; Evans, M.R. Humic Acid Seed and Substrate Treatments Promote Seedling Root Development. *HortScience* **2000**, *35*, 1231–1233. [CrossRef]
15. Türkmen, Ö.; Dursun, A.; Turan, M.; Erdinç, Ç. Calcium and Humic Acid affect Seed Germination, Growth, and Nutrient Content of Tomato (Lycopersicon esculentum L.) Seedlings under Saline Soil Conditions. *Acta Agric. Scand. Sect. B Plant Soil Sci.* **2004**, *54*, 168–174. [CrossRef]
16. Osman, A.S.; Rady, M.M. Effect of Humic Acid as an Additive to Growing Media to Enhance the Production of Eggplant and Tomato Transplants. *J. Hortic. Sci. Biotechnol.* **2014**, *89*, 237–244. [CrossRef]
17. Brooks, S.D.; DeMott, P.J.; Kreidenweis, S.M. Water Uptake by Particles Containing Humic Materials and Mixtures of Humic Materials with Ammonium Sulfate. *Atmos. Environ.* **2004**, *38*, 1859–1868. [CrossRef]
18. Zhang, J.; Yin, H.; Wang, H.; Xu, L.; Samuel, B.; Chang, J.; Liu, F.; Chen, H. Molecular Structure-Reactivity Correlations of Humic Acid and Humin Fractions from a Typical Black Soil for Hexavalent Chromium Reduction. *Sci. Total Environ.* **2019**, *651*, 2975–2984. [CrossRef]
19. Gerke, J. Concepts and Misconceptions of Humic Substances as the Stable Part of Soil Organic Matter: A Review. *Agronomy* **2018**, *8*, 76. [CrossRef]
20. Allen, R.G.; Pereira, L.S.; Raes, D.; Smith, M. *Crop Evapotranspiration-Guidelines for Computing Crop Water Requirements*; Food and Agriculture Organization: Rome, Italy, 1998.
21. Rueden, C.T.; Schindelin, J.; Hiner, M.C.; Dezonia, B.E.; Walter, A.E.; Arena, E.T.; Eliceiri, K.W. ImageJ2: ImageJ for the Next Generation of Scientific Image Data. *Bmc Bioinform.* **2017**, *18*, 529. [CrossRef]
22. R: A Language and Environment for Statistical Computing, R Foundation for Statistical Computing. Available online: https://www.R-project.org (accessed on 11 September 2019).
23. Zaller, J.G. Vermicompost in Seedling Potting Media can affect Germination, Biomass Allocation, Yields and Fruit Quality of Three Tomato Varieties. *Eur. J. Soil Boil.* **2007**, *43*, S332–S336. [CrossRef]
24. Finch-Savage, W.E.; Bassel, G.W. Seed Vigour and Crop Establishment: Extending Performance Beyond Adaptation. *J. Exp. Bot.* **2015**, *67*, 567–591. [CrossRef]
25. Luo, Y.; Liang, J.; Zeng, G.; Chen, M.; Mo, D.; Li, G.; Zhang, D. Seed Germination Test for Toxicity Evaluation of Compost: Its Roles, Problems and Prospects. *Waste Manag.* **2018**, *71*, 109–114. [CrossRef] [PubMed]
26. Ninnemann, H.; Fischer, K.; Brendler, E.; Liebner, M.; Rosenau, T.; Liebner, F. Characterisation of Humic Matter Fractions Isolated from Ammonoxidised Miocene Lignite. *J. Biobased Mater. Bioenergy* **2011**, *5*, 241–252. [CrossRef]
27. Zherebker, A.; Kostyukevich, Y.I.; Kononikhin, A.S.; Nikolaev, E.N.; Perminova, I.V. Molecular Compositions of Humic Acids Extracted from Leonardite and Lignite as Determined by Fourier Transform Ion Cyclotron Resonance Mass Spectrometry. *Mendeleev Commun.* **2016**, *26*, 446–448. [CrossRef]
28. Francioso, O.; Sanchez-Cortes, S.; Tugnoli, V.; Marzadori, C.; Ciavatta, C. Spectroscopic Study (DRIFT, SERS and 1H NMR) of Peat, Leonardite and Lignite Humic Substances. *J. Mol. Struct.* **2001**, *565*, 481–485. [CrossRef]
29. Houghton, J.; Thompson, K.; Rees, M. Does Seed Mass Drive the Differences in Relative Growth Rate Between Growth Forms? *Proc. R. Soc. B Boil. Sci.* **2013**, *280*, 20130921. [CrossRef] [PubMed]
30. Grime, J.P.; Hunt, R. Relative Growth-Rate: Its Range and Adaptive Significance in a Local Flora. *J. Ecol.* **1975**, *63*, 393. [CrossRef]
31. Poorter, H.; Remkes, C. Leaf Area Ratio and Net Assimilation Rate of 24 Wild Species Differing in Relative Growth Rate. *Oecologia* **1990**, *83*, 553–559. [CrossRef]

32. Shipley, B. Net Assimilation Rate, Specific Leaf Area and Leaf Mass Ratio: Which is Most Closely Correlated with Relative Growth Rate? A Meta-Analysis. *Funct. Ecol.* **2006**, *20*, 565–574. [CrossRef]
33. Medek, D.E.; Ball, M.; Schortemeyer, M. Relative Contributions of Leaf Area Ratio and Net Assimilation Rate to Change in Growth Rate Depend on Growth Temperature: Comparative Analysis of Subantarctic and Alpine Grasses. *New Phytol.* **2007**, *175*, 290–300. [CrossRef]
34. Marcelis, L.F.M.; Heuvelink, E.; Goudriaan, J. Modelling Biomass Production and Yield of Horticultural Crops: A Review. *Sci. Hortic.* **1998**, *74*, 83–111. [CrossRef]
35. Mašková, T.; Herben, T. Root:Shoot Ratio in Developing Seedlings: How Seedlings Change their Allocation in Response to Seed Mass and Ambient Nutrient Supply. *Ecol. Evol.* **2018**, *8*, 7143–7150. [CrossRef] [PubMed]
36. Kubota, C. Growth, Development, Transpiration and Translocation as Affected by Abiotic Environmental Factors. In *Plant Factory*; Elsevier BV: Amsterdam, The Netherlands, 2016; pp. 151–164.
37. Ryser, P. The Mysterious Root Length. *Plant Soil* **2006**, *286*, 1–6. [CrossRef]
38. Yano, K.; Kume, T. Root Morphological Plasticity for Heterogeneous Phosphorus Supply in *Zea mays* L. *Plant Prod. Sci.* **2005**, *8*, 427–432. [CrossRef]
39. Li, H.; Ma, Q.; Li, H.; Zhang, F.; Rengel, Z.; Shen, J. Root Morphological Responses to Localized Nutrient Supply Differ Among Crop Species with Contrasting Root Traits. *Plant Soil* **2013**, *376*, 151–163. [CrossRef]
40. Kramer-Walter, K.; Bellingham, P.; Millar, T.; Smissen, R.; Richardson, S.J.; Laughlin, D.C. Root Traits are Multidimensional: Specific Root Length is Independent from Root Tissue Density and the Plant Economic Spectrum. *J. Ecol.* **2016**, *104*, 1299–1310. [CrossRef]
41. Kawasaki, Y.; Yoneda, Y. Local Temperature Control in Greenhouse Vegetable Production. *Hortic. J.* **2019**, *88*, 305–314. [CrossRef]
42. Prasad, P.V.V.; Staggenborg, S.A.; Ristic, Z. Impacts of Drought and/or Heat Stress on Physiological, Developmental, Growth, and Yield Processes of Crop Plants. In *Response of Crops to Limited Water: Understanding and Modeling Water Stress Effects on Plant Growth Processes*; Ahuja, L.R., Reddy, V.R., Saseendran, S.A., Yu, Q., Eds.; American Society of Agronomy, Crop Science Society of America, Soil Science Society of America: Madison, WI, USA, 2008; pp. 301–355. [CrossRef]
43. Drobek, M.; Frąc, M.; Cybulska, J. Plant Biostimulants: Importance of the Quality and Yield of Horticultural Crops and the Improvement of Plant Tolerance to Abiotic Stress—A Review. *Agronomy* **2019**, *9*, 335. [CrossRef]
44. Lucini, L.; Rouphael, Y.; Cardarelli, M.; Bonini, P.; Baffi, C.; Colla, G. A Vegetal Biopolymer-Based Biostimulant Promoted Root Growth in Melon While Triggering Brassinosteroids and Stress-Related Compounds. *Front. Plant Sci.* **2018**, *9*, 11. [CrossRef]

Article

Developing Production Guidelines for Baby Leaf Hemp (*Cannabis sativa* L.) as an Edible Salad Green: Cultivar, Sowing Density and Seed Size

Renyuan Mi [1], Alan G. Taylor [2], Lawrence B. Smart [2] and Neil S. Mattson [1,*]

1 Horticulture Section, School of Integrative Plant Science, Cornell University, Ithaca, NY 14850, USA; rm974@cornell.edu

2 Horticulture Section, School of Integrative Plant Science, Cornell AgriTech, Cornell University, Geneva, NY 14456, USA; agt1@cornell.edu (A.G.T.); lbs33@cornell.edu (L.B.S.)

* Correspondence: nsm47@cornell.edu; Tel.: +1-607-255-0621

Received: 25 October 2020; Accepted: 4 December 2020; Published: 9 December 2020

Abstract: Scientific literature is lacking on cultural practices of baby leaf hemp production even though hemp (*Cannabis sativa* L.) is a widely grown crop for fiber and grain. The objective of this study was to develop a standard protocol to optimize yield and quality of baby leaf hemp production: cultivar screening, sowing density and seed size. Fresh weight (FW) and germination percentage was significantly affected by cultivars. Cultivars 'Picolo' and 'X-59' had a greater FW mainly due to greater germination percentage. In the sowing density experiment, 'Ferimon' and 'Katani' were evaluated at five seed densities, 0.65, 1.2, 1.75, 2.3 and 2.85 seeds·cm^{-2} (42 to 182 seeds per cell). The FW and FW per plant (FWPP) had a positive quadratic response and negative quadratic response, respectively. Regarding seed size, cultivars 'Anka,' 'Ferimon' and 'Picolo' had the largest percentage of seeds, 26% to 30%, within the medium width size between 3.18 and 3.37 mm. Using the largest sized seeds (3.77 mm) increased FW by 34%, 26% and 23% as compared to non-sorted 'Anka', 'Ferimon' and 'Picolo' seeds, respectively. Overall, a greater understanding of cultivar selection, sowing density and seed-size distribution can promote greater yield and quality of baby leaf hemp as an edible salad green.

Keywords: baby leaf hemp; cultivar selection; sowing density; seed-size distribution

---

1. Introduction

Cannabis sativa L. has long been used for fiber, nutritional grain and medicinal purposes with an established history and widespread utilization. Hemp was recorded as a textile fiber used in China 6000 years ago [1]. Hemp was also found in Middle Eastern and Central Asia used for medicinal purposes [2] in antiquity. Since the start of the deregulation of hemp in the US by the 2014 Farm Bill, a rapid increase in the hemp production area arose in the US [3]. Hemp production tripled from 2017 (25,713 acres) to 2018 (78,176 acres) [4]. Since the 2018 Farm Bill, hemp was officially removed from schedule I of the Controlled Substance Act and hemp was defined as the *C. sativa* plant containing a tetrahydrocannabinol (THC) level less or equal than 0.3% [5]. Most state hemp pilot programs report that production acreage is primarily for cannabidiol (CBD) followed by grain and fiber.

Leafy greens are valued by consumers because they can offer high nutritional phytochemicals, fiber and mineral elements that humans need and also provide an appealing appearance, aroma, taste and texture to include in the daily diet [6]. Recently, baby leaf greens have increased in popularity with consumers over mature leaf greens because of their special flavor, freshness, convenience, and their bioactive compounds [7]. In the study by Xiao et al. [8], most young leaves of baby greens had higher levels of phytochemicals than mature leaves. Currently, baby leaf hemp has been grown as a niche edible salad green. Anecdotally, we have observed a few producers in New York State that have been able to grow and sell their baby leaf hemp in New York City for up to $30 per pound. Based on the potential high-profit return, commercial growers are interested in baby leaf hemp production in controlled environment agriculture facilities (high tunnels, greenhouses and vertical farms). Growers usually harvest it at emergence of the third true leaf, which is approximately 12 to 18 d after sowing depending on the temperature and lighting condition. There is no published data on cultural requirements to support growers in achieving an efficient baby leaf hemp crop. After communicating with baby leaf hemp growers, it was found that their greatest concerns are the choice of cultivar and production methods.

Field hemp cultivars can be classified into four market classes: fiber, grain, medicinal and ornamentals [9], and breeding for improved cultivars has been conducted in all four market classes [10]. However, there is no scientific literature available on cultivar selection for baby leaf hemp. Information on the biomass yield and germination percentage of different cultivars is crucial for the success of commercial baby leaf hemp growers. Elite cultivars would not only offer maximum yield in a relative short period but would also provide uniformity and high quality which could bring higher profit for growers and lower the cost of production.

The effect of cultivar selection on yield is critical for successful production. For example, the length of the vegetative period of hemp was different between Ukrainian cultivars and French cultivars. In northern Europe, the advantage of early maturing cultivars is a shortened vegetative growth period, resulting in grain production before the onset of winter. However, for fiber production, northern Europe requires late maturing cultivars with longer vegetative growth period to maximize strong stem production [11]. By comparing nine cultivars from Ukraine, Hungary and France, the greatest single plot dry stem yield was collected from the Hungarian cultivars 'Kompolti,' 'Unico B' and French cultivar 'Futura 77' [12]. In the broader leafy greens literature, nine cultivars were evaluated by Grahn et al. [13] for determining the performance of extended season production in northwest Washington state. The greatest marketable yield was obtained by pak choi (*Brassica rapa* ssp. *chinensis* (L.) Hanelt.) 'Joi Choi' and mustard (*Brassica juncea* (L.) Czern.) 'Komatsuna'. For leaf lettuce (*Lactuca sativa* L. var. *crispa*), five cultivars ('Bergamo', 'Dubáček', 'Frisby', 'Lollo Rossa' and 'Redin') were investigated for yield. A significant effect of cultivar was exhibited in leaf head weight which varied from 164 to 502 g [14].

Field hemp is usually sown in high density to encourage rapid canopy closure and suppress weed growth [15]. High seeding rates also can promote yield and quality of hemp fiber by reducing branching and increasing the proportion of bast fiber content in the stem [16,17]. Bennett et al. [18] sowed field hemp in two densities, 0.015 and 0.03 seeds·cm^{-2}, and found greater yield and better weed control with the larger sowing density for all cultivars. However, no information is available on baby leaf hemp sowing density, leaving producers to question how to maximize yield. There is literature available on sowing density for common greenhouse vegetables such as microgreens or baby leaf greens. In a microgreen study, three microgreen cultivars were evaluated with five seed densities (1.1, 1.65, 2.2, 2.75 and 3.3 seeds·cm^{-2}), resulting in a quadratic increase in fresh weight (FW) and quadratic decrease in fresh weight per plant (FWPP) as sowing density increased from 1.1 to 3.3 seeds·cm^{-2} [19]. The FW and FWPP were inversely correlated in a microgreen experiment evaluating three seeding rates (0.81, 1.62 and 2.37 seeds·cm^{-2}) [20]. Therefore, due to a wide variation in results based on species and harvest stage, the optimal sowing density for baby leaf hemp production should be investigated in order to provide an effective method for commercial growers.

Factors such as seed size are highly likely to affect the germination percentage and biomass yield of plants because small seeds contain less nutrition, which may lead to reduced seedling growth. Hemp seeds used in baby leaf production have typically been industrial hemp seed lots with non-sorted seed due to seed availability and cost. Germination and emergence of switchgrass increased nonlinearly as the seed size increased from five treatments (40, 50, 60, 70 and 80° air valve settings of a South Dakota seed blower) [21]. There is additional evidence that the higher germination percentage was obtained with larger seeds compared to smaller seeds for *Erica vagans* L. A 90% germination percentage was attained from larger sized seeds with faster germination rate compared to smaller seeds [22]. A higher quality of seedlings was achieved with larger seed size of *Calluna vulgaris* L. [22]. The seed size of *Virola koschnyi* Aubl. 'Warb' did not significantly influence the germination percentage, but plant vigor was improved by larger seeds [23]. In general, a larger seed mass was demonstrated to produce more vigorous plants due to a more developed embryo and larger energy reserves [24].

In our work, we consider baby leaf hemp as a potential crop in the greenhouse environment grown on a soilless substrate and with liquid fertilizer. In this regard, baby leaf hemp could be grown year-round in a consistent manner. The objective of our work is to determine the impact of cultivar, seeding density and seed size distribution on the yield and quality (morphology, emergence) on hemp seedlings with a goal of optimizing procedures for commercial production in controlled environments.

2. Materials and Methods

Seed of nine hemp cultivars was obtained from those entered in Cornell University (Ithaca, NY, USA) 2018 field trials and grown under pilot program research authorization from NYS Dept of Agriculture and Markets. Of these nine cultivars, five were dual-purpose (D) and four were for dedicated grain cultivars (G). Nine cultivars were purchased from UNISeeds (Cobden, ON, Canada), Assocanapa USA (Lexington, KY, USA), HGI (Saskatoon, SK, Canada), Legacy Hemp (Hastings, MN, USA) and Parkland (Dauphin, MB, Canada) (Table 1). Prior to our experiments, there was no available information on flavor, aroma or other quality attributes of baby leaf hemp. Therefore, in consultation with commercial growers, the initial selection of these nine cultivars was based on cost (including seed price and shipping) as well as availability. At the time of our experiment, the lowest cost cultivar was 'Anka' ($5.83/kg) and the most expensive cultivars were 'Canda' and 'Joey' (each at $22.11/kg). For all experiments the following common methods were used. Plants were grown in a single layer glass greenhouse located at Cornell University in Ithaca, NY (42° N latitude) under ambient light and placed on a bench made by galvanized steel elevated 85 cm from the floor. Each experimental unit was planted in a polystyrene cell (8 × 8 × 6 cm; Dillen-ITML Greenhouse, Twinsburg, OH, USA) placed on the bench. For all experiments, the substrate mix was a custom seeding mix (Jiffy Group, Zwijndrecht, Zuid-Holland, The Netherlands), which was a blend of OMRI (Organic Materials Review Institute) approved coconut coir, peat moss from Jiffy Canada (Lorain, OH, USA) and dolomitic limestone. Substrate nutrient analysis was conducted (J.R. Peter's Inc., Allentown, PA, USA) with the following values: 0.34 ppm nitrate (NO_3-N), 2 ppm of ammonium (NH_4-N), 2.31 ppm phosphorus (P), 82 potassium (K), 6 ppm calcium (Ca), 10 ppm magnesium (Mg), 4 ppm sulfur (S), 0.04 ppm boron (B), 0.24 ppm iron (Fe), 0.01 ppm manganese (Mn), 0.03 ppm copper (Cu), 0.01 ppm zinc (Zn), 0.08 ppm molybdenum (Mo), 0.12 ppm aluminum (Al), 35 ppm sodium (Na) and 179 ppm chloride (Cl). The initial substrate pH was 5.58 and EC (Electrical Conductivity) was 0.63 $dS{\cdot}m^{-1}$. The substrate mix was prepared by mixing it with RO (reverse osmosis) water in a 2:1 ratio by volume to achieve adequate moisture. Cells were filled to a 5 cm substrate depth, seeds were sown and an additional 1 cm of the same substrate was covered on the top of the seeds. All treatments received the same access to irrigation water with water soluble fertilizer by sub-irrigation by filling a flat to a 3 cm level with a 150 $mg{\cdot}L^{-1}$ N nutrient solution (21 N-2.2 P-16.6 K Jack's All-Purpose Liquid Feed, J.R. Peter's Inc., Allentown, PA, USA) and allowing each cell to take up water for 90 s before removing. The germination period (time to seedling emergence) usually took 48 to 96 h depending on the temperature and lighting conditions. Plants were harvested when half of the seedlings reached the stage of emergence of the third true leaf. The time period from seed to harvest was around 13 to 18 d depending on the temperature and lighting conditions. Temperature and relative humidity during each experiment and crop cycle are listed in Table 2. In the experiments, an experimental unit was considered to be one 8 × 8 cm cell. Measurements were collected for germination percentage (number of seedlings emerged divided by total sown seeds), height (from the surface of the substrate to the tallest part of representative seedlings), and fresh weight (FW, using only the epicotyl, i.e., part of the plant above the cotyledons, based on commercial practice). After harvesting for FW, epicotyls from all plants in an experimental unit were bagged and placed in a 70 °C oven for 72 h to determine dry weight (DW). The fresh weight per plant (FWPP) was calculated as FW divided by number of seedling emergence for each treatment.

Table 1. Nine hemp cultivars used in this study with their sources and seed lot number.

Cultivars	Source	Seed Lot Number
'Anka'	UNISeeds	4371-6032
'Canda'	Parkland	828-17-04
'Ferimon'	UNISeeds	F1545X64110
'Joey'	Parkland	828-17-18S
'Katani'	HGI	15-DPKARE-01
'Picolo'	HGI	980-8-16KEFR-PICE-01
'USO-31'	Assocanapa USA	F1545R154001B
'Wojko'	Assocanapa USA	MH/2914/WOJ1
'X-59'	Legacy Hemp	2208.8

Table 2. Average Daily Temperature (ADT, °C) and Average Daily Relative Humidity (ADRH, %) in three cycles of each experiment.

Experiment	Crop Cycle	ADT (°C)	ADRH (%)
Cultivar selection	1	20	60
	2	19	66
	3	22	76
Sowing density	1	22	76
	2	23	80
	3	23	81
Seed size	1	19	41
	2	19	46
	3	19	37

2.1. Cultivar Selection Experiment

Cultivars 'Anka', 'Katani', 'Ferimon', 'Wojko', 'USO-31', 'X-59', 'Picolo', 'Canda' and 'Joey' were evaluated in this experiment (Figure 1). Seeds of these nine different cultivars were sown evenly on the substrate and covered with 1 cm of the same substrate with a sowing density at 1.2 seeds·cm^{-2} (i.e., 77 seeds per cell). For each crop cycle there were six experimental units. The time of the three crop cycles from seed sowing until harvest were 16th April to 29th April, 30th April to 13rd May and 18th June to 1st July, 2019, respectively. Plants were harvested when the third true leaf emerged for at least 50% of the plants, and we measured FW, DW, height, and germination percentage (as described above).

Figure 1. Typical appearance of nine baby leaf hemp at harvest stage (front view).

2.2. Sowing Density Experiment

Cultivars 'Ferimon' and 'Katani' were selected to evaluate sowing density of baby leaf hemp at five different seeding rates: 0.65, 1.2, 1.75, 2.3 and 2.85 seeds·cm^{-2} (i.e., 42, 77, 112, 147 and 182 seeds·cell^{-1}). The experiment was repeated over three crop cycles which took place: 19th June to 2nd July, 9th July to 20th July and 20th July to 31st July, 2019. Plants were harvested at the third true leaf stage and all parameters described above with FWPP were recorded for each experimental unit.

2.3. Seed Size Experiment

Cultivars 'Anka', 'Picolo' and 'Ferimon' were selected to evaluate the seed-size distribution and the effect of seed size on measured parameters. Seeds were sorted by size using eight different 23 cm × 23 cm hand screen sieves (Seedburo Equipment Company, Des Plaines, IL, USA) with round perforations for sorting by seed width. Seeds were sized through a series of stacked sieves from largest to smallest and seeds retained on a particular sieve were grouped: 3.77 mm (i.e., sieve size 9.5/64 inches), 3.57 mm (9/64 inches), 3.37 mm (8.5/64 inches), 3.18 mm (8/64 inches), 2.98 mm (7.5/64 inches), 2.78 mm (7/64 inches), 2.58 mm (6.5/64 inches) and 2.83 mm (6/64 inches). The distribution of 100 g of seed from each cultivar according to the sieve sizes was collected and analyzed (in terms of percent of seeds in each class by weight) with three replicates. In the evaluation of the effect of seed size on biomass/yield of baby leaf hemp, there were seven treatments according to seed width size: control (non-sorted seeds), sieve size 3.77, 3.57, 3.18, 2.98 and <2.98 mm. Sieved seeds of three entries from each treatment were sown evenly at a density 0.47 seeds·cm^{-2}. The experiment was repeated three times with seeding date and harvesting dates of: 8 December to 26 December 2019; 30 December 2019, to 17 January 2020; and 21 January to 8 February 2020, respectively. Plants were harvested at the third true leaf stage and all parameters mentioned above were recorded for each experimental unit.

2.4. Statistical Analysis

In cultivar selection experiment, for each crop cycle there were six blocks placed in a randomized complete block design where each block contained one experimental unit from each of the nine different cultivars. In the sowing density experiment, there were five experimental units per cultivar and sowing density treatments arranged in a randomized complete block design where each block consisted of one experimental unit from each cultivar at each of the five seed densities. Within a block, the 10 cells were completely randomized. For seed size experiment, each crop cycle had five blocks where each block consisted of one experimental unit per cultivar per seed size treatment. The experiment was also arranged in a randomized complete block design. The block was based on location in the greenhouse bench. All three experiments were replicated over time for a total of three crop cycles. Data were analyzed with R studio (Version 1.2.1335, RStudio, Inc., Boston, MA, USA), using a mixed model including linear and quadratic regression (when treatments followed a quantitative independent variable, i.e., sowing density), Analysis of Variance (ANOVA), and mean separation comparison by Tukey's honestly significant difference test (alpha = 0.05) with the following packages in R studio: library(ggplot2), library(multcomp), library(emmeans), library(lsmeans), library(lme4).

3. Results

3.1. Cultivar Selection Experiment

There was a significant difference for all measured parameters in response to cultivar, crop cycle, and cultivar × crop cycle interaction but there was no significant difference due to block except for fresh weight (Table 3). There was no significant cultivar × block interaction, except dry weight (Table 3).

Table 3. Analysis of Variance on measured parameters for baby leaf hemp.

ANOVA Table					
Main Effects & Interactions	**Germination**	**Height**	**Fresh Weight**	**Dry Weight**	**Fresh Weight Per Plant**
Cultivar	***	***	***	***	***
Block	NS	NS	*	NS	NS
Crop Cycle	***	***	***	***	***
Cultivar × Block	NS	NS	NS	*	NS
Cultivar × Cycle	***	**	**	**	*

NS, *, **, *** Nonsignificant or significant at $p \leq 0.05$, 0.01 or 0.001, respectively.

There was a significant difference observed between the nine cultivars for germination percentage (Table 3), which ranged from 51% ('Wojko') to 81% ('Picolo') (Table 4). 'Picolo', 'X-59' and 'Ferimon' had a significantly greater germination percentage than 'USO-31', 'Canda', 'Wojko' and 'Joey'. There were no significant differences between 'Anka' and 'Katani' for germination percentage, but they had a significantly lower germination percentage than 'Picolo' and 'X-59'. Therefore, regarding germination, 'Picolo' and 'X-59' had good germination capacity among these nine cultivars for the seed lots tested in this experiment.

Table 4. Growth parameters of nine baby leaf hemp cultivars. Data represent means of 18 experimental units (3 crop cycles each with 6 experimental units per treatment).

Cultivars	Germination Percentage (%)	Height (cm)	Fresh Weight (g·cell^{-1})	Dry Weight (g·cell^{-1})	Fresh Weight Per Plant (g·plant^{-1})
'Anka'	67 ^{c,d}	12.4 ^{b}	8.6 ^{b,c}	1.0 ^{c,d}	0.17 ^{b}
'Canda'	58 ^{e,f}	11.8 ^{b}	8.7 ^{b,c}	1.1 ^{b,c}	0.20 ^{a}
'Ferimon'	72 ^{b,c}	12.3 ^{b}	9.7 ^{a,b}	1.1 $^{a–c}$	0.18 ^{a,b}
'Joey'	53 ^{f}	11.5 ^{b}	6.9 ^{d}	0.8 ^{e}	0.17 ^{a,b}
'Katani'	70 $^{b–d}$	13.7 ^{a}	9.7 ^{a,b}	1.2 ^{a,b}	0.18 ^{a,b}
'Picolo'	81 ^{a}	12.5 ^{b}	10.3 ^{a}	1.3 ^{a}	0.16 ^{b}
'USO-31'	64 ^{d,e}	12.5 ^{b}	8.9 ^{b}	1.0 ^{b,c}	0.19 ^{a,b}
'Wojko'	51 ^{f}	11.8 ^{b}	7.5 ^{c,d}	0.8 ^{d,e}	0.19 ^{a}
'X-59'	75 ^{a,b}	12.0 ^{b}	9.8 ^{a,b}	1.8 ^{a,b}	0.17 ^{b}

Letters represent mean separation comparison using Tukey's HSD (alpha = 0.05).

There was no significant difference between the nine cultivars for height except 'Katani', which was significantly taller than other cultivars. The range of height of these nine cultivars was between 11.5 and 13.7 cm (Table 4). A significant difference between the nine cultivars was observed for FW, which ranged from 6.9 g·cell^{-1} to 10.3 g·cell^{-1} (Table 4). 'Picolo' had the greatest FW and was 33% larger than 'Joey', which had the smallest FW. 'Picolo' had a significantly greater FW than five cultivars ('USO-31', 'Canda', 'Ank', 'Wojko' and 'Joey'). 'X-59,' 'Katani', 'Ferimon' and 'USO-31' had a significantly larger FW than 'Wojko' or 'Joey'. The FW (per cell) was closely related to germination percentage, so germination percentage seems to be an important attribute for obtaining a good yield.

The DW results closely followed FW. Dry weight varied from 0.78 to 1.27 g·cell^{-1} between the nine cultivars (Table 4). The greatest DW was for 'Picolo', which was 39% greater than the lowest cultivar, 'Joey.' 'Picolo', 'X-59' and 'Katani' exhibited the greatest DW and these were significantly greater than 'Anka', 'Wojko' and 'Joey'.

The range of FWPP among nine cultivars was from 0.163 to 0.196 g·plant^{-1} (Table 4). 'Canda' and 'Wojko' had a significantly larger FWPP than 'X-59', 'Anka' and 'Picolo'. No significant difference in FWPP was observed between 'USO-31', 'Katani', 'Ferimon' and 'Joey'.

3.2. Sowing Density Experiment

In the investigation of optimal sowing density on baby leaf hemp production, no significant effect of sowing density was observed on the germination of 'Ferimon' or 'Katani'. In general, 'Ferimon' had a significantly greater germination percentage than 'Katani'. The mean germination percentage of 'Ferimon' and 'Katani' was 69% and 63%, respectively.

There was no significant effect of sowing density on height of 'Katani' or 'Ferimon'. Overall, 'Katani' was slightly taller (mean of 14.7 cm across treatments) than 'Ferimon' (mean of 13.3 cm across treatments) (Table 5). FW of both 'Katani' and 'Ferimon' showed a positive response to sowing density, whereby FW increased quadratically as sowing density increased from 0.65 to 2.85 seeds·cm^{-2} (Figure 2; Table 6). There was no significant interaction between cultivar and sowing density (Table 5). The increasing FW response to sowing density began to plateau at the two greatest densities (Figure 2). For example, FW of 'Katani' increased 28% as density increased from 0.65 to 1.2 seeds·cm^{-2}, but increased by only 11% as density increased from 2.3 to 2.85 seeds·cm^{-2}. Similarly, for 'Ferimon', FW increased 27% as density increased from 0.65 to 1.2 seeds·cm^{-2}, but increased by only 7% as density increased from 2.3 to 2.85 seeds·cm^{-2}.

Table 5. Analysis of Variance for sowing density parameters and model-accounted variability R^2.

ANOVA Table					
Main Effect and Interaction	**Germination**	**Plant Height**	**Fresh Weight**	**Dry Weight**	**Fresh Weight Per Plant**
Density	NS	NS	***	***	***
Density^2	NS	NS	**	NS	**
Cultivar	*	*	NS	NS	**
Cultivar × Density	NS	NS	NS	NS	NS
R^2	0.193	0.843	0.926	0.888	0.869

NS, *, **, *** Nonsignificant or significant at $p \leq 0.05$, 0.01 or 0.001, respectively. R^2 represented model-accounted variability for each parameter.

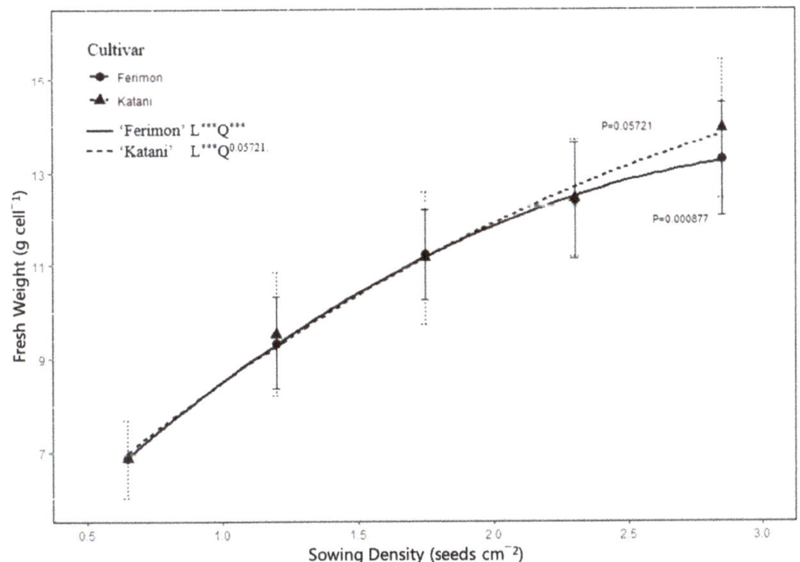

Figure 2. Fresh weight per cell of baby leaf hemp 'Ferimon' and 'Katani' in response to increasing sowing density. Data represent mean ± SE (Standard Error) of 15 experimental units (3 crop cycles each with 5 experimental units per treatment per cultivar). Significance of linear (L) and quadratic (Q) regression represented as ***, significant at $p \leq 0.001$.

Table 6. Regression values for growth parameters by sowing density for 'Katani' and 'Ferimon', including squared terms for a quadratic model.

Coefficients	Estimate [A]	Std. Error [A]	p-Value [B]
Fresh Weight			
'Katani'			
Intercept	11.6	1.10	***
Density	5.3	1.18	***
Density^2	−0.641	0.330	0.057
'Ferimon'			
Intercept	11.0	0.876	***
Density	6.11	0.933	***
Density^2	−0.919	0.262	***
Dry Weight			
'Katani'			
Intercept	1.26	0.149	***
Density	0.641	0.158	***
Density^2	−0.068	0.044	NS
'Ferimon'			
Intercept	1.21	0.133	***
Density	0.669	0.142	***
Density^2	−0.085	0.040	*
Fresh Weight Per Plant			
'Katani'			
Intercept	0.501	0.028	***
Density	−0.195	0.030	***
Density^2	0.035	0.008	***
'Ferimon'			
Intercept	0.436	0.022	***
Density	−0.129	0.023	***
Density^2	0.020	0.007	***

NS, *, **, *** Nonsignificant or significant at $p \leq 0.05$, 0.01 or 0.001, respectively. [A] Estimated response when all other treatment effects are equal to zero [B] Significance when treatment effects are at their average value.

For DW, a similar pattern was found as in FW, in which there was also a positive response to sowing density. 'Ferimon' displayed both significant linear and quadratic response to DW as the sowing density increased (Figure 3; Table 6). The increase of DW as sowing density increased also plateaued at the maximum sowing density. For example, DW of 'Ferimon' increased 37% as sowing density increased from 0.65 to 1.2 seeds·cm^{-2}, but DW only had a 9% increase as sowing density increased from 2.3 to 2.85 seeds·cm^{-2}. However, for 'Katani', the linear regression represented a better fit than a quadratic regression (Figure 3). For 'Katani', DW increased from 0.85 to 1.79 g as sowing density increased from 0.65 to 2.85 seeds·cm^{-2}.

Both cultivars had a significant response of FWPP to sowing density. No significant interaction between cultivar and sowing density occurred, but there was a significant difference between cultivars (Table 5). 'Katani' had a significantly larger FWPP than 'Ferimon'. There was a significant quadratic decrease in FWPP of both 'Katani' and 'Ferimon' as the sowing density increased from 0.65 to 2.85 seeds·cm^{-2} (Figure 4; Table 6). For example, for 'Katani', the FWPP decreased 29%, 22%, 17% and 11% when the sowing density increased from 0.65 to 1.2, 1.2 to 1.75, 1.75 to 2.3 and 2.3 to 2.85 seeds·cm^{-2}, respectively. For 'Ferimon', the FWPP decreased 25%, 17%, 16% and 15% as the sowing density increased from 0.65 to 1.2, 1.2 to 1.75, 1.75 to 2.3 and 2.3 to 2.85 seeds·cm^{-2}, respectively.

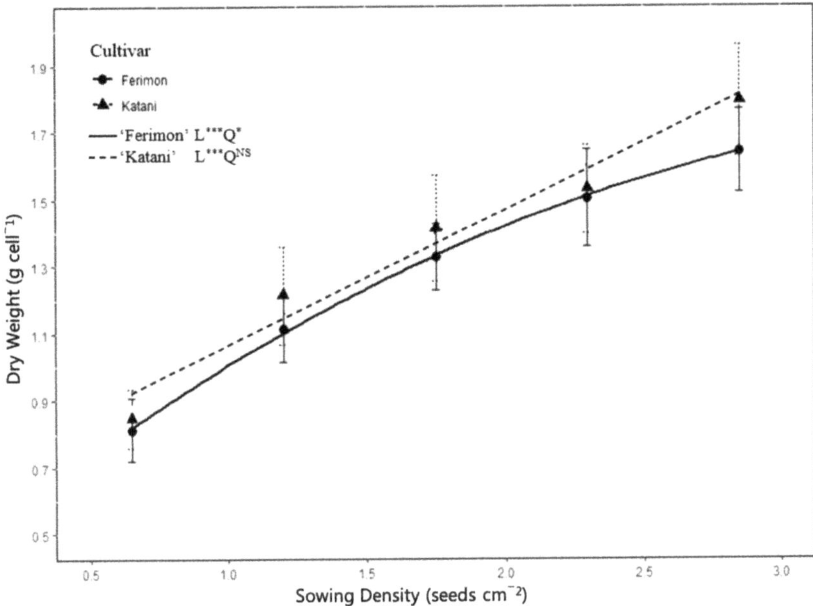

Figure 3. Dry weight per cell of baby leaf hemp 'Ferimon' and 'Katani' in response to increasing sowing density. Data represent mean ± SE of 15 experimental units (3 crop cycles each with 5 experimental units per treatment per cultivar). Significance of linear (L) and quadratic (Q) regression represented as NS, *, ***, Nonsignificant or significant at $p \leq 0.05$ or 0.001, respectively.

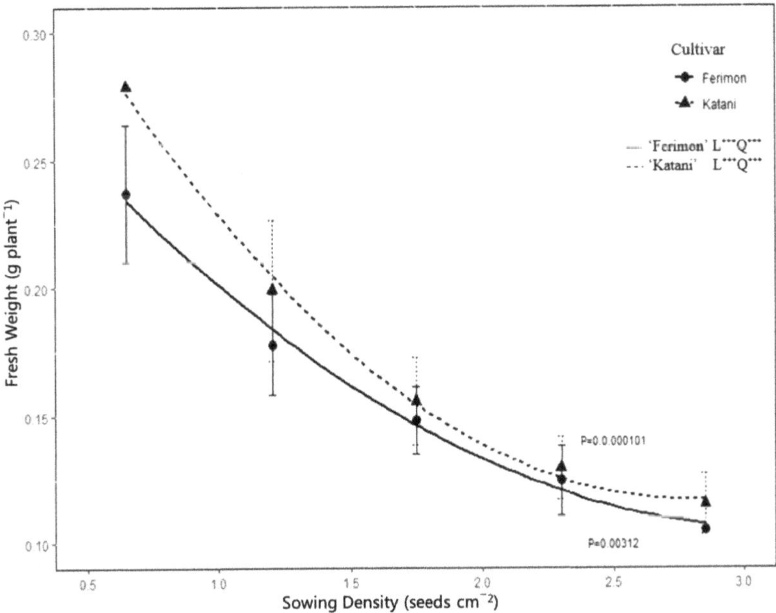

Figure 4. Fresh weight per plant of baby leaf hemp 'Ferimon' and 'Katani' in response to increasing sowing density. Data represent mean ± SE of 15 experimental units (3 crop cycles each with 5 experimental units per treatment per cultivar). Significance of linear (L) and quadratic (Q) regression represented as ***, significant at $p \leq 0.001$.

3.3. Seed Size Experiment

To determine the effect of seed size and distribution on yield and quality of baby leaf hemp seedlings, we evaluated the seed-size distribution. It exhibited a similar pattern for all three cultivars studied. The greatest percentage of seeds (by weight) were in the category of sieve size 3.18 and the lowest percentage of seeds were in the category of sieve size <2.38 mm width (Figure 5). Seeds with sieve size greater than 2.78 mm width represented 90% to 94% of all seeds.

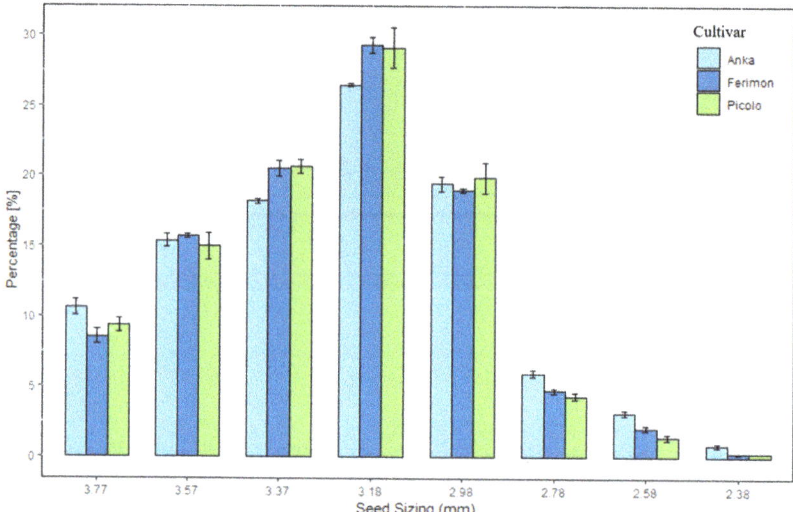

Figure 5. Hemp seed-size distribution for 'Anka', 'Ferimon' and 'Picolo'. Data are means ± SE of three seed lots (each consisting of 100 g) per cultivar.

A significant effect of seed size treatment and cultivar was observed for all growth parameters (Table 7). There were significant interactions between seed size and cultivar for FW and Height. There were also significant effects of block and cycle for all parameters except germination.

Table 7. Analysis of Variance for seed size parameters.

ANOVA Table					
Main Effects & Interactions	**Germination**	**Height**	**Fresh Weight**	**Dry Weight**	**Fresh Weight Per Plant**
Treatment	***	***	***	***	***
Cultivars	***	***	***	***	***
Block	NS	*	***	***	***
Cycle	NS	***	***	***	***
Treatment × Cultivars	NS	**	*	NS	NS
Treatment × Block	NS	NS	NS	NS	NS
Treatment × Cycle	NS	NS	NS	NS	*

NS, *, **, *** Nonsignificant or significant at $p \leq 0.05$, 0.01 or 0.001, respectively.

The effect of seed size was shown for each parameter of each cultivar [25]. Averaged for the three cultivars, seed size <2.98 mm had the lowest germination compared to all other sizes and not the non-sorted control (Figure 6). The largest sized seed fractions had greater dry weight and fresh weight per plant, while the smallest size fractions had lower dry weight and fresh weight per plant compared to the control (Figure 6). For height, a significant interaction between seed size treatments and cultivars was found, and the height of 'Anka', 'Ferimon' and 'Picolo' ranged from 7.07 to 10.0, 8.6 to 10.3 and 8.5 to 10.4 cm, respectively (Table 7; Figure 7). For 'Anka' and 'Ferimon', the largest seed size of 3.77 mm was significantly taller than non-sorted treatment, but not for 'Picolo'. Only seed size less than 2.98 mm had a shorter height than non-sorted treatment among three cultivars (Figure 7).

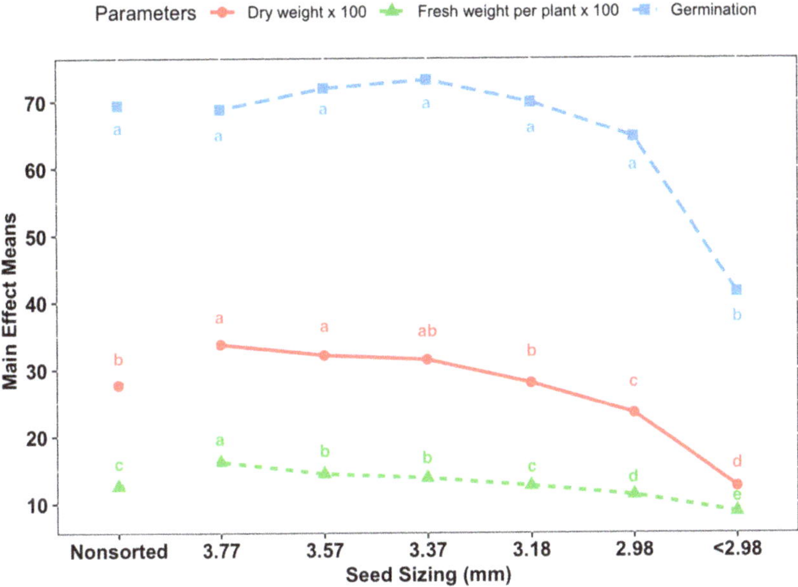

Figure 6. Main effect means of growth parameters for 'Anka', 'Ferimon' and 'Picolo' in response to seed size treatment. Data represent mean of 15 experimental units (3 crop cycles each with 5 experimental units per treatment). Letters represent mean separation comparison using Tukey's HSD (alpha = 0.05).

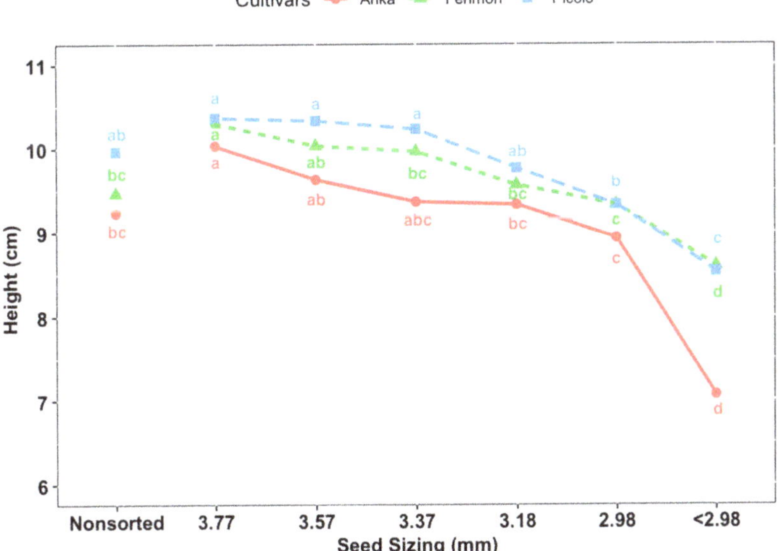

Figure 7. Height of baby leaf hemp cultivars 'Anka', 'Ferimon' and 'Picolo' in response to seed size treatment. Data represent mean of 15 experimental units (3 crop cycles each with 5 experimental units per treatment). Letters represent mean separation comparison using Tukey's HSD (alpha = 0.05).

Overall, there was a pattern of declining FW as seed size decreased. 'Ferimon' had the greatest FW among three cultivars. 'Ferimon' and 'Picolo' showed a consistent trend in FW with respect to seed size effect (Figure 8). However, 'Anka' had an unexpectedly lower FW for seed size 3.57 mm. Seed size 3.77 and 3.57 mm had a significantly larger FW than non-sorted seeds for 'Ferimon' and 'Picolo', but for 'Anka', only seed size 3.77 mm had a significantly larger FW than the non-sorted treatment. The largest-sized seeds (3.77 mm) had a FW that was 35%, 27% and 23% greater than non-sorted seeds for 'Anka', 'Ferimon' and 'Picolo,' respectively (Figure 8). For 'Ferimon' and 'Picolo', seed size less than 2.98 mm had a significantly smaller FW than non-sorted treatment, but for 'Anka', seed size of <2.98 or 2.98 mm had a significantly smaller FW than non-sorted treatment. 'Anka,' 'Ferimon' and 'Picolo' had FW reductions of 71%, 55% and 49% for the smallest sized seeds, respectively, relative to non-sorted seeds.

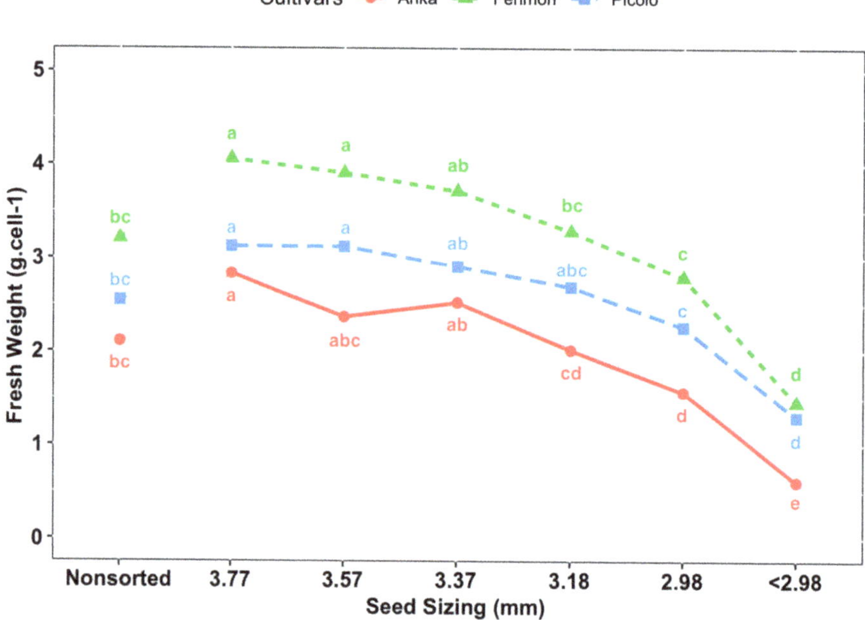

Figure 8. Fresh weight of baby leaf hemp cultivars 'Anka', 'Ferimon' and 'Picolo' in response to seed size treatment. Data represent mean of 15 experimental units (3 crop cycles each with 5 experimental units per treatment). Letters represent mean separation comparison using Tukey's HSD (alpha = 0.05).

4. Discussion

4.1. Cultivar Selection Experiment

The objective of cultivar selection was to determine which field hemp dual or grain cultivars produce the greatest quality (germination and FW) when grown as an edible salad green. 'Anka' was used as an industry standard cultivar for a number of field hemp experiments by our colleagues at Cornell University due to seed availability and cost, but the seed lot we used showed poor performance in germination. In our study, we found that germination percentage was highly correlated with FW ($p = 0.0002$, $R^2 = 0.858$) and DW ($p = 0.0005$, $R^2 = 0.816$) among the nine cultivars for the seed lots tested. Because the baby leaf hemp production cycle is short (to reach emergence of the third true leaf), the yield performance was highly related to the germination percentage of each cultivar. We propose that each sown seed producing a vigorous seedling contributes more to the final yield than other factors such as height of the plant. Therefore, for commercial baby leaf hemp growers, selection of cultivars with good germination is critical for success. In our study, 'Picolo', 'X-59', 'Ferimon' and 'Katani' were good performing cultivars based on germination and subsequent yield. Regarding cultivar selection, while there is no previous research with baby leaf hemp, other studies with baby leaf greens or microgreens illustrated that cultivar selection has a large effect on yield enhancement [13,14]. In an examination of 10 microgreen species, there was a large variability in germination time with 1 to >14 d required to reach 75% germination as well as a large range of germination percentage between 10% to 98% under 12 h light and dark, respectively. In another example, the yield of perennial wall rockets (*Diplotaxis tenuifolia* (L.) DC.) was affected by cultivars, which illustrates the importance of cultivar studies for optimizing marketable weight [11]. In fiber hemp, Lisson and Mendham [12] determined cultivars for the greatest single plot dry stem yield. Except for cultivar selection for yield, the success of growing fiber hemp also depended on the selection of cultivars less sensitive to photoperiod and cultivation on well drained sites [12]. In our study, selecting elite cultivars with high quality of germination and fresh weight yield can contribute to optimization of baby leaf hemp production practices. While we determined response in terms of yield and germination, other attributes such as sensory preferences and nutritional values were impacted by cultivar [25]. Therefore, both best production practices and consumer preferences need to be considered to achieve success in the baby greens market. There could be a variation of quality between seed lots in one cultivar. In my experiment, the limitation of time only allowed me to repeat the experiment three times with the same seed lot for each cultivar. In future work, more evaluation of these cultivars with different seed lots should be conducted to further confirm the results.

4.2. Sowing Density Experiment

Usually commercial seed companies provide recommended seeding rates for growers, but because baby leaf hemp is a new niche crop, there is currently no information on this subject. Although the research by Bennett et al. [18] used seeding rates for fiber hemp in low and high seed rates, 0.015 and 0.03 seeds·cm^{-2}, respectively, our seeding rates were 22 (0.65 seeds·cm^{-2}) to 95 (2.85 seeds·cm^{-2}) times greater. Therefore, we adapted information from other baby leaf green species such as arugula, mizuna (*Brassica rapa nipposinica*—(L.H. Bailey.) Hanelt.) and mustard, which were sown at a seeding rate from 1.1 to 3.3 seeds·cm^{-2}. The FW of these microgreens also demonstrated a quadratic increase in yield as sowing density increased [19] with a similar pattern whereby diminishing increases in FW/DW were observed at the highest densities. In the evaluation of the microgreen table beet (*Beta vulgaris* L.), the commercially recommended sowing density 201 g·m^{-2} could lead to a greater shoot fresh weight per m^2 than treatments with lower seeding rates [20]. However, a negative aspect found by Murphy et al. [20] was that at higher sowing density resulted into a lower biomass shoots, which is similar to our finding of lower FWPP as density increases. In our experiment, we found that density did not affect height, but since it affected FWPP, at higher density we have thinner and lighter plants that are more prone to lodging. Three microgreen species in a culinary assessment study were utilized at a sowing density of 3 seeds·cm^{-2} [26]. Because the hemp seeds were slightly larger than these Brassicaceae microgreen seeds, we used sowing density from 0.65 to 2.85 seeds·cm^{-2}. While 2.85 seeds·cm^{-2} led to the greatest FW/DW in our study (Figures 2 and 3), we did observe an elevated susceptibility of disease (gray-colored mold on the surface of substrate) in the maximum seeding rate which was not reflected in the data. Hemp is very susceptible to diseases such as gray mold (*Botrytis cinerea*), hemp canker (*Fusarium* spp.) and damping off (mostly caused by *Pythium* spp.) [27]. Therefore, based on yield results of our experiment, the recommended sowing density for commercial growers would be 2.3 to 2.85 seeds·cm^{-2}. We observed that increased sowing density extended the time period to harvest. We harvested all plants at the same time during the third true leaf emergence for the average of plants in all treatments, but it was obvious that with a lower sowing density, plants tended to grow faster and thicker with more true leaves formed. Future work should examine the effect of sowing density on seeding development rate, further measures of quality (stem thickness, disease incidence) and the interaction with other cultural factors (for example, at higher density greater air circulation may be required to prevent foliar-borne pathogens).

4.3. Seed Size Experiment

For conventional baby leaf and microgreens crops, seed companies typically sell seed lots of uniform size for commercial growers. However, little seed sizing was observed in the hemp seed lots we used, possibly as this is a new product niche. Based on our observation of previous experiments, the performance of baby leaf hemp including biomass yield and germination was impacted by different sized seeds. Smaller seeds also tend to produce more abnormal seedlings or result in inferior germination. Abnormal seedlings exhibited an absence of lateral roots and true leaf formation (Figure 9). After depletion of nutrients in cotyledons, cotyledons tended to be shriveled. We found that larger-sized seeds performed better (germination, FW, FWPP) than non-sorted seed lots or lower-sized seeds. Other researchers had also linked seed size to crop performance, with factors such as germination and seedling quality [21,22]. Lettuce seeds were reported to have a positive relationship between seed size and germination percentage and seedling vigor [28]. Based on our findings, we also believe that hemp seed lots with more uniform size distribution would have greater yield and quality than non-sorted seed lots. This is because seeds of different size tend to have more variable germination which would then lead to disparities in size and performance, such as larger seeds emerging earlier and shading smaller seeds. For commercial growers, we recommend sieving and discarding the smallest portion of seeds (less than 2.98 mm width) in order to maximize the yield. A limitation of seed lot effect should be considered in the future research. We used the same seed lot with three crop cycles in the seed size experiment, so the effect of seed size on plant performance could also be affected by the seed lot effect. Further studies should be conducted with the hemp seed size and seedlings to understand the effect of seed size on seedling quality (leaf area or stem thickness), nutrient level (phytochemical or mineral nutrients) and interaction with fertilizers (whether large-sized seeds may require less fertilizer to obtain the optimal yield).

(a) (b)

Figure 9. Images of abnormal 'Picolo' seedling (**a**) and as compared to a healthy seedling (**b**).

5. Conclusions

The success of baby leaf hemp production requires optimal environmental conditions and cultural management. Based on the results of our studies, and the seed lots we have access to, we recommend growers consider 'Picolo', 'Ferimon', 'X-59', and 'Katani' as productive baby leaf hemp cultivars with high germination percentage that exhibit relatively high yield potential. Similarly, other field hemp cultivars available on the market may be suitable for baby leaf production if they are disease-free and have a high germination rate. Although the sensory analysis and nutrition assessment has not been studied yet, these cultivars could bring the optimal production (fresh weight) for growers to ensure their profit. For commercial sowing density, we recommend 2.3 to 2.85 seeds·cm^{-2} when harvested at the stage of emergence of the third true leaf. If harvested at an earlier or later development stage, alternative seed densities would need to be studied. Although there was a slight yield benefit with greater than 2.3 seeds·cm^{-2}, there was increased incidence of disease observed at the highest density, especially in some growing conditions (high relative humidity or poor airflow). Finally, we found that seed size affected germination and yield in baby leaf hemp production. Around 3% to 25% yield increase could be achieved by sieving and discarding the smallest portion of seeds (<2.98 mm), depending on cultivar (Figure 8). In this research, cultural production methods of baby leaf hemp were developed as a foundational study, but the understanding of baby leaf hemp growing method is still limited. The effect of other management techniques, including light quantity, light quality, CO_2 enrichment and fertilizer practices, should be studied for baby leaf hemp in future work.

Author Contributions: Investigation, experimental design, data collection and analysis, writing and original draft preparation, R.M.; conceptualization on all aspects of the project, data interpretation and editing the manuscript, N.S.M.; conceptualization on seed size study, data interpretation and editing the manuscript, A.G.T.; conceptualization on variety selection, data interpretation and editing the manuscript, L.B.S. All authors have read and agreed to the published version of the manuscript.

Funding: This research received no external funding.

Acknowledgments: This material is based upon work that is supported by the United States Hatch Funds under Multi-state Project W-4168 under accession number 1007938 to the second author. This work was supported by New York State Department of Agriculture and Markets through grants AC477 and AC483 from Empire State Development Corporation.

Conflicts of Interest: The authors declare no conflict of interest.

References

1. Li, H.L. An archaeological and historical account of cannabis in China. *Econ. Bot.* **1974**, *28*, 437–448. [CrossRef]
2. Russo, E.B. History of cannabis and its preparations in saga, science, and sobriquet. *Chem. Biodivers.* **2007**, *4*, 1614–1648. [CrossRef]
3. Hemp in the Farm Bill, What Does It Mean? Available online: https://www.votehemp.com/hemp-news/hemp-in-the-farm-bill-what-does-it-mean/ (accessed on 12 August 2020).
4. U.S. Hemp Crop Report. 2018. Available online: https://www.votehemp.com/wp-content/uploads/2019/01/Vote-Hemp-Crop-Report-2018_nobleed.pdf (accessed on 12 August 2020).
5. McConnell, M.; Wyden, R.; Merkley, J.; Paul, R. *Hemp Farming Act of 2018*; United States Congress: Washington, DC, USA, 2018.
6. Liu, S.; Manson, J.E.; Lee, I.M.; Cole, S.R.; Hennekens, C.H.; Willett, W.C.; Buring, J.E. Fruit and vegetable intake and risk of cardiovascular disease: The Women's Health Study. *Am. J. Clin. Nutr.* **2000**, *72*, 922–928. [CrossRef] [PubMed]
7. Martínez-Sánchez, A.; Luna, M.C.; Selma, M.V.; Tudela, J.A.; Abad, J.; Gil, M.I. Baby-leaf and multi-leaf of green and red lettuces are suitable raw materials for the fresh-cut industry. *Postharvest Biol. Technol.* **2012**, *63*, 1–10. [CrossRef]
8. Xiao, Z.; Lester, G.E.; Luo, Y.; Wang, Q. Assessment of vitamin and carotenoid concentrations of emerging food products: Edible microgreens. *J. Agric. Food Chem.* **2012**, *60*, 7644–7651. [CrossRef] [PubMed]
9. De Meijer, E.P.M. Fibre hemp cultivars: A survey of origin, ancestry, availability and brief agronomic characteristics. *J. Int. Hemp Assoc.* **1995**, *2*, 66–73.
10. Salentijn, E.M.; Zhang, Q.; Amaducci, S.; Yang, M.; Trindade, L.M. New developments in fiber hemp (*Cannabis sativa* L.) breeding. *Ind. Crops Prod.* **2015**, *68*, 32–41. [CrossRef]
11. Senevirathne, G.I.; Gama-Arachchige, N.S.; Karunaratne, A.M. Germination, harvesting stage, antioxidant activity and consumer acceptance of ten microgreens. *Ceylon J. Sci.* **2019**, *48*, 91–96. [CrossRef]
12. Lisson, S.N.; Mendham, N.J. Cultivar, sowing date and plant density studies of fibre hemp (*Cannabis sativa* L.) in Tasmania. *Aust. J. Exp. Agric.* **2000**, *40*, 975–986. [CrossRef]
13. Grahn, C.M.; Benedict, C.; Thornton, T.; Miles, C. Production of baby-leaf salad greens in the spring and fall seasons of northwest Washington. *HortScience* **2015**, *50*, 1467–1471. [CrossRef]

14. Koudela, M.; Petříková, K. Nutrients content and yield in selected cultivars of leaf lettuce (*Lactuca sativa* L. var. *crispa*). *Hortic. Sci.* **2008**, *35*, 99–106. [CrossRef]
15. Ontario Ministry of Agriculture and Food. Available online: http://www.omafra.gov.on.ca/english/crops/facts/00-067.htm#economics (accessed on 2 October 2020).
16. Struik, P.C.; Amaducci, S.; Bullard, M.J.; Stutterheim, N.C.; Venturi, G.; Cromack, H.T.H. Agronomy of fibre hemp (*Cannabis sativa* L.) in Europe. *Ind. Crops Prod.* **2000**, *11*, 107–118. [CrossRef]
17. Van der Werf, H.M.; Wijlhuizen, M.; De Schutter, J.A.A. Plant density and self-thinning affect yield and quality of fibre hemp (*Cannabis sativa* L.). *Field Crops Res.* **1995**, *40*, 153–164. [CrossRef]
18. Bennett, S.J.; Snell, R.; Wright, D. Effect of variety, seed rate and time of cutting on fibre yield of dew-retted hemp. *Ind. Crops Prod.* **2006**, *24*, 79–86. [CrossRef]
19. Allred, J.A. Environmental and Cultural Practices to Optimize the Growth and Development of Three Microgreen Species. Master's Thesis, Cornell University, Ithaca, NY, USA, 2017.
20. Murphy, C.J.; Llort, K.F.; Pill, W.G. Factors affecting the growth of microgreen table beet. *Int. J. Veg. Sci.* **2010**, *16*, 253–266. [CrossRef]
21. Aiken, G.E.; Springer, T.L. Seed size distribution, germination, and emergence of 6 switchgrass cultivars. *J. Range Manag.* **1995**, *48*, 455–458. [CrossRef]
22. Vera, M.L. Effects of altitude and seed size on germination and seedling survival of heathland plants in north Spain. *Plant Ecol.* **1997**, *133*, 101–106. [CrossRef]
23. Gonzalez, E.J. Effect of seed size on germination and seedling vigor of *Virola koschnyi* Warb. *For. Ecol. Manag.* **1993**, *57*, 275–281. [CrossRef]
24. Hartmann, H.T.; Kester, D.E.; Davies, F.T.; Geneve, R. *Hartmann & Kester's Plant Propagation: Principles and Practices*; Pearson Education UK: Harlow, UK, 2013.
25. Mi, R. Cultural Management, Production and Consumer Sensory Evaluation of Baby Leaf Hemp (*Cannabis sativa* L.) as an Edible Salad Green. Master's Thesis, Cornell University, Ithaca, NY, USA, 2020.
26. Renna, M.; Di Gioia, F.; Leoni, B.; Mininni, C.; Santamaria, P. Culinary Assessment of Self-Produced Microgreens as Basic Ingredients in Sweet and Savory Dishes. *J. Culin. Sci. Technol.* **2017**, *15*, 126–142. [CrossRef]
27. McPartland, J.M. A review of Cannabis diseases. *J. Int. Hemp Assoc.* **1996**, *3*, 19–23.
28. Sharples, G.C. The Effects of Seed Size on Lettuce Germination & Growth. *Prog. Agric. (Arizona)* **1970**, *22*, 10–11.

Review

Modern Seed Technology: Seed Coating Delivery Systems for Enhancing Seed and Crop Performance

Irfan Afzal [1], Talha Javed [1], Masoume Amirkhani [2] and Alan G. Taylor [2,*]

1. Seed Physiology Lab, Department of Agronomy, University of Agriculture, Faisalabad 38040, Pakistan; iafzal@uaf.edu.pk (I.A.); talhajaved54321@gmail.com (T.J.)
2. Cornell AgriTech, School of Integrative Plant Science, Horticulture Section, Cornell University, Geneva, New York, NY 14850, USA; ma862@cornell.edu

* Correspondence: agt1@cornell.edu

Received: 9 October 2020; Accepted: 3 November 2020; Published: 5 November 2020

Abstract: The objective of modern seed-coating technology is to uniformly apply a wide range of active components (ingredients) onto crop seeds at desired dosages so as to facilitate sowing and enhance crop performance. There are three major types of seed treating/coating equipment: dry powder applicator, rotary pan, and pelleting pan with the provisions to apply dry powders, liquids, or a combination of both. Additional terms for coatings produced from these types of equipment include dry coating, seed dressing, film coating, encrustments, and seed pelleting. The seed weight increases for these different coating methods ranges from <0.05% to >5000% (>100,000-fold range). Modern coating technology provides a delivery system for many other materials including biostimulants, nutrients, and plant protectants. This review summarizes seed coating technologies and their potential benefits to enhance seed performance, improve crop establishment, and provide early season pest management for sustainable agricultural systems.

Keywords: seed enhancement; seed treatment; seed dressing; seed coating; film coat; pellet; organic agriculture

---

1. Introduction

High seed quality is always demanded by farmers and may result in up to a 30% increase in crop yields [1,2]. Sowing high-quality seeds is essential, but their use does not guarantee successful stand establishment. The difference in time between sowing and stand establishment is a crucial period. Seeds may be exposed to a wide range of biotic and abiotic stresses resulting in decreased stand performance [3]. However, judicious use of chemical, biochemical, and biological seed treatments can protect and enhance establishment, growth and potential productivity [4]. In this review, seed treatments refer to materials that are active components, while seed dressings are the minimal coating that results after the application of seed treatments onto seeds. Seed treatments are most effective when they are objective oriented and crop specific to ensure optimal stand establishment and enhance yields under changing climatic conditions [5].

Seed treatments may be applied commercially by the seed industry or in some cases "on farm" for crop protection and enhanced seedling growth [2,6]. There is also a growing trend for the development and use of organically approved treatments for sustainable agriculture. Collectively, innovative seed coating technologies are needed as delivery systems for the application of active ingredients at effective dosages to crop seeds [7,8].

A brief history of seed treatments for plant protection illustrates the practical need for better delivery systems and improved ability to sow seeds [9]. Copper sulphate was found to be an effective seed treatment for bunt on cereals in the 1800s when applied as a soak. However, treating large quantities of seed required subsequent drying that made the process cumbersome and time consuming. The soaking process was replaced by the "heap" or "barn floor" method where a small amount of liquid was sprinkled over the seed and then mixed [9]. The soaking (also known as steeping) method is still in use for sugar beet seed using the method described by Halmer (2000) [10].

In 1866, a technique was developed to improve sowing of cotton seed using a paste of wheat flour to form a pellet [11]. During the mid-20th century, many coating technologies for improved agricultural productivity were developed and reviewed by Jeffs (1986) [9]. Seed coating technology continued to advance through the 1970s to 1990s and reviewed by Taylor and Harman (1990), Scott (1989) and Hill (1999) [7,12,13]. More recent reviews focus on seed enhancements and seed coating equipment in the 21st century by Taylor (2003), Pedrini et al. (2017), Halmer (2000), and Pedrini et al., (2020) [6,8,10,14].

Seed enhancements may be defined as post-harvest treatments that improve germination or seedling growth or facilitate the delivery of seeds and other materials required at time of sowing [15]. Seed coating is used for the application of biostimulants, plant nutrients, (including inoculants) and other products that will ameliorate biotic and abiotic stresses encountered after sowing [11,16].

The global market for seed coating materials (colorants, polymers, fillers and other additives) in 2019 was US $1.8 billion and is forecasted to reach $3.0 billion by 2025 [17]. The major group of active ingredients are chemical seed treatments estimated between $3 to $5 billion in 2020, and accounts for at least 2/3 of the total seed treatment market [18]. The biological seed treatment market includes a wide range of biologicals including biofertlizers, biopesticides and biostimulants [19]. The biological seed treatment market is estimated between $1 to $1.5 billion in 2020, and bioinoculants are the dominant group with about 70% of total [18].

The focus of this review is the use of selected seed coating components, including liquids and solid particulates, with designated seed coating equipment and technology for uniform delivery of treatments over seeds uniformly. Applications of selected seed treatment and coatings are presented as biostimulants, nutrients, and in management of abiotic and biotic stress. Seed coating technologies described may be applied to a wide range of crop seeds: grains, oilseed, vegetable, ornamentals, and other seed species [20].

There is considerable research and development by industry in the broader field of seed treatments, and much of this technology is proprietary. Many biological seeds treatments are being developed and marketed for pest management and as biostimulants. However, it is beyond the scope of this publication to critically review the merits and efficacy of these biologicals, though they are used commercially. Therefore, this review focuses on published papers and most are from refereed journals. This paper contains 112 references with 97 published papers or book chapters, 6 patents, 6 websites and 3 personal communications cited. Moreover, to provide relevancy to the seed coating industry, eight companies were acknowledged to provide valuable input in preparation of this review.

2. Seed Treatment Active Components and Other Coating Materials

A wide range of materials is used in seed treatments and coatings. These materials were categorized by their composition and origin as synthetic chemicals (SYN), natural products or derivatives from natural products (NP), biological agents (BIO) and minerals mined from the earth (MIN) (Table 1). Among these categories, particular materials may be used for organic use and labelling, and the US Organic Materials Review Institute (OMRI) [21] approved materials were noted as organic (*OR*). Seed treatment and coatings are further characterized by function, as active components, liquids or solid particulates.

Table 1. Seed treatment and coating materials grouped as active components, liquids and solid particulates. Each group of material is further classified by function and composition. Abbreviations for material source/origin: Synthetic Chemicals—SYN, Natural products or derivatives—NP, Biologicals—BIO, Mineral—MIN, substances may be Organically approved—*OR*.

Active Components	Liquids	Solid Particulates
Biostimulants	**Water Colorants**	**Binders**
• SYN, NP, BIO *(OR)*	• SYN, NP *(OR)*	• Also, under Liquids • Soy flour: NP *(OR)*
Plant nutrients	**Adjuvants**	**Fillers**
• SYN, MIN *(OR)*	• SYN *(OR)*	• Diatomaceous earth (DE): MIN *(OR)* • Limestone: MIN *(OR)* • Gypsum: MIN *(OR)* • Bentonite: MIN *(OR)* • Vermiculite: MIN *(OR)* • Talc: MIN *(OR)* • Zeolite: MIN *(OR)* • Silica: MIN *(OR)* • BaSO4: MIN
Abiotic stress: Drought and Salinity	**Binders**	
• SYN, BIO *(OR)*	• Polyvinyl alcohol (PVOH) and • Polyvinyl acetate (PVAc): SYN • Methyl cellulose: SYN • Carboxymethyl cellulose (CMC): SYN • Plant starches: NP *(OR)* • Gum Arabic: NP *(OR)*	
Plant Protectants		
• SYN, NP, BIO, MIN *(OR)*		
Inoculants		
• BIO *(OR)*		

2.1. Active Components

The purpose of active ingredients is aimed at protecting and enhancing seed and seedling performance in terms of germination, growth and development. The mode of action of the active ingredient dictates its role for protection and/or enhancement [16]. Active ingredients discussed in this paper include biostimulants, plant nutrients, protectants from abiotic and biotic stress, and inoculants (Table 1). Seed protectants are the most widely used group of ingredients for controlling pathogens and pests at the time of sowing. Fungicides, insecticides, nematicides, and bactericides are grouped as protectants [22]. Selected fungal and/or bacterial microorganisms are used commercially for plant protection, and as inoculants for nitrogen fixation [22,23]. Abiotic stresses due to saline soil conditions or drought stress may occur after sowing and selected biological and synthetic seed treatments may be applied in the seed coating to alleviate these stresses. Elicitors are being investigated as active components for pest management [24–26], and drought stress [27]. There is increased interest and demand for biostimulant- and nutrient based seed treatments [8]

2.2. Liquids

Active components must be applied to seeds so that they adhere onto seeds throughout storage until planted. In addition, seeds treated with pesticides must easily be recognized as treated. Colorants are commonly used to indicate that seeds are treated and constitute about 60% of coating ingredient components, and in the case of seed pelleting are applied at the end of coating process [8]. Colorants also provide a visual of assessment of application uniformity, and cosmetic appearance. Water is the universal carrier of liquids that are atomized onto seeds during the coating process, and atomization is best achieved with low viscosity liquids. The proportion of water in the applied liquid is adjusted to maintain low solution viscosity. Adjuvants are used [20] as most chemical seed treatment active ingredients have limited water solubility, so surfactants are needed to produce aqueous seed treatment formulations. Surfactants may serve as an active component, and a seed coating technology with surfactants was documented to enhance germination and stand establishment when sown in water repellent soils [28].

Seed coating binders act as adhesives to adhere treatments to seeds. The binder provides the coating integrity during and after drying. They prevent cracking and dusting off during handling and sowing [2]. Commonly used binders (Table 1) for maintaining physical integrity of seeds are: polyvinyl alcohol [29], polyvinyl acetate [30], methyl cellulose [31], and carboxymethyl cellulose [32]. For organic seed coatings plant starches (maltodextrins) [33] and gum Arabic [34] are commonly used.

Most binders are commonly referred to as polymers [35]. In preparing binders in water, solution viscosity must be low for complete atomization of the liquid onto seeds, based on the fourth author's experience preferably <100 centipoise (cP), or <0.1 pascal-second (Pa-s).

2.3. Solid Particulates

Solid particulates are the bulking materials used in seed coating technologies and form the physical coating after drying [7,30]. Solid particulates may also be binders. Solid particulate binders are applied as fine powders and become hydrolyzed as water is applied during the coating process. Fillers are also fine powders and can be mixed with the solid particulate binders to produce a seed-coating blend. Successful seed pelleting depends upon the optimization and selection of the most appropriate filler materials that do not interfere with germination [32].

Filler materials are generally inexpensive, non-toxic, easily available, and produce a uniform coating surface texture that should not impede radicle emergence [6]. Several filler materials are used for seed pelleting including diatomaceous earth [36], limestone, gypsum [32], bentonite [34], vermiculite [37], talc [38], zeolite [32], silica sand [39] and barium sulphate [40] (Table 1). These fillers are generally mineral materials that are mined from the earth with minimal modification except for grinding to obtain a fine powder size used in seed coating. Particle size should pass through a 200-mesh sieve (<75 µm) for uniform distribution over the seed surface based on the fourth author's experience.

3. Seed Coating Equipment and Methods

The seed treatment and coating materials described in Section 2 provides an extensive list of potential ingredients. The next step in the seed coating process is when selected ingredients are applied with appropriate equipment to produce the final coated product. The selection of seed coating equipment and coating method is determined primarily by the dosage of actives, liquids and solid components applied per unit of seed. There are three major types of seed coating equipment used today: dry coating, rotary pan and pelleting pan (Figure 1). This coating equipment used singly or in some cases in tandem is paired with five coating methods: dry powder, seed dressing, film coating, encrusting and pelleting [6,8,10]. The overall goal of all coating equipment and methods is to achieve good application uniformity and adherence. Processes should not cause mechanical injury to seeds during coating [35].

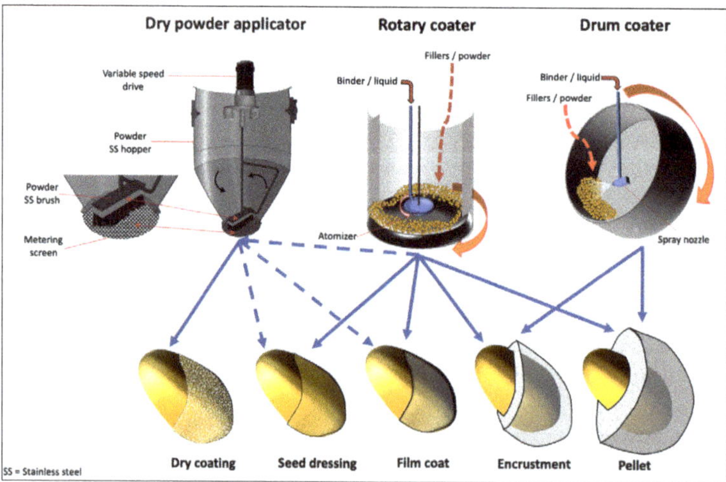

Figure 1. The three major types of seed coating equipment: dry powder applicator, rotary coater and drum coater used to produce five seed coatings: dry coating, seed dressing, film coat, entrustment and seed pellet.

3.1. Dry Powder Coating

Dry powder application is a seed coating method used for mixing seeds with a dry powder. The older term for this application method is "planter box" treatment [6]. Dry powders, also known as dusts, [20] are used for fungal or bacterial treatments followed by drying (hydration/dehydration) and seeds can have a shorter shelf-life after application [6]. This technology can be conducted on-farm for the application of labeled treatments for the control of a pests [9].

Dry powder application equipment and technology has evolved to allow for more precise loading of material onto seeds. As can be seen in Figure 1 [41,42] a rotating stainless-steel brush sifts a powder material through a metering screen (Figure 1). The equipment is calibrated on a weight basis to deliver powder to a given weight of seed. The seed is not shown in the illustration, but would be moving underneath the dry powder applicator via most delivery systems (auger, conveyor, seed tender, etc.) [https://www.ctapplicators.com] [43]. This equipment is used for stand-alone dry powder application, or for the application of finishing powders after seed dressing or film coating (described in Sections 3.2 and 3.3). Another dry powder feeder equipment uses a computer-controlled auger with hopper vibrator to deliver coating powders, finishing powders or dry powder actives to seed by volumetric or weight basis [44]. Dry powder carriers may act as lubricants to improve seed flowability by reducing seed-to-seed friction in the planter [6]. The most common dry powders are talc and graphite [45], and recent research revealed that soy-based protein is an environmentally friendly and cost-effective seed lubricant that improves flow and singulation during planting without creating dust [46]. Thus, the use of soy-based protein has the potential to reduce the risk of negative impact on pollinators and people.

The dosage of dry coating powders applied to seeds is limited by their adherence onto seeds, and ranges from 0.06 to 1.0% of seed weight (Figure 2). This loading rate is inversely proportional to seed size, and the amount of powder retained increases as seed size decreases due to the increase in seed surface area of smaller seeds [45].

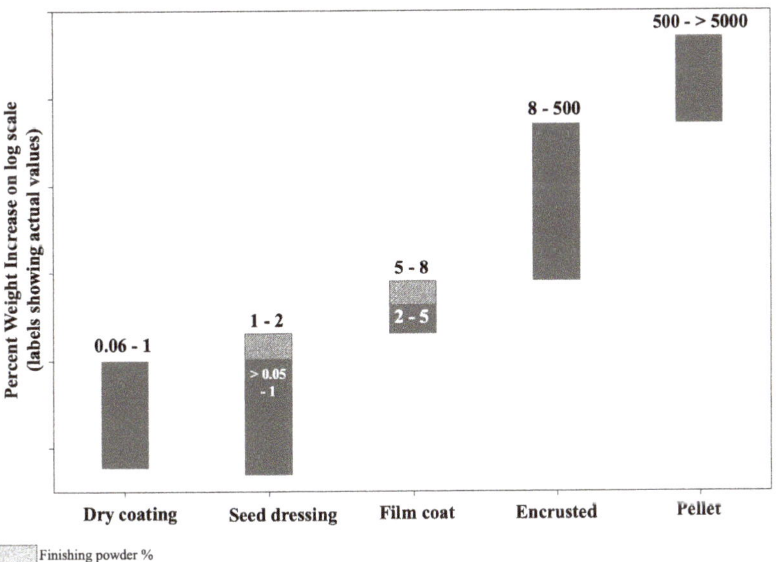

Figure 2. Percent weight increase after dry coating, seed dressing, film coat, entrustment and seed pellet technologies. The grey shaded bar for seed dressing and film coat is addition of a finishing powder during the coating process. The percent weight increase shown on a log scale to aid comparison between technologies.

3.2. Seed Dressing

Seed dressing is the most widely used method for low dosages of active components onto seeds [33]. Although there are many types of equipment used for coating [9], the most commonly used device is the rotary coater (Figure 1). Liquids are applied onto a spinning disc and atomized onto seeds that are spinning inside a metal cylinder, then the freshly treated seeds are discharged. A wide range of active materials especially chemical plant protectants can be applied with this method.

The dosage of liquid seed treatment formulations typically ranges from <0.05 to 1.0% by weight (Figure 2). For higher loading rates of chemical seed treatment, in particular insecticides, finishing powders or fluency powders are added immediately after the liquid application to absorb excess liquid [45]. The dry finishing powders can be added into the rotary coater during operation or applied immediately downstream with the dry seed coating equipment (Figure 1).

3.3. Film Coating

Film coating originally developed for the pharmaceutical and confectionary industries was adapted as a seed coating method [6]. Film coating consists of producing a continuous thin layer over the seed surface. The rotary coater is the primary seed coating equipment used for film coating (Figure 1). Film coating polymers (liquid components) are formulated to dissolve/dispense active ingredient prior to application on seeds. Film coating resulted in 90% application recovery [7], with little modification of shape and size during this process [7,8]. Film coating has gained in use and is the most adaptable among all seed applied technologies. The performance of film-coated seed is evaluated on the basis of germination and dust control. Film coating improves flow-ability of seed during treating/processing and sowing operations. This value-added treatment is preferred over conventional methods due to excellent delivery of protectants on value seeds and have a cosmetic appearance [6].

The weight increases for film-coated seed, ranges from 2 to 5% of seed weight (Figure 2) [16]. Seed weight build-up greater than 5% requires other seed coating equipment with drying capability during coating, primarily a ventilated pan and fluidized bed seed coating facilitate concurrent treating and drying [10]. However, both side-ventilated or perforated pan and fluidized bed are used much less in commercial practice than the rotary pan technology. As described for seed treatment (Section 3.2), dry finishing powders can be added into the rotary coater, or with the dry seed coating equipment to increase loading from 5–8% (Figure 2).

The choice of film forming polymers is important for success in field sowing [5,13] and in the protection of the environment. Corn seeds coated with a proprietary film-forming polymer, PolySeed CF (Rigrantec, Porto Alegre, RS, Brazil), improved precision seed placement compared to graphite treated or non-coated seeds with significant reduction in dust formation and leaching of applied insecticides [47]. Further, the film coating polymer had good seed treatment adhesion resulting in less dust-off into the environment [47].

3.4. Encrusting

Encrusting is a seed coating method with the addition of liquids and solid particulates that results in a coated seed that is completely covered, but the original seed shape is retained [16]. Encrusted seeds can be referred to as mini-pellets [6] or sometimes as coated seeds. The primary coating methods to produce encrusted seed are the rotary coater or coating pan (Figure 1). The addition of large amounts of water during encrusting requires that the freshly coated seed be dried to back to its original seed moisture content prior to packaging and storing. The weight increase after encrusting can range from 8 to 500% (Figure 2).

Encrusted seeds have been shown to improve seedling emergence. Significantly higher germination of fescue seeds was measured when seeds were encrusted before storage compared to encrusting after storage or non-treated seeds [48]. The seed coating thickness or percent build-up may impact germination rate, and encrusted seed requires more time to germinate as compared to film-coated seed [49]. The amount of binder used in producing encrusted coatings changes mechanical properties including integrity, compressive strength and time to disintegrate after soaking [50].

3.5. Pelleting and Agglomeration

Seed pelleting is a continuation of the encrusting coating process resulting in even greater build-up so that the original size or shape of the coated crop seed is not visible [8,16]. The materials and techniques used for this purpose are proprietary [8], but common mineral materials cited in the literature and in patents are presented (Table 1). The binders may be liquid or formulated as dry powders (Table 1). Dry powder binders are mixed with filler materials to produce a coating blend [51], only requiring water applied during the coating process as the liquid. The percent weight increase after pelleting and drying ranges from 500 to >5000 percent (Figure 1). It is common that the percent weight increase is expressed as a ratio of seed weight to dried pellet weight, so a 500% weight increase is a 1:5 build-up of seed to coating.

The selection of liquids paired with fillers (Table 1) is essential to ensure that the pelleted seed will germinate unimpeded by the pellet matrix [16]. The pelleting seed industry has conducted tremendous research and development on optimizing commercial pelleting products for growers. The demand of pelleted seed continues to grow among growers so seeds can be planted with precision. Precise seed spacing achieved with pelleted seed reduces the need for thinning operations. Pelleted seeds are commonly used for growing transplants. Pelleting is frequently performed on high-value, small-seeded horticultural crops (e.g., onion, lettuce, carrot, tobacco, and tomato [6,32,34,36].

Material properties for successful pelleting include particle size distribution, porosity, water absorbing and holding capacity and lack of toxicity [32]. For tobacco seed pelleting, a combination of bentonite and talc [38] or pumice [52] was highly recommended. Similarly, diatomaceous earth and a combination of gypsum and calcium carbonate were found to be effective in broccoli [53] and lettuce [32], respectively. Calcium peroxide was added as a seed coating component [12] after sowing in a water-saturated soil with limited oxygen availability, the calcium peroxide releases oxygen gas to the germinating seed. Calcium peroxide applied in a seed pellet improved emergence and crop establishment of rice under submerged conditions [54].

Pelleting requires the most time and expertise compared to other coating technologies due to extensive application of active components, liquids, and solid particulates (Table 2). the pellet should not cause any restriction to germination when sown in the field. pellet integrity is dependent on the selection of material (fillers and binders) and appropriate technology [7].

Table 2. Comparison of amount of coating components and time needed for the dry coating, seed dressing, film coat, entrustment and seed pellet technologies. The (+) for seed dressing and film coating is the addition solid particulates as finishing powders. Relative comparisons are noted with number '+'.

Coating Technology	Active Components	Liquids	Solid Particulates	Time Needed to Treat/Coat
Dry powder	+	0	+	+
Seed dressing	+	+	0 (+)	+
Film coating	++	+	0 (+)	+
Encrusting	+++	+++	+++	+++
Pelleting	++++	++++	++++	++++

The objective of all the described coating methods thus far is for each seed to be singulated during the coating process to avoid doubles or agglomerates (two or more seeds in one coated propagule). However, it may be needed in certain cases to have more than one seed in a pellet. Seed agglomeration is an alternative coating technology in which multiple seeds are pooled into a single delivery unit [36]. The purpose of this technology is to sow multiple seeds of the same seed lot, different varieties of the same crop or multiple seed species. Seed agglomerates may be produced with a pan coater or rotary coater (Figure 2). Other agglomeration technologies use extrusion equipment [14] and molding technology [48]. Moreover, producing "seed balls" is a pelleting technique that utilizes materials, seeds and supporting additives in small amounts such as mineral fertilizer [55]. Both seed agglomeration technologies are used for improving handling and sowing of small-seeded species for arid land restoration [56,57].

3.6. Comparison of Seed Treatment and Coating Technologies

Five seed treatment and coating technologies were discussed in Sections 3.1–3.5, and now each technology can be compared to provide relative differences. The range of weight increase after treatment/coating is shown for the five methods and is expressed on a log scale to better visualize percent weight increase or build-up (Figure 2). The coating technologies cover from <0.05% to >5000% weight increase (>100,000-fold range) that accommodates all crop seed specific treatment and coating needs and applications. Additional comparisons of the five coating methods are illustrated with respect to weight increase after coating, and the relative amounts of active components, liquids and solid particulates applied, and the time required to treat or coat a batch of seeds (Table 2). All coating technologies can apply active components, but the potential amount per unit seed is limited by coating technology. No water or liquids are applied with the dry powder method, while with the other coating methods the amount of water/liquids increase is proportional to the percent weight increase (Figure 2). Solid particulates may be added with seed dressing and film coating as the amount of water increases resulting in "stickiness" during seed treating and inadvertent agglomeration. The solid particulates are termed drying powders [20], finishing powders or fluency agents that help absorb excess moisture applied during coating. As stated previously, these drying powders can also serve as seed lubricants to reduce friction as seed flows through the seed treater or planter [20]. There is a clear distinction in choice of seed coating technology with respect to the amount of water applied. Seed dressing and film coating as described do not require further drying after treatment, while encrusting and pelleting require post-coating drying to remove excess water and to dry seeds to their original seed moisture content. Finally, each seed dressing/coating method requires time, and longer processing times are needed as the amount of coating materials increases.

Many factors affect the final coated seed properties including the rotator and atomizing disc rpm, the solid particulate particle size, porosity, water holding capacity, and the binder adhesion properties [49]. The success of coating process and uniform distribution of active components requires time for mixing in the coating equipment [35] and for accurate adherence of binder and powder to seeds [34].

There are two types of seed treatment/coating equipment systems: batch treater and continuous flow treaters [20]. A batch treater matches a known amount of seed with seed treatment and coating material at one time, while the continuous flow treats a known amount of seed with seed treatment and coating material at a given flow rate [35]. Dry powder applicator or rotary coater may be either a batch or continuous flow based in equipment design, while most drum coater technology used for small-seeded vegetable crop seeds is performed on a batch basis. Each seed coating method (Figure 1) requires precise metering to deliver the target dosage onto seeds. Seed treatment equipment is needed to proportion an accurate amount of material to the seed. Computer technology is often used to monitor seed flow and seed treatment application, known as proportion control [35]. There are two stages of seed treatment application to achieve uniformity of application from seed to seed: primary and secondary application [35]. Primary application is the direct application of liquids onto seeds, for example the atomizer (atomizing disk) in the rotary coater disperses liquids directly onto seeds (Figure 1). Secondary application is the seed-to-seed transfer of the applied material during mixing while in the seed coating equipment [35]. Dosage can be expressed on a weight basis, for example g/100 kg seed or quantity per seed, for example mg ai/seed (ai—active ingredient) [35].

4. Efficacy of Seed Treatments and Coatings

4.1. Biostimulants

There has been considerable effort over many decades on applying chemicals to seeds to improve germination and seedling growth. The term "biostimulants" was adopted in the 21st century and provides a better definition and grouping of materials that serve to enhance plant performance. Biostimulants may be defined as natural compounds that trigger physiological and molecular processes modulating crop yield and quality. There are several categories of plant biostimulants and these materials are natural products or biologicals. A review of biostimulants applied as seed coatings is summarized by category (Table 3): beneficial bacteria and fungi [58–60], plant and animal-derived proteins, protein hydrolysates and amino acids [50,51,53,61], carbohydrate derivatives [62,63], seaweed [64] and herbal extracts [65]. There are no seed applied references for other biostimulant categories including vitamins, humic and fulvic acids. All these compounds may enhance plant metabolism when applied in small quantity, but their mode of action is only partially understood [66,67].

Table 3. Review of biostimulants applied as seed treatments on seed germination, seedling growth and other measured parameters.

Active Components (Source *)	Crop	Application Mode/Type of Experiment	Main Findings	Reference
Paraburkholderia phytofirmans **PsJN Strain: BIO**	Wheat	Seed coating—10 g seeds required 40 µL of the coating product Agicote Rouge T17 and PsJN inoculum (10^8 CFU mL^{-1})	Increased straw yield (55–100%), grain yield (43–100%) and thousand kernels weight (19–58%) compared with the inoculated control in all 3 seasons	[58]
Arbuscular Mycorrhizal Fungi Inoculum: BIO	Cowpea	Seed coating—*Rhizophagus irregularis*	No effect on seed yield, but 66% increase on shoot dry weight compared to the control	[59]
Plant growth promoting bacteria: BIO	Cowpea	Seed coating—*Pseudomonas libanensis*	Plant biomass and seed yield significantly enhanced 101% and 52% compared to the control	[59]
Arbuscular Mycorrhizal Fungi: BIO	Chickpea	Seed coating—a mixture of equal proportions of five *R. irregularis* isolates	Increased pod (160%), seed numbers (148%), and grain yield (140%) in field compared to the control	[60]
Soy flour: NP	Broccoli	Seed coating—Application of plant-based protein to the seeds	All treatments with >30% soy flour in the coating had greater fresh and dry weight, leaf area compared with the control	[53]
Soy flour: NP Vermicompost: NP, (OR)	Broccoli	Seed coating—Co-application of vermicompost and plant-based protein to the seeds	Seedling growth improved, increased shoot length (up to 114%), Shoot dry weight (42%) and root dry weight (51.5%) compared to the non-treated control seeds	[51]
Soy flour: NP Vermicompost: NP, (OR)	Red clover-Ryegrass	Seed coating—Co-application of vermicompost and plant-based protein to the seeds	All treatments showed a 40 to 60% increase in seedling dry weight and the seedling vigor indexes were 15% to 27% higher than control for red clover and 40% for ryegrass	[50]

Table 3. *Cont.*

Active Components (Source *)	Crop	Application Mode/Type of Experiment	Main Findings	Reference
Amino acid mixtures: NP	Cucumber	Seed coating—Application of 5 different amino acid mixtures to seeds	Total leaf area and dry weight were 35–50% and 26–30% higher for all amino acid mixtures (containing proline, hydroxyproline or their combination, amino acid mixture without proline and/or hydroxyproline) in comparison with no amino acid in coating	[61]
Chitosan nanoparticles: NP	Chilli	Seed coating—20 and 100 ppm chitosan using top-spray fluidized bed coating equipment	Chitosan treatments enhanced germination (6–7%) and decreased seed fungal infection (12–28%) compared with the intact control seeds	[62]
Chitosan: NP	Artichoke	Seed coating—10 mL of 3% or 4% (*w*/*v*) chitosans solution applied to 100 g seeds	4% (*w*/*v*) Chitosan B enhanced seedling growth (20%) compared with non-treated seeds	[63]
Seaweeds: NP, (*OR*)	Radish	Seed coating—Algal homogenate 50 mg/g seeds	Seedlings' length was 23% higher than in the control and seedling dry weight was 26% higher than in the control	[64]

* Source of material: Natural products or derivatives—NP, Biologicals—BIO, substances may be Organically approved—*OR*.

The application of biostimulant components has not been widely integrated as seed treatments in agriculture. Biostimulants applied as seed treatments and coatings are more cost effective and provide great potential to enhance stand establishment compared to foliar and soil application methods [51,59]. The global market for biostimulants applied as seed treatments in 2015 and 2019 was USD 112 million and USD 181 million, respectively and is forecasted to reach USD 338 million by 2025 [68].

The studies summarized in Table 3, reports on the beneficial effects of biostimulants applied as seed treatments and coatings on germination enhancement and growth stimulation on several crop species. For example, Amirkhani et al. [51,53] reported that seed coating with plant-derived protein enhanced germination indices and seedling uniformity, as well as the vigor index of broccoli, compared to non-coated seeds under optimum conditions. Moreover, the co-application of plant-derived protein and a nutrient-rich micronized vermicompost as a dry seed-coating binder and biostimulant significantly enhanced plant biometric parameters in germination and greenhouse studies [51,53]. In another study, Qiu et al. [50] reported enhancement in the percent germination and germination rate in red clover, and root enhancement in ryegrass, in response to biostimulant seed coating. The above studies suggest that biostimulant seed treatment practices enhanced uptake of soil-media nitrogen. The application of nitrogen in the seed coating accounted for less than 5% of the total nitrogen taken up by the roots. Therefore, the biostimulant was not merely a nitrogen fertilizer, but acted as a biostimulant to enhance nutrient uptake [51,53].

4.2. Nutrient Coating

Adequate nutrient availability is very important starting at the early stages of plant growth. Seed coating with appropriate amounts of macro- and preferentially micro-nutrients can reduce nutrient losses by placement on the seed, and also reduce competition from weeds. However, germination and seedling growth can also be hindered by macronutrient coatings due to phytotoxicity. To prevent such toxicity, direct contact of nutrients should be avoided with seeds by including the initial layer or boundary layer followed by the nutrient coating.

Several investigations conducted on plant nutrients applied as seed coatings were summarized by Scott (1989), Farooq (2012) and Masuthi (2009) [12,69,70]. Successful coating of phosphorus on oats improved early plant growth [71]. In rice seeds, boron (2 g/kg seed) was applied as seed coating and significantly increased grain yield and boron contents over a control [72]. Losses of nutrients by seed coating reduced the cost of production as compared to soil applications [69]. Conventional broadcasting of fertilizers exhibited higher cost and losses, while coating with an equivalent rate of nutrients significantly produced higher yield of cereal crops [69,73]. Slow release nutrient (N-P-K) coating on maize seeds resulted in improved emergence and yield attributes as compared to conventional compound fertilizer application in the field [74].

The effects of applied nutrients to a wide range of field and vegetable crop seeds with pre-defined quantity of nutrients are summarized, and plant improvements in germination, emergence, plant growth and yield were cited (Table 4). All fertilizers were synthetic chemicals, but several are available as organically approved. Zinc oxide [75,76] and zinc sulphate [70,77–80] are the most promising micronutrients used in seed coating of cereal crops and pulses. Wiatrak [81,82] evaluated the effect of polymer coating with manganese, copper and zinc on wheat and soybean crops and found a cost-effective technique for the enhancement of plant growth and ultimate yield of both crops [81,82]. In another study, coating with a range of micronutrients (Zn, B, K, Mo, Fe, Mg, Mn) increased productivity of cotton, chickpea, groundnut and pigeon pea with minimum expenditure and higher returns [83].

Table 4. Review of plant nutrients applied as seed treatments on seed germination, seedling growth, yield and other measured parameters.

Active Components (Source *)	Crop	Application mode/Type of Experiment	Main Findings	Reference
Boric acid (H_3BO_3) SYN (*OR*)	Rice	Application of H_3BO_3 at 1, 1.5, 2, 2.5 and 3 g B/kg by seed coating	2 g B/kg significantly decreased panicle sterility, increased 1000-kernal weight (7%), grain yield (20%) and B contents (24%) over control	[72]
Calcium oxide (CaO) SYN	Tomato	Application of CaO, bentonite and talc combination by pelleting	Increased final emergence (23%) and seedling growth over control. Better storability of pelleted seeds after 5 months	[34]
Monopotassium phosphate (KH_2PO_4) SYN	Pearl millet	Application of KH_2PO_4 at a rate of 400 g P ha^{-1} by seed priming and seed coating	Seed coating increased vegetative biomass over 400% at early stages and panicle yield (50%) compared to control. The time to flowering (10–14 days) reduced by seed coating	[84]
Micronutrients SYN	Wheat	Application of mixture of manganese, copper and zinc micronutrients by polymer coating	Seed coating with 395 mL 100 kg $seeds^{-1}$ improved dry matter yield (23%), N uptake (25%), P uptake (23%) and grain yield (2%) over control	[81]
Micronutrients SYN	Soybean	Application of mixture of manganese, copper and zinc micronutrients by polymer coating	Compared to control, increased grain yield (14%) and plant Normalized Difference Vegetation Index (10.5%) with polymer seed coating at 395 mL 100 kg $seeds^{-1}$	[82]
Micronutrients SYN	Cotton Pigeon pea Chickpea Groundnut	Seed polymer coating with various micronutrients	Increased yield to the extent of 17% in cotton, 20% in pigeon pea, 16% in chickpea and 14% in groundnut over control	[83]

Table 4. *Cont.*

Active Components (Source *)	Crop	Application mode/Type of Experiment	Main Findings	Reference
Zinc oxide (ZnO) SYN	Maize Soybean Pigeon pea	Application of ZnO at 25 and 50 mg Zn/g seeds by seed coating	Significant increased germination (93–100%) compared to control (80%). Improved growth and hormonal activity	[75]
Zinc oxide (ZnO) SYN	Rice	Application of ZnO coated urea at 2.0% (*w*/*w*) by seed coating	Significant increase in yield (28%) and micronutrients (40%) over control	[76]
Zinc sulfate and Boric acid ($ZnSO_4 + H_3BO_3$) SYN (*OR*)	Soybean	Seed coating with the dose 0.8 kg of H_3BO_3 + 0.8 kg of $ZnSO_4$/kg seeds.	Significantly improved growth and reduced shoot dry matter production	[77]
Zinc sulfate and Zinc chloride ($ZnSO_4 + ZnCl_2$) $ZnSO_4$ SYN (*OR*), $ZnCl_2$ SYN	Wheat	Application of $ZnSO_4$ + $ZnCl_2$ at 1.25 g Zn/kg by seed coating	Improved chlorophyll a and b contents. Enhanced grain yield and Zn contents	[78]
Zinc sulfate ($ZnSO_4$) SYN (*OR*) Borax SYN (*OR*) Arappu leaf powder (*OR*)	Cowpea	Application of $ZnSO_4$, Borax and arappu leaf powder at 250 mg, 100 mg and 250 g/kg seed respectively by seed pelleting	Increased grain yield by 32% over control. Seed pelleting with arappu leaf powder alone and in combination with $ZnSO_4$ improved yield parameters	[70]
Zinc sulfate and Boric acid ($ZnSO_4 + H_3BO_3$) SYN (*OR*)	Stylosanthes	Application of $ZnSO_4$ with 90 g and H_3BO_3 with 120 g per kg seed	Significantly improved growth, development and modulation	[79]

* Source of material: Synthetic Chemicals—SYN, Natural products or derivatives—NP, Biologicals—BIO, Mineral—MIN, substances may or may not be Organically approved—*OR*.

4.3. Abiotic Stress

Abiotic stresses may occur in the field and have a deleterious effect on germination and stand establishment. Abiotic stresses may be caused by drought stress or salinity stress. Both chemical and biological seed treatments and coatings have the potential to ameliorate deleterious effects of transient abiotic stress [4,85]. Superabsorbent polymers (SAPs) are hydrophilic polymers that can absorb over one hundred times their weight in water and have a long history of use in agriculture [86]. Seed coating technologies were developed to incorporate SAPs with filler materials to produce encrusted or pelleted seeds [87,88]. Hydro-absorbers and SAP improved germination potential by early and rapid completion of imbibition and active metabolism phases by improving water availability around the sown seed [49]. SAP supplies sufficient moisture and ensures oxygen availability to germinating seed under normal and stressful conditions [89]. SAP seed coatings were shown to increase germination and stand establishment at substantially lower application rates than soil-applied SAPs [90–92].

Salinity stress reduces soil water availability and results in an excess of sodium ions in the soil. Biological seed treatments may partially ameliorate the deleterious influence of salinity on plant growth. A commercial seed treatment formulation of *Trichoderma harziannum* was applied onto squash (*Cucurbita pepo*) seeds and studied in pot experiments in the greenhouse [93]. Pots were irrigated with 50 and 100 mM NaCl solutions and plant weight and leaf mineral content analyzed. The biological seed treatment increased plant growth at both salinity levels compared to the non-treated control. Moreover, the biological increased the leaf potassium to sodium ratio suggesting that one mechanism of a beneficial biological was altered mineral uptake. In another study, seed treatment with *T. harzianum* alleviated biotic, abiotic, and physiological stresses in germinating seeds and seedlings [94]. A recent review on beneficial microbes applied as seed coatings stated several plant beneficial microbes (PBMs) enhanced drought or salinity tolerance [22].

4.4. Plant Protectants and Inoculants

Management of biotic stresses in agriculture is synonymous with plant protectants applied as seed treatments and coatings. These seed treatments may be fungicides, insecticides, bactericides, and nematicides [20]. In agriculture, control of these pests should be considered if damage exceeds an economic threshold [35]. Plant protectants are applied in anticipation that economic damage will occur from soil-borne or air-borne pathogens and/or pests. Therefore, seed treatments provide insurance from potential biotic stresses either singularly or in combination, as in the case with soil-borne pathogens and insect pests. A wide range of active components may serve as plant protectants including: synthetic chemicals, natural products, and biologicals (Table 1). Some of these plant protectants may be organically approved for use in crop protection. Based on the seed-treatment active component and its formulation, dosage and other attributes, these actives may be applied with specific paring of equipment. Methods include: dry powder applicator, rotary coater or drum coater to apply dry coating, seed dressing, film coat, and encrustment or pellet (Figure 1).

The literature on seed treatments as plant protectants is beyond the scope of this review. However, selected papers are highlighted on seed treatments and coatings as seed enhancements. Herbicide safeners are seed treatments that negate the potential herbicidal effect of selective herbicide chemistries on crop plants. Thus, herbicide safeners are tools for specialty crops and other plant species that lack chemical weed control options. The herbicide safener, fluxofenim was effective on field soil treated with the herbicide, metolachlor on switchgrass (*Panicum virgatum*) [95]. Biologicals also known as plant beneficial microbes (PBMs) [22] may provide inconsistent pest management under a wide range of field conditions encountered at time of sowing. Synthetic chemical seed treatments provide more reliable pest control for conventional agriculture but are prohibited for organic crop production. Biopesticides that are derived from natural products or microbes and are organically approved have potential for pest management comparable to synthetic chemical seed treatments. Spinosad, a biopesticide for foliar application, was investigated as an onion seed treatment at Cornell University [96]. Spinosad seed treatment was comparable in efficacy to chemical seed treatments in the control of onion maggot (*Delia antiqua*). An organic formulation of spinosad was also effective for control of onion maggot and seed-corn maggot (*Delia platura*) when used in combination with other seed treatments [97]. Collectively, seed-coating technology as described in this paper provides a delivery platform for many other active components for improved pest management that are environmentally friendly for sustained systems. In addition, new generation biochemical, bio-pesticides reduces the reliance on synthetic agrochemical seed treatments [97,98]. Greater efficacy of fungicides has been achieved with good treatment adhesion resulting in less dusting [98].

The use of plant extracts as seed treatments can improve seed quality and reduce infestation of microbial pathogens [99]. Such plant extracts have antibiotic and antimicrobial properties that help in alleviation of biotic and abiotic stresses during seed emergence in the soil [100]. Natural occurring plant extracts are readily available, less expensive, and have promising effects on germination, plant growth, and yield as compared to traditional chemical fungicide treatments [99,101].

Seed pelleting was effective in sowing sesame seed. Pelleting significantly enhanced plant height, lateral branches and number of capsules per plant as compared to non-pelleted seeds [102]. Damping-off disease incidence was significantly reduced by pelleting of sesame seeds with the plant growth promoting microbe (strain E681) [29]. Pelleting does not normally affect shelf-life. An investigation on the storage of pelleted seeds revealed that quality of tobacco seed after pelleting was maintained up to 720 days when properly stored in aluminum cans [103].

Microbial seed coating is a method of coating seeds with plant beneficial microorganisms such as plant growth promoting bacteria (PGPB), rhizobia, and fungi to increase crop growth and yield through improvement in nutrition and protection against diseases and pathogens [22,23]. Coating seeds with beneficial microbes is an efficient delivery system for application of beneficial microbes and is a promising tool for inoculation of different crop seeds with a reduced use of inoculum as compared to traditional seed treatments [7,12,37]. A typical inoculant formulation is based on the selection of the microorganism, a suitable carrier, and related additives [22,104]. Combination of carbon source materials with rhizobia not only aids in the survival of bacterial strains as a food source but also provides protection from the external environment [105]. In addition to seed coating as a carrier for food bases, pH can be adjusted for optimum growth of beneficial microbes [7]. Lime pelleting was shown to be helpful for rhizobia survival by neutralizing fertilizer acidity close to the seed [4]. Application of compatible rhizosphere microbes to chickpea seeds was effective to alleviate biotic stress through enhanced stand establishment, growth, and molecular attributes. Peat and biochar were effective for providing protection to the rhizobia by tightly absorbing it and preventing direct exposure to the external environment [106]. After seed coating with bacterial strains, rapid desiccation should be avoided, by selection of appropriate filler materials. Microbial survival on coated seeds may be attenuated, and generally old chemistry seed treatment fungicides including captain, thiram and carboxin are not recommended with Rhizobium inoculants [107]. Therefore, compatibility of new seed treatments should be tested to ensure efficacy of the biological.

4.5. Other Coatings

Different marker substances including visible dyes, fluorescent tracers and magnetic powders were incorporated into coatings to trace the seed in the supply chain and protect the true seeds from fake seeds in the market [8,52]. Color-coding is the most widely used marker system in coating processes for identification of a specific variety or seed treatment [23]. Colored seed is an indication of a seed coat treatment with appropriate fungicide or pesticide and is used to reduce the risk of livestock or human consumption [8]. Natural colorants can be used for storage of soybean seeds without loss of vigor [108]. Additionally, researchers have also evaluated the efficacy of fluorescein, rhodamine, and magnetic powder as anti-counterfeiting labels in tobacco seeds in order to enhance seed security in the supply chain [52]. Riboflavin is a natural fluorescent compound and was used for marking cucumber seeds for authentication [109]. Riboflavin was not phytotoxic after application nor after seed storage compared to non-treated seeds, and riboflavin fluorescence was not diminished after 10 months' storage [109].

5. Conclusions and Future Prospects

Seed coating technologies have many virtues including protecting seeds from pests and diseases at the time of sowing and improving flowability for precision seeding [15]. Improved stand establishment and seedling vigor under biotic and abiotic stresses can be achieved by using appropriate seed coating equipment, methods, and materials. The growing demand for coated seeds is documented with many small and large companies in the market. Despite the extensive information on natural or synthetic active components, coating methods and polymers, the seed industry in many developing countries is not adopting this technology. Farmers in these countries are not utilizing seed treatments due to lack of resources as compared to 100% adoption in developed countries [4]. Usually, economical treatments are preferred if cost is not exceeding USD 20 per planted hectare [7]. Therefore, the success of seed coating technology depends upon the selection of inexpensive and readily available coating agents with low cost. Collectively, cost effective, simple materials and methods are needed for use in third world countries.

There is limited information available on the shelf life of treated and coated seeds. Specifically, it would be helpful to know if seed treatment phytotoxicity increases with the loss of seed vigor in storage [110,111]. Is there a reduction in seed treatment efficacy after seed storage, particularly for seeds treated with biologicals [7]? Investigations are needed on how to better integrate seed coating technologies with weed management exploiting herbicide safeners [95], or herbicide seed treatments [112]. Additional research and development are needed for new biochemical, bio-pesticide plant protectants [96,97] that can be used for organic or conventional crop production for sustainable agricultural systems. Lastly, the knowledge of seed treatment and coating technologies should be directed for reliable and consistent stand establishment under changing climatic conditions. To accomplish these goals will require the development of new active components, with complimentary coating equipment, and coating technologies. This can best be achieved by continued efforts from multidisciplinary teams of seed scientists, agronomists, chemists, pest management specialists and engineers. These achievements may be accomplished through a partnership of academia with industry for the development of cost-effective materials and methods for wide-scale adoption in developed and third-world countries.

Author Contributions: I.A. Conceptualization, writing and original draft preparation, T.J. Writing and original draft preparation. M.A. Writing, preparation of both figures and three tables, A.G.T. Conceptualization and editing the manuscript. All authors have read and agreed to the published version of the manuscript.

Funding: This material is based upon work that is supported by the United States Hatch Funds under Multi-state Project W-4168 under accession number 1007938 to the fourth author, and Pakistan Science Foundation under a research project PSF/NSLP-489 to the first author.

Acknowledgments: The authors thank Hilary Mayton for critically reviewing this manuscript, and helpful suggestions were provided by Simone Pedrini for developing the illustration in Figure 1. We thank valuable input from seed treatment and coating industries, and seed companies: ABM, Aginnovations, BASF, Beck's Hybrids, CTApplicators, Germains, Incotec and Syngenta.

Conflicts of Interest: The authors declare no conflict of interest.

References

1. Ellis, R.H. Seed and seedling vigor in relation to crop growth and yield. *J. Plant Growth Regul.* **2004**, *11*, 249–255. [CrossRef]
2. Afzal, I.; Rehman, H.U.; Naveed, M.; Basra, S.M.A. Recent advances in seed enhancements. In *New Challenges in Seed Biology-Basic and Translational Research Driving Seed Technology*; InTechOpen: London, UK, 2016; pp. 47–74.
3. Zinsmeister, J.; Leprince, O.; Buitink, J. Molecular and environmental factors regulating seed longevity. *Biochem. J.* **2020**, *477*, 305–323. [CrossRef] [PubMed]
4. Sharma, K.K.; Singh, U.S.; Sharma, P.; Kumar, A.; Sharma, L. Seed treatments for sustainable agriculture—A review. *J. Appl. Nat. Sci.* **2015**, *7*, 521–539. [CrossRef]
5. Halmer, P. Seed technology and seed enhancement. *Acta Hort.* **2008**, *771*, 17–26. [CrossRef]
6. Taylor, A.G. Seed treatments. In *Encyclopedia of Applied Plant Sciences*; Thomas, B.D.J., Murphy, B.G., Eds.; Elsevier Academic Press: Cambridge, UK, 2003; pp. 1291–1298.
7. Taylor, A.G.; Harman, G.E. Concepts and technologies of selected seed treatments. *Annu. Rev. Phytopathol.* **1990**, *28*, 321–339. [CrossRef]
8. Pedrini, S.; Merritt, D.J.; Stevens, J.; Dixon, K. Seed coating: Science or marketing spin? *Trends Plant Sci.* **2017**, *22*, 106–116. [CrossRef] [PubMed]
9. Jeffs, K.A. *Seed Treatment*, 2nd ed.; The British Crop Protection Council (BCPC) Publication: Surrey, UK, 1986; p. 332.
10. Halmer, P. Commercial seed treatment technology. In *Seed Technology and Its Biological Basis*; Black, M., Bewley, J.D., Eds.; Sheffield Academic Press: Sheffield, UK, 2000; pp. 257–286.
11. Porter, F.E.; Scott, J.M. Seed coating methods and Purposes: A status report. In *Proceedings of the Short Course for Seedsmen*; Mississippi Agricultural and Forestry Experiment Station: Prairie, MS, USA, 1979.
12. Scott, J.M. Seed coatings and treatments and their effects on plant establishment. *Adv. Agron.* **1989**, *42*, 43–83.
13. Hill, H.J. Recent developments in seed technology. *J. New Seeds* **1999**, *1*, 105–112. [CrossRef]
14. Pedrini, S.; Balestrazzi, A.; Madsen, M.; Bhalsing, K.; Hardegree, S.; Dixon, K.W.; Kildisheva, O.A. Seed enhancement: Getting seeds restoration-ready. *Restor. Ecol.* **2020**, *28*, S266–S275. [CrossRef]
15. Taylor, A.G.; Allen, P.S.; Bennett, M.A.; Bradford, K.J.; Burris, J.S.; Misra, M.K. Seed enhancements. *Seed Sci. Res.* **1998**, *8*, 245–256. [CrossRef]
16. Taylor, A.G. Seed storage, germination, quality and enhancements. In *The Physiology of Vegetable Crops*, 2nd ed.; Wien, H.C., Stutzel, H., Eds.; CAB International: Wallingford, UK, 2020; pp. 1–30.

17. Seed Coating Materials Market by Type (Polymers, Colorants, Pellets, Minerals/Pumice, and Other Additives), Crop Type (Cereals & Grains, Oilseeds & Pulses, Fruits & Vegetables, Flowers & Ornamentals), & by Region—Global Trends & Forecasts to 2020. Available online: https://www.marketsandmarkets.com/Market-Reports/seed-coating-materials-market-149045530.html (accessed on 18 September 2020).
18. Taylor, A.G.; Trimmer, M.; DunhamTrimmer International Bio Intelligence, Lakewood Ranch, Florida. Global chemical and biological seed treatments market. Personal communication, 2020.
19. Biological Products Markets around the World. Available online: http://www.bpia.org/wp-content/uploads/2018/03/Biological-Products-Markets-Around-The-World.pdf (accessed on 30 October 2020).
20. Buffington, B.; Beegle, D.; Lindholm, C. *Seed Treatment a National Pesticide Applicator Manual*; Pesticide Educational Resources Collaborative (PERC), University of California Davis: Davis, CA, USA, 2018.
21. Organic Materials Review Institute. Available online: https://www.omri.org/ (accessed on 14 September 2020).
22. Rocha, I.D.S.; Ma, Y.; Souza-Alonso, P.; Vosátka, M.; Freitas, H.; Oliveira, R.S. Seed coating: A tool for delivering beneficial microbes to agricultural crops. *Front. Plant Sci.* **2019**, *10*, 1357. [CrossRef]
23. Ma, Y. Seed coating with beneficial microorganisms for precision agriculture. *Biotechnol. Adv.* **2019**, *37*, 107423. [CrossRef]
24. Kalaivani, K.; Kalaiselvi, M.M.; Senthil-Nathan, S. Effect of methyl salicylate (MeSA), an elicitor on growth, physiology and pathology of resistant and susceptible rice varieties. *Sci. Rep.* **2016**, *6*, 34498. [CrossRef]
25. Klessig, D.F.; Manohar, M.; Baby, S.; Koch, A.; Danquah, W.B.; Luna, E.; Park, H.J.; Kolkman, J.M.; Turgeon, B.G.; Nelson, R.; et al. Nematode ascaroside enhances resistance in a broad spectrum of plant–pathogen systems. *J. Phytopathol.* **2019**, *167*, 1–8. [CrossRef]
26. Lee, M.W.; Huffaker, A.; Crippen, D.; Robbins, R.T.; Goggin, F.L. Plant elicitor peptides promote plant defences against nematodes in soybean. *Mol. Plant Pathol.* **2018**, *19*, 858–869. [CrossRef]
27. Tayyab, N.; Naz, R.; Yasmin, H.; Nosheen, A.; Keyani, R.; Sajjad, M.; Hassan, M.N.; Roberts, T.H. Combined seed and foliar pre-treatments with exogenous methyl jasmonate and salicylic acid mitigate drought induced stress in maize. *PLoS ONE* **2020**, *15*, e0232269. [CrossRef]
28. Madsen, M.D.; Petersen, S.; Taylor, A.G. Seed Coating Compositions and Methods for Applying Soil Surfactants to Water-Repellent Soil. U.S. Patent 9,554,502 B2, 31 January 2017.
29. Ryu, C.M.; Kim, J.; Choi, O.; Kim, S.H.; Park, C.S. Improvement of biological control capacity of *Paenibacillus polymyxa* E681 by seed pelleting on sesame. *Biol. Control* **2006**, *39*, 282–289. [CrossRef]
30. Chen, Y.; Turnblad, K.M. Insecticidal Seed Coating. U.S. Patent 0,177,526 A1, 28 November 2002.
31. Lopisso, D.T.; Kühlmann, V.; Siebold, M. Potential of soil-derived fungal biocontrol agents applied as a soil amendment and a seed coating to control *Verticillium wilt* of sugar beet. *Biocontrol Sci. Technol.* **2017**, *27*, 1019–1037. [CrossRef]
32. Kangsopa, J.; Hynes, R.K.; Siri, B. Lettuce seeds pelleting: A new bilayer matrix for lettuce (*Lactuca sativa*) seeds. *Seed Sci. Technol.* **2018**, *46*, 521–531. [CrossRef]
33. Kimmelshue, C.; Goggi, A.S.; Cademartiri, R. The use of biological seed coatings based on bacteriophages and polymers against *Clavibacter michiganensis subsp. nebraskensis* in maize seeds. *Sci. Rep.* **2019**, *9*, 17950. [CrossRef]
34. Javed, T.; Afzal, I. Impact of seed pelleting on germination potential, seedling growth and storage of tomato seed. *Acta Hortic.* **2020**, *1273*, 417–424. [CrossRef]
35. Danielson, B.; Gaul, A. *Category 4, Seed Treatment—Iowa Commercial Pesticide Applicator Manual, CS16*; Iowa State University Extension and Outreach: Ames, IA, USA, 2011; p. 40.
36. Sikhao, P.; Taylor, A.G.; Marino, E.T.; Catranis, C.M.; Siri, B. Development of seed agglomeration technology using lettuce and tomato as model vegetable crop seeds. *Sci. Hortic.* **2015**, *184*, 85–92. [CrossRef]
37. Cho, S.; Seo, H.; Oh, Y.; Lee, E.; Choi, I.; Jang, Y.; Song, Y.; Min, T. Selection of coating materials and binders for pelleting onion (*Allium cepa* L.) seed. *J. Korean Soc. Hortic. Sci.* **2000**, *41*, 593–597.
38. Guan, Y.J.; Wang, J.C.; Hu, J.; Tian, Y.X.; Hu, W.M.; Zhu, S.J. A novel fluorescent dual-labeling method for anti-counterfeiting pelleted tobacco seeds. *Seed Sci. Technol.* **2013**, *41*, 158–163. [CrossRef]
39. Sooter, C.A.; Millier, W.F. The effect of pellet coatings on the seedling emergence from lettuce seeds. *Trans. Am. Soc. Agric. Eng.* **1978**, *21*, 1034–1039. [CrossRef]
40. Vereenigde, M.H.; Wittlaan, J.; Haag, D.J.R. Seed Coating Composition. European Patent No. Eur. Patent 2229808 A1, 22 September 2010.
41. Hirsch, G.W. Powdered Seed Treatment Applicator. U.S. Patent 7,487,892, 10 February 2009.

42. Hirsch, G.W. Powder Dispenser Assembly. U.S. Patent 8,556,129, 15 October 2013.
43. Changing Times, LLC. Available online: https://www.ctapplicators.com/ (accessed on 10 September 2020).
44. Marks, P.; Aginnovation LLC, Lodi, CA, USA; Taylor, A.G. Dry powder seed coating equipment. Personal communication, 2020.
45. Anderson, D. Talc and Graphite: What You Need to Know before you Plant Machinery. *AgWeb J.* **2014**. Available online: https://www.agweb.com/article/talc_and_graphite_what_you_need_to_know_before_you_plant_NAA_Dan_Anderson (accessed on 14 September 2020).
46. Badua, S.A.; Sharda, S.; Strasser, R.; Cockerlin, K.; Ciampitti, I. Comparison of soy protein based and commercially available seed lubricants for seed flowability in row crop planters. *Appl. Eng. Agric.* **2019**, *35*, 593–600. [CrossRef]
47. Avelar, S.A.G.; Sousa, F.V.D.; Fiss, G.; Baudet, L.; Peske, S.T. The use of film coating on the performance of treated corn seed. *Rev. Bras. Sementes* **2012**, *34*, 186–192. [CrossRef]
48. Olivera, M.E.; Ferrari, L.; Araoz, S.; Postulka, E.B. Improvements on physiological seed quality of *Festuca arundinacea* schreb by encrusting technology: Products and storage effects. *Science* **2017**, *10*, 33–37. [CrossRef]
49. Gorim, L.; Asch, F. Effects of composition and share of seed coatings on the mobilization efficiency of cereal seeds during germination. *J. Agron. Crop Sci.* **2012**, *198*, 81–91. [CrossRef]
50. Qiu, Y.; Amirkhani, M.; Mayton, H.; Chen, Z.; Taylor, A.G. Biostimulant seed coating treatments to improve cover crop germination and seedling growth. *Agronomy* **2020**, *10*, 154. [CrossRef]
51. Amirkhani, M.; Mayton, H.S.; Netravali, A.N.; Taylor, A.G. A seed coating delivery system for bio-based biostimulants to enhance plant growth. *Sustainability* **2019**, *11*, 5304. [CrossRef]
52. Guan, Y.; Wang, J.; Tian, Y.; Hu, W.; Zhu, L.; Zhu, S.; Hu, J. The novel approach to enhance seed security: Dual anti-counterfeiting methods applied on tobacco pelleted seeds. *PLoS ONE* **2013**, *8*, e57274. [CrossRef]
53. Amirkhani, M.; Netravali, A.; Huang, W.; Taylor, A.G. Investigation of soy protein–based biostimulant seed coating for broccoli seedling and plant growth enhancement. *Hortic. Sci.* **2016**, *51*, 1121–1126. [CrossRef]
54. Mei, J.; Wang, W.; Peng, S.; Nie, L. Seed pelleting with calcium peroxide improves crop establishment of direct-seeded rice under waterlogging conditions. *Sci. Rep.* **2017**, *7*, 1–12. [CrossRef] [PubMed]
55. Nwankwo, C.I.; Blaser, S.R.G.A.; Vetterlein, D.; Neumann, G.; Herrmann, L. Seed ball-induced changes of root growth and physico-chemical properties—A case study with pearl millet. *J. Plant Nutr. Soil Sci.* **2018**, *181*, 768–776. [CrossRef]
56. Madsen, M.D.; Davies, K.W.; Williams, C.J.; Svejcar, T.J. Agglomerating seeds to enhance native seedling emergence and growth. *J. Appl. Ecol.* **2012**, *49*, 431–438. [CrossRef]
57. Gornish, E.; Arnold, H.; Fehmi, J. Review of seed pelletizing strategies for arid land restoration. *Restor. Ecol.* **2019**, *27*, 1206–1211. [CrossRef]
58. Ben-Jabeur, M.; Kthiri, Z.; Harbaoui, K.; Belguesmi, K.; Serret, M.D.; Araus, J.L.; Hamada, W. Seed coating with thyme essential oil or *Paraburkholderia phytofirmans* PsJN strain: Conferring septoria leaf blotch resistance and promotion of yield and grain isotopic composition in wheat. *Agronomy* **2019**, *9*, 586. [CrossRef]
59. Ma, Y.; Látr, A.; Rocha, I.; Freitas, H.; Vosátka, M.; Oliveira, R.S. Delivery of inoculum of *Rhizophagus irregularis* via seed coating in combination with *Pseudomonas libanensis* for cowpea production. *Agronomy* **2019**, *9*, 33. [CrossRef]
60. Rocha, I.; Duarte, I.; Ma, Y.; Souza-Alonso, P.; Látr, A.; Vosátka, M.; Freitas, H.; Oliveira, R.S. Seed coating with arbuscular mycorrhizal fungi for improved field production of chickpea. *Agronomy* **2019**, *9*, 471. [CrossRef]
61. Wilson, H.T.; Amirkhani, M.; Taylor, A.G. Evaluation of gelatin as a biostimulant seed treatment to improve plant performance. *Front. Plant Sci.* **2018**, *9*, 1006. [CrossRef] [PubMed]
62. Chookhongkha, N.; Sopondilok, T.; Photchanachai, S. Effect of chitosan and chitosan nanoparticles on fungal growth and chilli seed quality. In Proceedings of the International Conference on Post-harvest Pest and Disease Management in Exporting Horticultural Crops-PPDM2012 973, Bangkok, Thailand, 21 February 2012; pp. 231–237.
63. Ziani, K.; Beatriz, U.; Juan, I.M. Application of bioactive coatings based on chitosan for artichoke seed protection. *Crop Prot.* **2010**, *29*, 853–859. [CrossRef]
64. Michalak, I.; Dmytryk, A.; Schroeder, G.; Chojnacka, K. The application of homogenate and filtrate from baltic seaweeds in seedling growth tests. *Appl. Sci.* **2017**, *7*, 230. [CrossRef]

65. Ben-Jabeur, M.; Vicente, R.; López-Cristoffanini, C.; Alesami, N.; Djébali, N.; Gracia-Romero, A.; Serret, M.D.; López-Carbonell, M.; Araus, J.L.; Hamada, W. A novel aspect of essential oils: Coating seeds with thyme essential oil induces drought resistance in wheat. *Plants* **2019**, *8*, 371. [CrossRef]
66. Jardin, D.P. Plant biostimulants: Definition, Concept, Main Categories and Regulation. *Sci. Hortic.* **2015**, *196*, 3–14. [CrossRef]
67. Calvo, P.; Nelson, L.; Kloepper, J.W. Agricultural uses of plant biostimulants. *Plant Soil* **2014**, *383*, 3–41. [CrossRef]
68. Taylor, A.G.; Trimmer, M.; DunhamTrimmer International Bio Intelligence, Lakewood Ranch, Florida. Global market value of biostimulant seed treatments. Personal communication, 2020.
69. Farooq, M.; Wahid, A.; Siddique, K.H. Micronutrient application through seed treatments: A review. *J. Soil Sci. Plant Nutr.* **2012**, *12*, 125–142. [CrossRef]
70. Masuthi, D.A.; Vyakaranahal, B.S.; Deshpande, V.K. Influence of pelleting with micronutrients and botanical on growth, seed yield and quality of vegetable cowpea. *Karnataka J. Agric. Sci.* **2009**, *22*, 898–900.
71. Peltonen, S.P.; Kontturi, M.; Peltonen, J. Phosphorus seed coating enhancement on early growth and yield components in oat. *Agron. J.* **2006**, *98*, 206–211. [CrossRef]
72. Rehman, A.U.; Farooq, M. Boron application through seed coating improves the water relations, panicle fertility, kernel yield, and biofortification of fine grain aromatic rice. *Acta Physiol. Plant.* **2013**, *35*, 411–418. [CrossRef]
73. Guan, Y.; Song, C.; Gan, Y.; Li, F.M. Increased maize yield using slow-release attapulgite-coated fertilizers. *Agron. Sustain. Dev.* **2014**, *34*, 657–665. [CrossRef]
74. Dong, Y.J.; He, M.R.; Wang, Z.L.; Chen, W.F.; Hou, J.; Qiu, X.K.; Zhang, J.W. Effects of new coated release fertilizer on the growth of maize. *J. Soil Sci. Plant Nutr.* **2016**, *16*, 637–649. [CrossRef]
75. Adhikari, T.; Kundu, S.; Rao, A.S. Zinc delivery to plants through seed coating with nano-zinc oxide particles. *J. Plant Nutr.* **2016**, *39*, 136–146. [CrossRef]
76. Shivay, Y.S.; Kumar, D.; Prasad, R.; Ahlawat, L.P.S. Relative yield and zinc uptake by rice from zinc sulphate and zinc oxide coatings onto urea. *Nutr. Cycl. Agroecosyst.* **2008**, *80*, 181–188. [CrossRef]
77. Acha, A.J.; Vieira, H.D.; Freitas, M.S.M. Perennial soybean seeds coated with high doses of boron and zinc. *Afr. J. Biotechnol.* **2006**, *15*, 1998–2005.
78. Rehman, A.; Farooq, M. Zinc seed coating improves the growth, grain yield and grain biofortification of bread wheat. *Acta Physiol. Plant.* **2006**, *38*, 238. [CrossRef]
79. Xavier, P.B.; Vieira, H.D.; Amorim, M.M. Physiological potential of Stylosanthes spp. seeds cv. Campo Grande in response to coating with zinc and boron. *J. Seed Sci.* **2016**, *38*, 314–321. [CrossRef]
80. Ullah, A.; Farooq, M.; Hussain, M.; Ahmad, R.; Wakeel, A. Zinc seed coating improves emergence and seedling growth in desi and kabuli chickpea types but shows toxicity at higher concentration. *Int. J. Agric. Biol.* **2019**, *21*, 553–559.
81. Wiatrak, P. Influence of seed coating with micronutrients on growth and yield of winter wheat in Southeastern Coastal Plains. *Am. J. Agric. Biol. Sci.* **2013**, *8*, 230.
82. Wiatrak, P. Effect of polymer seed coating with micronutrients on soybeans in Southeastern Coastal Plains. *Am. J. Agric. Biol. Sci.* **2013**, *8*, 302–308. [CrossRef]
83. Vasudevan, S.N.; Doddagoudar, S.R.; Sangeeta, I.M.; Shakuntala, N.M.; Patil, S.B. Augmenting productivity of major crop through seed polymer coating with micronutrients and foliar spray. *J. Adv. Agric. Technol.* **2016**, *3*, 150–154. [CrossRef]
84. Karanam, P.V.; Vabez, V. Phosphorus coating on pearl millet seed in low P Alfisol improves plant establishment and increases stover more the seed yield. *Exp. Agric.* **2010**, *46*, 457–469. [CrossRef]
85. Chandrika, K.P.; Prasad, R.D.; Godbole, V. Development of chitosan-PEG blended films using *Trichoderma*: Enhancement of antimicrobial activity and seed quality. *Int. J. Boil. Macromol.* **2019**, *126*, 282–290. [CrossRef]
86. Quastel, J.H. 'Krilium' and synthetic soil conditioners. *Nature* **1953**, *171*, 7–10. [CrossRef]
87. Tsujimoto, T.; Sato, H.; Matsushita, S. Hydration of Seeds with Partially Hydrated Super Absorbent Polymer Particles. U.S. Patent 5,930,949, 3 August 1999.
88. Leinauer, B.; Serena, M.; Singh, D. Seed coating and seeding rate effects on turfgrass germination and establishment. *HortTechnology* **2010**, *20*, 179–185. [CrossRef]

89. Gorim, L.; Asch, F. Seed coating with hydro-absorbers as potential mitigation of early season drought in sorghum (*Sorghum bicolor* L. Moench). *Biology* **2017**, *6*, 33. [CrossRef] [PubMed]
90. Berdahl, J.D.; Barker, R.E. Germination and emergence of Russian wildrye seeds coated with hydrophilic materials. *Agron. J.* **1980**, *72*, 1006–1008. [CrossRef]
91. Chen, H.W.; Jiang, S.T.; Zhou, J.Q.; Zhao, Y.Y.; Wang, J.H. Super absorbent polymer seed coating and its effect on physiological features of corn seed. *J. Hefei Univ. Technol. (Nat. Sci.)* **2004**, *27*, 242–246.
92. Su, L.Q.; Li, J.G.; Xue, H.; Wang, X.F. Super absorbent polymer seed coatings promote seed germination and seedling growth of *Caragana korshinskii* in drought. *J. Zhejiang Univ. Sci.* **2017**, *18*, 696–706. [CrossRef]
93. Yildirim, E.; Taylor, A.G.; Spittler, T.D. Ameliorative effects of biological treatments on growth of squash plants under salt stress. *HortTechnology* **2006**, *111*, 1–6. [CrossRef]
94. Mastouri, F.; Björkman, T.; Harman, G.E. Seed treatment with *Trichoderma harzianum* alleviates biotic, abiotic, and physiological stresses in germinating seeds and seedlings. *Phytopathology* **2010**, *100*, 1213–1221. [CrossRef]
95. Rushing, J.B.; Baldwin, B.S.; Taylor, A.G.; Owens, V.N.; Fike, H.J.; Moore, K.J. Seed safening from herbicidal injury in switchgrass establishment. *Crop Sci.* **2013**, *53*, 1650–1657. [CrossRef]
96. Nault, B.A.; Straub, R.W.; Taylor, A.G. Performance of novel insecticide seed treatments for managing onion maggot in onion fields. *Crop Prot.* **2006**, *25*, 58–65. [CrossRef]
97. Wilson, R.G.; Orloff, S.B.; Taylor, A.G. Evaluation of insecticides and application methods to protect onions from onion maggot, Delia antiqua and seedcorn maggot, Delia platura, damage. *Crop Prot.* **2014**, *67*, 102–108. [CrossRef]
98. Moretti, E.; Nault, B.A. Onion maggot control in onion, 2019. *Arthropod Manag. Tests* **2020**, *45*, tsaa007. [CrossRef]
99. Mbega, E.R.; Mortensen, C.N.; Mabagala, R.B.; Wulff, E.G. The effect of plant extracts as seed treatments to control bacterial leaf spot of tomato in Tanzania. *J. Gen. Plant Pathol.* **2012**, *78*, 277–286. [CrossRef]
100. Findura, P.; Hara, P.; Szparaga, A.; Kocira, S.; Czerwińska, E.; Bartoš, P.; Treder, K. Evaluation of the effects of allelopathic aqueous plant extracts, as potential preparations for seed dressing, on the modulation of cauliflower seed germination. *Agriculture* **2020**, *10*, 122. [CrossRef]
101. Mancini, V.; Romanazzi, G. Seed treatments to control seedborne fungal pathogens of vegetable crops. *Pest. Manag. Sci.* **2014**, *70*, 860–868. [CrossRef]
102. Dogan, T.; Zeybek, A. Improving the traditional sesame seed planting with seed pelleting. *Afr. J. Biotechnol.* **2009**, *8*, 6120–6126.
103. Carvalho, M.L.M.; Lopes, C.A.; Ribeiro, A.M.P.; Vasconcelos, M.C. Could packing and pelleting keep the quality of tobacco seeds during storage? *J. Seed Sci.* **2018**, *40*, 296–303. [CrossRef]
104. Deaker, R.; Roughley, R.J.; Kennedy, I.R. Legume seed inoculation technology—A review. *Soil Biol. Biochem.* **2004**, *36*, 1275–1288. [CrossRef]
105. Tufail, M.S.; Krebs, G.L.; Ahmad, J.; Southwel, A. The effect of Rhizobium seed inoculation on yields and quality of forage and seed of berseem clover (*Trifolium alexandrinum* L.) and its impact on soil fertility and smallholder farmer's income. *J. Anim. Plant Sci.* **2018**, *28*, 1493–1500.
106. Głodowska, M.; Schwinghamer, T.; Husk, B.; Smith, D. Biochar based inoculants improve soybean growth and nodulation. *Agric. Sci.* **2017**, *8*, 1048–1064. [CrossRef]
107. Belles, D.; Seed Care, Syngenta Crop Protection, Phoenix, Arizona; Taylor, A.G. Rhizobia compatibility. Personal communication, 2020.
108. Tripathi, B.; Pandey, A.; Bhatia, R.; Walia, S.; Yadav, A.K. Improving soybean seed performance with natural colorant-based novel seed-coats. *J. Crop Improv.* **2015**, *29*, 301–318. [CrossRef]
109. Sikhao, P.; Teeraponchaisit, P.; Taylor, A.G.; Siri, B. Seed coating with riboflavin, a natural fluorescent compound, for authentification of cucumber seeds. *Seed Sci. Technol.* **2014**, *42*, 171–179. [CrossRef]
110. Taylor, A.G.; Salanenka, Y.A. Seed treatments: Phytotoxicity amelioration and tracer uptake. *Seed Sci. Res.* **2012**, *22*, S86–S90. [CrossRef]
111. Taylor, A.G.; Eckenrode, C.J.; Straub, R.W. Seed treatments for onions: Challenges and progress. *HortTechnology* **2001**, *36*, 199–205.

112. Kanampiu, F.K.; Kabambe, V.; Massawe, C.; Jasi, L.; Friesen, D.; Ransom, J.K.; Gressel, J. Multi-site, multi-season field tests demonstrate that herbicide seed-coating herbicide-resistance maize controls Striga spp. and increases yields in several African countries. *Crop Prot.* **2003**, *22*, 697–706. [CrossRef]

Publisher's Note: MDPI stays neutral with regard to jurisdictional claims in published maps and institutional affiliations.

Article

Evaluating Genotypes and Seed Treatments to Increase Field Emergence of Low Phytic Acid Soybeans

Benjamin J. Averitt [1], Gregory E. Welbaum [2], Xiaoying Li [2], Elizabeth Prenger [3], Jun Qin [4] and Bo Zhang [2,*]

1 Department of Crop and Soil Sciences, University of Georgia, Athens, GA 30602, USA; ben.averitt@uga.edu
2 School of Plant and Environmental Sciences, Virginia Tech, Blacksburg, VA 24060, USA; welbaum@vt.edu (G.E.W.); xiaoying@vt.edu (X.L.)
3 Department of Plant Science, University of Missouri, Columbia, MO 65211, USA; eprenger@missouri.edu
4 Hebei Academy of Agricultural and Forestry Sciences, Shijiazhuang 050051, China; junqin@soybreeding.com
* Correspondence: bozhang@vt.edu

Received: 10 September 2020; Accepted: 29 October 2020; Published: 30 October 2020

Abstract: Low phytic acid (LPA) soybean [*Glycine max* (L.) Merr] genotypes reduce indigestible PA in soybean seeds in order to improve feeding efficiency of mono- and agastric animals, but often exhibit low field emergence, resulting in reduced yield. In this study, four LPA soybean varieties with two different genetic backgrounds were studied to assess their emergence and yield characters under 12 seed treatment combinations including two broad-spectrum, preplant fungicides (i.e., ApronMaxx (mefenoxam: (R,S)-2-[(2,6-dimethylphenyl)-methoxyacetylamino]-propionic acid methyl ester; fludioxonil: 4-(2,2-difluoro-1,3-benzodioxol-4-yl)-1H-pyrrole-3-carbonitrile) and Rancona Summit (ipconazole: 2-[(4-chlorophenyl)methyl]-5-(1-methylethyl)-1-(1H-1,2,4-triazol-1-ylmethyl) cyclopentanol; metalaxyl: N-(methooxyacetyl)-N-(2,6-xylyl)-DL-alaninate)), osmotic priming, and MicroCel-E coating. Two normal-PA (NPA) varieties served as controls. Both irrigated and non-irrigated plots were planted in Blacksburg and Orange, Virginia, USA in 2014 and 2015. Results revealed that three seed treatments (fungicides Rancona Summit and ApronMaxx, as well as Priming + Rancona) significantly improved field emergence by 6.4–11.6% across all genotypes, compared with untreated seeds. Seed priming was negatively associated with emergence across LPA genotypes. Seed treatments did not increase the yield of any genotype. LPA genotypes containing *mips* or *lpa1/lpa2* mutations, produced satisfactory emergence similar to NPA under certain soil and environmental conditions due to the interaction of genotype and environment. Effective seed treatments applied to LPA soybeans along with the successful development of LPA germplasm by soybean breeding programs, will increase use of LPA varieties by commercial soybean growers, ultimately improving animal nutrition while easing environmental impact.

Keywords: field emergence; low phytic acid; seed treatment; soybean

1. Introduction

Grain soybean [*Glycine max* (L.) Merr] is one of the most important crops for animal feed in the United States due to its high protein content and wide adaptability. Seventy-five percent of phosphorus (P) in soybean seeds is in the form of phytic acid (PA), myo-inositol-1,2,3,4,5,6-hexakisphosphate, which is indigestible for agastric and monogastric animals such as swine, poultry, and most aquatic animals, leading to low feeding efficiency [1]. In addition, other essential minerals, such as calcium, iron, manganese, and zinc, are bound by phytic acid, forming insoluble phytate salts, that render them unavailable, resulting in nutrient deficiencies in monogastric animals [2]. Furthermore, these nondigestible phytate salts are excreted by animals and become an important source of P pollution detrimental to the environment causing massive algal blooms and fish death [3,4].

Although animal producers have long added synthetic phytase to animal feed to improve PA digestibility, a much more effective method would be the utilization of low-PA (LPA) seeds developed from mutant lines. Three mutant alleles have been reported to create soybean LPA varieties [5]. The first two, *lpa1* and *lpa2*, were both discovered in mutant line CX-1834. These alleles lower phytate by producing a truncated ABC transporter responsible for partitioning PA into seeds [6]. The third mutant allele, *mips1*, is responsible for the first step in PA biosynthesis, catalyzing the NADH-dependent conversion of glucose-6-phosphate to myo-inositol-3-phosphate [7]. However, these mutations not only reduce the seed phytic acid levels in soybean, but also affect the pathways associated with seed development, leading to reduced seed germinability and ultimately low emergence [8,9]. Recent studies showed that many transcriptional genes in biological processes, such as those related to phytic acid metabolism and seed dormancy were involved in this process and the expression diversification of antioxidation-related and hormone-related genes were reported to strongly contribute to variations of emergence rate of LPA soybean lines [8–10]. So far, the mechanism of seed emergence in LPA soybean lines remains unclear and requires further exploration.

Poor field emergence has greatly hindered the use of LPA germplasm in soybean breeding programs [10]. Many attempts have been made to improve emergence in the past few decades. Previous studies showed that soybean seeds produced in temperate environments exhibited higher field emergence than those from tropical/subtropical environments, which illustrates the importance of seed production environment on LPA cultivars for commercial production [11–13]. Maupin and Rainey (2011) reported some seeds derived from the LPA genotype (*mips1*) had field emergence (above 85%) similar to normal-PA (NPA) soybean lines, indicating the potential to develop high emerging LPA soybean lines from natural variations within some LPA mutants [12]. Recently, several new LPA soybean lines (such as 56CX-1273), which display rapid emergence and good agronomic performance, have been developed using traditional crossbreeding methods as well as transgenic technologies [2,14].

Seed treatments improve field emergence of a broad range of crops including soybean. Fungicide treatment is one of the most commonly used to increase soybean stand establishment because it protects seed/seedling from seed- and soil-borne diseases, such as seed rot and damping-off caused by *Phytophthora* spp. [15,16]. Seed priming increases soybean seed vigor, and consequently improves seedling emergence under normal or stressful conditions [17]. Priming involves a controlled hydration procedure followed by redrying applied preplant that allows initial metabolic processes required for seed germination to occur prior to planting resulting in faster germination and uniform field establishment [18]. Additionally, mineral nutrients have been applied preplant as seed coating treatments to improve seedling growth [19,20]. Micro-Cel E, a synthetic calcium silicate, produced by the hydrothermal reaction of diatomaceous silica and high purity lime, can supply plant-essential nutrients and has pesticidal properties. Micro-Cel can be applied as a seed coating and may improve soybean stand establishment.

However, no seed treatment consistently increases the field emergence of LPA soybean lines. The purpose of this study was to apply twelve combinations of four seed treatments: two fungicide treatments (ApronMaxx and Rancona Summit) reported to greatly improve soybean emergence previously [21,22], osmotic priming with potassium phosphate solution and seed coating using Micro-Cel E, to increase field emergence of LPA soybeans. The objective was to: (1) evaluate the seed and seedling vigor of four newly developed LPA soybean varieties (56CX-1283, MD 03-5453, V12-4557, and V12-BB144) with two different genetic backgrounds (i.e., 56CX-1283 and MD 03-5453 having both the *lpa1* and *lpa2* alleles, while V12-4557 and V12-BB144 have the *mips1* allele), and (2) establish a preplant seed enhancement treatment that can effectively improve field emergence and establishment of LPA soybeans.

2. Materials and Methods

2.1. Plant Materials

Six maturity group V soybean varieties were studied: four LPA and two NPA (Table 1). The four LPA genotypes were 56CX-1283, MD 03-5453, V12-4557 and V12-BB144. 56CX-1283 and MD 03-5453 were developed by the USDA-ARS-Purdue University and University of Maryland, respectively, and contain both the *lpa1* and *lpa2* alleles. V12-4557 and V12-BB144 were developed at Virginia Tech and have the *mips1* allele. Seeds of all six varieties were grown in the same field at the Virginia Tech Kentland Research Farm near Blacksburg, VA using identical agronomic practices the previous year. Seeds of all six genotypes were dried to 9% moisture (dwt basis) after harvest and stored in sealed paper bags maintained in dark in a room maintained at 21 °C until planted the following growing season. The LPA varieties' PA content ranged from 2132 to 4421 ppm. AG 5632 (Bayer, Pittsburgh, PA) and 5002T [23] are both NPA commercial varieties. Their PA content ranged from 5887 to 6116 ppm. MD 03-5453 and V12-4557 have a history of poor field emergence and were not tested in 2014 but were added into the study in 2015.

Table 1. The phytic acid (PA) content, genetic source of the low-PA (LPA) trait, and the years planted for each soybean genotype in this study.

Genotype	LPA Gene	Years Planted	PA Content (ppm)
5002T	N/A	2014, 2015	6116.10
AG 5632	N/A	2014, 2015	5886.72
56CX-1283	*lpa1/lpa2*	2014, 2015	2486.03
MD 03-5453	*lpa1/lpa2*	2015	2131.68
V12-4557	*mips1*	2015	4060.80
V12-BB144	*mips1*	2014, 2015	4420.50

2.2. Field Plot Design and Trait Measurement

The experimental design was a triplicated split plot generalized randomized complete block design (GRCBD) wherein the main plots were blocked by the two locations (VT Kentland Farm, Blacksburg and VT Northern Piedmont Research Station, Orange, VA) and split into irrigated and non-irrigated subplots. The Blacksburg location has Hayter loam fine-loamy, mixed, active, mesic Ultic Hapludalfs soil type. The Northern Piedmont site near Orange, VA has Davidson clay loam fine, kaolinitic, thermic Rhodic Kandiudults soil type [24]. Plots were irrigated shortly after planting and kept wet until emergence to simulate damp spring planting conditions typically encountered. All plots were planted using a small-plot mechanical seeder in the last week of May. Orange usually a little warmer than Blacksburg (about 4 °C higher in average in 2014 and 2015) and gets more precipitation (32 more mm in average in 2014 and 2015) than Blacksburg in late May. Each plot was planted in two 3.05 m-long rows spaced 0.82 m apart with 80 seeds per row at a density of 26 seeds per meter. Stand counts were taken at the V1 stage (one set of unfolded trifoliolate leaves) [25]. The plots were once-over destructively harvested in late October (Orange) and early November (Blacksburg). Grain weight and moisture content were recorded for each plot and converted to yield (kg ha^{-1}) at 13% moisture on dry weight basis. Phytic acid was measured by high-throughput indirect Fe colorimetry [26].

2.3. Seed Treatments

Twelve seed treatment combinations (Table 2) were tested in 2014: MicroCel-E (synthetic calcium silicate, $CaSiO_3$; Manville, Denver, CO); osmotic priming; two fungicides, ApronMaxx (mefenoxam: (R,S)-2-[(2,6-dimethylphenyl)-methoxyacetylamino]-propionic acid methyl ester; fludioxonil: 4-(2,2-difluoro-1,3-benzodioxol-4-yl)-1H-pyrrole-3-carbonitrile; Syngenta Crop Protection, Greensboro, NC) and Rancona Summit (ipconazole:

2-[(4-chlorophenyl)methyl]-5-(1-methylethyl)-1-(1H-1,2,4-triazol-1-ylmethyl) cyclopentanol; metalaxyl: N-(methooxyacetyl)-N-(2,6-xylyl)-DL-alaninate; Valent USA, Walnut Creek, CA, USA); all possible two and three way combinations; and an untreated control. The specific treatments were selected based on prior unpublished results of germination tests. MicroCel-E was ineffective in 2014, so it was excluded in the 2015 trials (Table 2).

Table 2. Seed treatments used in this study.

Treatment	Years Used	Use
Control	2014, 2015	Untreated control
ApronMaxx	2014, 2015	Broad spectrum fungicide
MicroCel-E	2014	Weak fertilizer
Priming	2014, 2015	Post-harvest preplant controlled hydration treatment followed by redrying prior to planting
Rancona Summit	2014, 2015	Broad spectrum fungicide
Priming + Rancona	2014, 2015	
Priming + ApronMaxx	2014, 2015	
Priming + MicroCel-E	2014	
Priming + MicroCel-E + Rancona	2014	
Priming + MicroCel-E + ApronMaxx	2014	
MicroCel-E + Rancona	2014	
MicroCel-E + ApronMaxx	2014	

MicroCel-E was applied to seeds in seed coating. Polyvinyl acetate based-adhesive, $(C_4H_6O_2)_n$ (Elmer's Glue-All, Elmer's Products, Westerville, OH, USA) was diluted 10 times with tap water and misted on seeds (2.5 mL/1000 seeds) in a rotating bowl. Powedered MicroCel-E was slowly added by hand to coat seeds with a thin layer (2.5 mg MicroCel-E/1000 seeds). Seeds were immediately dried at 32 °C in a forced-air dryer for 24 h.

Before seeds were incubated in osmoticum for priming, they were surface sanitized with 30% bleach (8.25%, sodium hypoclorite, NaOCl) solution for 4 min then rinsed in double distilled water (ddH_2O). A single layer of seeds was osmotically primed in 3% monopotassium phosphate (KH_2PO_4) solution in ddH_2O on two thicknesses of germination blotter paper (Anchor Paper Co., St. Paul, MN, USA, 9.5 × 9.5 cm) saturated with 20 mL of solution. Seeds were sealed in square clear plastic boxes (10.1 × 10.1 × 3.5 cm OD). Seeds were placed in an incubator at 16 °C for 72 h in dark and force-air dried to their original moisture content.

Seeds were briefly soaked with two aqueous broad-spectrum fungicides, ApronMaxx and Rancona Summit according to label instructions. Twelve mL ApronMaxx was mixed with 10 mL red dye and 78 mL water, and 26 mL Rancona was mixed with 10 mL dye and 64 mL water, respectively. Seeds were treated by applying 2.25 mL of fungicide solution per 1000 seeds in a rotating drum. The seeds were force-air dried to their original moisture content after treated. For treatments with both fungicides and MicroCel-E, solutions were modified to contain either 8 mL Rancona, 7.5 mL dye, and 34.5 mL water; or 6 mL ApronMaxx, 7.5 mL dye, and 36.5 mL water. Three mL of fungicide solutions were applied per 1000 seeds. Once treated, the seeds were dried in a 32 °C dryer for 24 h. All untreated controls were also dried as previously described, so moisture contents of all treatments ranged from 7.5 to 9.5% (dwt basis, 17 h 103 °C).

2.4. Statistical Analysis

Correlation analysis of linear lines was calculated using JMP 11 software (SAS Inc, Raleigh, NC, USA) and R software package "corrplot" [27]. A split-split plot analysis of variance was performed using R software with packages "lattice" [28], "car" [29], and "agricolae" [30]. Analysis of variance effects included treatment, genotype, location, irrigation, and replication. The AOV function was used to perform an ANOVA using the following formula for each year, 2014 and 2015, separately. Means of significant *F*-test were separated by Tukey's Honestly Significant Difference ($HSD_{\alpha = 0.05}$).

3. Results

3.1. Effects of Genetic and Environmental Factors on Field Emergence

In 2014, an ANOVA revealed significant variation among treatments, varieties, and irrigation regimes. Significant interactions included treatment × genotype, treatment × location, line × location, treatment × irrigation, line × irrigation, treatment × line × location, treatment × location × irrigation, and line × location × irrigation ($p < 0.05$). In 2015, an ANOVA revealed significant variation among treatments and genotype. Significant interactions occurred for treatment × line, treatment × location, line × location, line × irrigation, and treatment × line × irrigation ($p < 0.05$). Field emergence data were averaged separately for each irrigation regime, location in 2014 and 2015 for all soybean genotypes (Table 3). The average field emergence of NPA AG 5632 was 82.6% in 2014, higher than LPA varieties 56CX-1283 (72.1%) and V12-BB144 (70.3%), both of which had significantly higher emergence than NPA genotype 5002T (68.4%). In 2015, 79.5% of 56CX-1283 emerged, followed by NPA genotypes AG-5632 and 5002T without significant differences. For LPA genotype V12-4557, 71.9% of seeds emerged not significantly different from 5002T. Both V12-BB144 (61.8%) and LPA genotype MD 03-5453 (46.7%) were significantly lower than all other genotypes grown in 2015. Across treatments in both years, LPA genotype 56CX-1283 seeds emerged to similar percentages compared to NPA varieties, while the LPA genotype V12-BB144 had variable performance relative to the NPA varieties in Northern Piedmont (Table 3). MD03-5453 and V12-4557 had lower field emergence than the NPA varieties in 2015 with MD03-5453 being the lowest.

For all varieties, field emergence was 4.2% higher in 2014 compared to 2015, probably partly due to the introduction of MD 03-5453 into the study. An overall trend of higher mean emergence in non-irrigated trials compared to irrigated trials occurred across years and locations, indicated that excessive water may have been applied, irrigation increased disease pressure, or some genotypes were sensitive to moist soils. Field emergence percentage varied by location from 77.9% in Orange in 2014 to 68.8% in Blacksburg for 2014. The 2015 mean emergence in Blacksburg was 75.2%, and 63.0% for Orange (Table 3).

3.2. General Effects of Seed Treatments on Field Emergence

Seeds treated by Rancona Summit displayed the highest emergence across locations and irrigation regimes in 2014 with an average of 82.1% (Table 4). Rancona Summit was followed in descending order by ApronMaxx (81.9%), the control (80.2%), MicroCel-E + ApronMaxx (78.4%), MicroCel-E + Rancona Summit (76.8%), and Priming + Rancona Summit (76.3%). The untreated control was not significantly different from any seed treatment. Untreated seeds emerged to higher percentages than MicroCel-E, Priming + MicroCel-E + Rancona Summit, Priming, Priming + MicroCel-E, Priming + MicroCel-E + ApronMaxx, and Priming + ApronMaxx.

Emergence data were collected on fewer treatments in 2015 after ineffective treatments were identified in 2014. Untreated seed emergence was 67.4%, significantly lower compared with the three most effective seed treatments. Rancona Summit (79.0%) and ApronMaxx (76.6%) producing the highest emergence (Table 4). The Priming + Rancona Summit treatment performed better compared to the control in 2015 with emergence of 73.8%. Untreated control emergence was significantly higher than Priming and Priming + ApronMaxx treatments.

Table 3. Field emergence of two normal-PA (NPA) and four LPA soybean varieties grown at Blacksburg and Orange in 2014 and 2015.

Line	Irrigation	Phytate	Emergence %					
			2014 Only	2015 Only	2014 BB [1]	2014 O	2015 BB	2015 O
5002T	I [2]	Normal	64.6 *	71.7 *	65.4d	71.4c	76.5bc	74.7cd
	N		72.1	79.5				
	Mean		68.4c [3]	75.6ab				
AG-5632	I	Normal	79.2 *	77.4	76.8b	88.4a	85.5a	72.2cd
	N		86.0	79.9				
	Mean		82.6a	78.7a				
56CX-1283	I	*lpa1/lpa2*	69.7 *	79.2	67.6d	76.5b	82.4ab	76.7bc
	N		74.4	79.9				
	Mean		72.1b	79.5a				
V12-BB144	I	*mips1*	66.0 *	56.5 *	65.2d	75.3b	77.1bc	46.8f
	N		74.6	66.9				
	Mean		70.3b	61.8c				
MD 03-5453	I	*lpa1/lpa2*	-	44.8	-	-	54.0e	39.6g
	N		-	48.7				
	Mean		-	46.7d				
V12-4557	I	*mips1*	-	68.1 *	-	-	75.8bc	68.2d
	N		-	76.0				
	Mean		-	71.9b				
	Mean		73.3	69.1	68.8	77.9	75.2	63.0

[1] BB = Blacksburg; O = Orange. [2] I = Irrigated; N = Non-irrigated. [3] Values within a column with values for a single year (2014 only; 2015 only or within a set of columns with values for a single year (2014 BB and O; 2015 BB and O) followed by the same letter are not significantly different based on Tukey's HSD ($\alpha = 0.05$). * Indicates a significant difference ($p \leq 0.05$) in emergence between irrigation regimes for the corresponding genotype.

Table 4. Average field emergence and Tukey's separation of means for 12 seed treatment combinations in 2014 and 2015.

Treatment	Irrigation	Emergence %					
		2014 Only	2015 Only	2014 BB [1]	2014 O	2015 BB	2015 O
Control (untreated)	I [2]	77.8	64.4				
	N	82.6	70.3	74.0fghij	86.3a	71.0b	63.8cd
	Mean	80.2ab [4]	67.4c				
ApronMaxx	I	77.9 *	73.7				
	N	85.8	79.5	78.3cdefg	85.5ab	83.4a	69.8bc
	Mean	81.9a	76.6ab				
Rancona	I	78.6 *	76.2				
	N	85.6	81.7	76.8defgh	87.4a	86.3a	72.0b
	Mean	82.1a	79.0a				
Microcel	I	72.2	-				
	N	76.7	-	69.7ijk	79.2bcdef	-	-
	Mean	74.5c	-				
Priming	I	63.7 *	55.7				
	N	73.0	58.8	63.4lmn	73.3fghijk	62.6d	52.0e
	Mean	68.4de	57.3d				
MxA [3]	I	76.6	-				
	N	80.2	-	74.9efghi	81.8abcd	-	-
	Mean	78.4abc	-				
MxR	I	73.0 *	-				
	N	80.7	-	70.2ijk	83.5abc	-	-
	Mean	76.8bc	-				
PA	I	61.0	58.0				
	N	63.9	63.5	56.9o	68.0jklm	68.9bcd	52.6e
	Mean	62.4f	60.7d				
PR	I	71.5 *	70.3 *				
	N	81.3	77.4	71.1hijk	81.2abcde	79.5a	68.5bcd
	Mean	76.3bc	73.8b				
PM	I	60.2 *	-				
	N	70.9	-	62.2mno	68.9ijkl	-	-
	Mean	65.5ef	-				
PxMxA	I	59.3 *	-				
	N	68.1	-	59.8no	67.6klm	-	-
	Mean	63.7f	-				
PxMxR	I	66.8	-				
	N	73.0	-	67.9jklm	71.9ghijk	-	-
	Mean	69.9d	-				
	Mean	73.3	69.1	68.8	77.9	75.3	63.1

[1] BB = Blacksburg; O = Orange; [2] I = Irrigated; N = Non-irrigated; [3] M = MicroCel-E; A = ApronMaxx; R = Rancona; P = Priming; [4] Values within a column with values for a single year (2014 only; 2015 only or within a set of columns with values for a single year (2014 BB & O; 2015 BB & O) followed by the same letter are not significantly different based on Tukey's HSD ($\alpha = 0.05$). * Indicates a significant difference ($p \leq 0.05$) in emergence between irrigation regimes for the corresponding.

3.3. *Effects of Seed Treatments on Field Emergence by PA Phenotype*

Analysis of the two NPA and four LPA soybean varieties and six treatments showed patterns of field emergence among phytic acid types in 2015 (Figure 1). The untreated control treatment average field emergence for the two *mips1* genotypes was 70.5%, while untreated seed emergence for the two *lpa1*/*lpa2* genotypes was 59.3%. Seeds treated with Rancona Summit, ApronMaxx, and Priming + Rancona had slightly higher average field emergence compared to untreated seeds.

Priming and Priming + ApronMaxx tended to decrease emergence relative to untreated seeds. The overall mean of untreated emergence of the LPA varieties was 64.0%, while emergence of the two NPA varieties was 72.2%. This confirms that the LPA varieties may have lower inherent emergence than NPA varieties, although this could be largely due to the inclusion of the low-emerging genotype MD 03-5453 and the likelihood of greater disease incidence in wet soils. ApronMaxx and Rancona fungicide treatments, however, improved the emergence of LPA varieties by between 12.9% and 14.1%. This improvement suggests that fungicide treatments have the potential, when optimized, to improve LPA seed emergence to essentially the same percentages as NPA varieties.

3.4. *Effect of Seed Treatments on Field Emergence of LPA and NPA Genotypes*

The application of some seed treatments to the six different soybean LPA and NPA genotypes significantly improved field emergence, but effects were specific to each line (Table 5). The treatment × genotype combinations produced significant variations in both 2014 and 2015. In 2014, no treatment × genotype combination significantly improved emergence over untreated seeds of the same genotype. ApronMaxx, Rancona Summit, MicoCel-E, MicroCel-E + ApronMaxx, MicroCel-E + Rancona Summit, and Priming + Rancona Summit treatments increased emergence relative to untreated controls for at least one of four genotypes grown in 2014 (Table 5). For LPA genotypes, nearly all treatments decreased emergence relative to the control by 2–37%. The exceptions to this decrease was line V12-BB144, which ApronMaxx, Rancona Summit, and MicroCel-E + ApronMaxx treatments slightly increased (0.2–1.4%) or had no effect on emergence. MicroCel-E, Priming, Priming + ApronMaxx, Priming + MicroCel-E, Priming + MicroCel-E + ApronMaxx, and Priming + MicroCel-E + Rancona Summit each had a significant decrease in emergence relative to the control for at least one of four genotypes (Table 5).

In 2015, several treatment × genotype combinations improved emergence compared to untreated seeds of the same genotype. ApronMaxx and Rancona Summit treatments increased emergence over untreated seeds of LPA line MD 03-5453, and the highest emergence was 69.5% (Table 5). Rancona Summit as well as Priming + Rancona Summit increased emergence for NPA line 5002T compared to the control. ApronMaxx, Rancona Summit, and Priming + Rancona Summit treatments increased emergence for most genotypes relative to untreated, but the increases were not significant except as described previously. Both Priming and Priming + ApronMaxx treatments decreased emergence for five out of six genotypes from 0.7–27% compared to the control (Table 5).

3.5. *Effect of Treatments, Genotypes, and Treatment × Line Interactions on Yield*

No seed treatment significantly increased yield of any genotype in either year (Table 6). However, significant differences in yield existed among genotypes in both years. In 2014, NPA line AG-5632 had the highest mean yield across all treatments at 5145 kg ha^{-1}, followed by LPA line 56CX-1283 (4849 kg ha^{-1}), NPA line 5002T (4768 kg ha^{-1}), and V12-BB144 (4479 kg ha^{-1}). Only 56CX-1283 and 5002T were not significantly different. In 2015, the yield of 56CX-1283 and AG-5632 were not significantly different. Genotypes 5002T, V12-BB144, and V12-4557 yielded significantly less than either 56CX-1283 or AG 5632. The lowest yielding line was MD 03-5453 at 1211 kg ha^{-1}, significantly lower than all other genotypes. No seed treatment increased yield compared to the control for each respective genotype. Seed treatment differences existed among genotypes, but not for treatments applied to the same genotype.

A

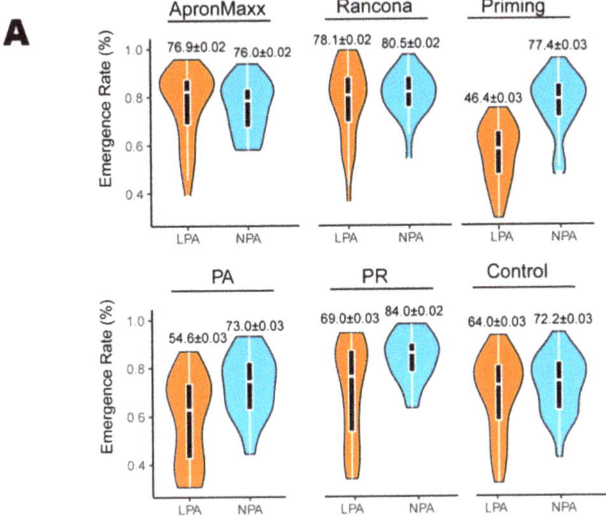

B

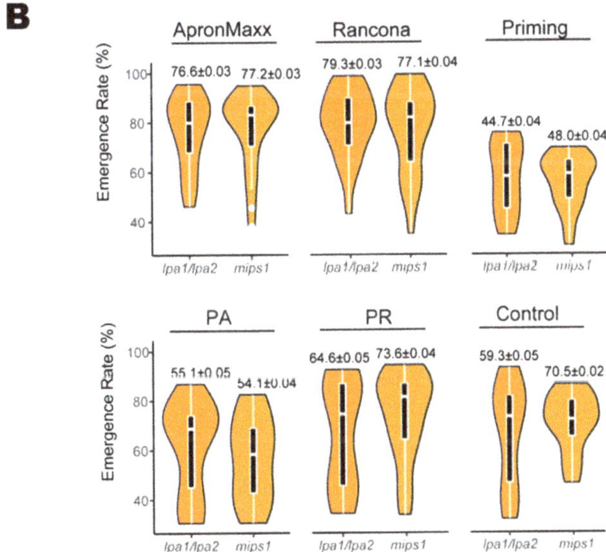

Figure 1. Field emergence for six seed treatments across different soybean varieties grown in 2015. Emergence rates of four low phytic acid (LPA) and two normal phytic acid (NPA) soybean varieties (**A**) as well as two LPA varieties with *lpa1* and *lpa2* alleles and two other LPA varieties with the *mips1* allele (**B**) were calculated. White horizontal lines at the center of each box show median values. The bounds of each black box show the quartiles, and the upper and lower bars show the maximum and minimum values, respectively. This image was drawn using ggplot2 package. Numbers above the violin plot indicate the means ± standard deviations. PA: Priming + Rancona; PR: Priming + ApronMaxx.

Table 5. Effects of seed treatments on field emergence in NPA and LPA soybeans and Tukey's separation of means in 2014 and 2015.

Line	Phytate	Emergence %												
		C [1]	A	R	M	P	MxA	MxR	PA	PR	PM	PxMxA	PxMxR	Mean
2014														
5002T	NPA [2]	77.6	84.4	81.3	70.4	66.0 *	72.1	75	53.3 *	70.7	51.5 *	50.8 *	67.1 *	68.4c [3]
AG-5632	NPA	81.8	87.3	87.3	82	79.4	87.7	82.5	80.3	84.5	80	77.6	80.8	82.6a
56CX-1283	LPA	85.3	79.9	83.6	73.9 *	69.4 *	76.4	77.9	48.4 *	79.1	71.0 *	58.4 *	61.3 *	72.1b
V12-BB144	LPA	75.9	75.9	76.1	71.6	58.5 *	77.3	72	67.7	70.2	59.6 *	68	70.5	70.3b
Mean	-	80.2ab	81.9a	82.1a	74.5c	68.4de	78.4abc	76.8bc	62.4f	76.3bc	65.5ef	63.7f	69.9d	-
2015														
5002T	NPA	65.1	77.8	78.8 *		76.3			71.6	84.0 *				75.6ab
AG-5632	NPA	79.3	74.3	82.3		78.6			74.3	84.1				78.7a
56CX-1283	LPA	80.4	85.4	88.6		63.2 *			74.6	85.1				79.5a
V12-BB144	LPA	67.9	72.5	71.6		41.3 *			47.9 *	69				61.8c
MD 03-5453	LPA	36.3	67.7 *	69.5 *		26.2			35.5	44.1				46.7d
V12-4557	LPA	72.9	81.8	82.6		56.7 *			60.4	78.6				71.9b
Mean	-	67.4c	76.6ab	79.0a		57.3d			60.7d	73.8b				-

[1] C = control; A = ApronMaxx; R = Rancona Summit; M = MicroCel-E; P = Priming; [2] NPA = normal phytic acid; LPA = low phytic acid; [3] Values followed by the same letter within bordered columns or rows are not significantly different based on Tukey's HSD ($\alpha = 0.05$). * Indicates a treatment is significantly different from the control treatment for the corresponding genotype.

Table 6. Effects of seed treatments on yield of six soybean lines grown at Blacksburg and Orange and Tukey's separation of means in 2014 and 2015.

Line	Phytate	C [1]	A	R	M	P	MxA	MxR	PA	PR	PM	PxMxA	PxMxR	Mean
		2014												
		Yield kg ha^{-1}												
5002T	NPA [2]	4782	4983	4909	4782	4459	4842	5057	4815	4701	4519	4573	4768	4768b [3]
AG-5632	NPA	4936	5151	5151	5091	4956	5077	5387	5185	5219	5185	5151	5205	5144a
56CX-1283	LPA	4882	4882	4922	4829	4882	4936	5010	4546	5098	4808	4707	4721	4849b
V12-BB144	LPA	4304	4431	4566	4734	4465	4425	4492	4580	4526	4156	4539	4499	4479c
Mean	-	4728ab	4862ab	4889ab	4356ab	4687ab	4822ab	4990a	4782ab	4896ab	4667b	4741ab	4795ab	-
		2015												
5002T	NPA	3477	3618	3551		3558			3564	3511				3544b
AG-5632	NPA	4284	3827	3867		4270			3867	3901				4008a
56CX-1283	LPA	4001	3921	4304		4479			4102	4055				4143a
V12-BB144	LPA	3685	3894	3732		3141			3268	3470				3531b
MD 03-5453	LPA	1110	1506	1076		659			1446	1439				1211c
V12-4557	LPA	3571	3800	3531		3477			3443	3289				3517b
Mean	-	3416a	3430a	3336a		3302a			3309a	3255a				-

[1] C = control; A = ApronMaxx; R = Rancona Summit; M = MicroCel-E; P = Priming [2] NPA = normal phytic acid; LPA = low phytic acid [3] Values followed by the same letter within bordered columns or rows are not significantly different based on Tukey's HSD ($\alpha = 0.05$).

4. Discussion

A major use of grain soybean is animal feed because of its high protein content. However, high levels of PA in soybean seeds may lead to animal mineral and protein malnutrition. In addition, phytic acid phosphorus excreted by monogastric animals such as poultry, swine, and fish can become a pollutant. These problems have provided plant geneticists with an incentive to develop LPA soybean varieties [31]. However, PA is also important for the growth and development of soybean seedlings because it is a primary storage reserve of phosphate in seeds. Phosphate is an essential component of adenosine triphosphate (ATP) that provides energy necessary for seedling growth and development. Thus, phosphorus is essential for the general health and vigor of developing seedlings. Reducing seed phytate by re-engineering synthesis pathway often has the unintended consequence of reducing seedling vigor and harming crop establishment [32]. Unfortunately, LPA soybeans often exhibit lower field emergence, making them problematic to grow particularly during stressful growing conditions.

This study included MD 03-5453 and 56CX-1283 expressing *lpa1*/*lpa2* homologs responsible for a low phytic acid phenotype (Table 1). In combination, *lpa1*/*lpa2* lower the PA content to about 25% of NPA genotypes while the remaining 75% phosphorus is inorganic [1,33]. This study also included V12-4557 and V12-BB144 genotypes, expressing *mips1*, another allele responsible for LPA soybeans (Table 1). Compared with *lpa* mutants, *mips1* mutants have higher seed PA content where it usually accounts for 50% of total phosphorus. However, *mips1* mutants increase feed efficiency for mono-and a-gastric animals with the added benefit of a modified, beneficial sugar profile. Since they have higher PA than *lpa* mutants, germination would be predicted to be similar to wild-type soybeans.

The emergence data for LPA genotypes were inconsistent between genotypes and years compared to NPA genotypes. The *mips1* LPA genotype V12-BB144 showed higher emergence than NPA 5002T in 2014 but significantly lower emergence in 2015, while emergence of *mips1* LPA genotype V12-4557 was essentially the same as 5002T in 2015, the only year it was grown. LPA genotype MD 03-5453, containing *lpa1*/*lpa2*, emerged to lower percentages than all others in 2015. However, except for MD 03-5453 and V12-BB144 in 2015, the other two LPA genotypes exhibited average field emergence of around 70% or greater. This suggests that LPA genotypes containing *mips* or *lpa1*/*lpa2* mutations, can produce satisfactory emergence if seeds are carefully produced and stored properly prior to planting. Other studies have shown that LPA genotypes, *lpa1*/*lpa2* as well as *lpa1*/*lpa2* with GmIPK2 silenced, produced satisfactory germination or field emergence [14,34]. Maupin and Rainey (2011) reported average emergence of between 74–84% for varieties with *mips* or *lpa1*/*lpa2* mutations tested across 12 unique environments [12]. Anderson and Fehr (2008) reported up to 81.0% field emergence for *lpa1*/*lpa2* mutants from various seed sources [11].

Final grain yield was only loosely correlated with field emergence. Grain yield was not significantly affected by seed treatments. Soybean plants compensate by producing more pods per plant at wider spacings, so when emergence is slightly reduced, as was the case for most treatments in this study, yield was not affected [35].

Inconsistencies in emergence data among genotypes, treatments, and years were influenced by several important seed quality factors irrespective of the genetic-controlling phytate accumulation. This study was conducted at a cooler location at 650 m elevation (Blacksburg) and a warmer climate (Orange) at lower elevation in the Virginia Piedmont with vastly different soil types. Edaphic differences between locations such as soil microbes, soil texture, water holding capacity, etc., likely contributed to variation in emergence among genotypes and treatments complicating the conclusions about the role of seed phytate on emergence.

Environmental factors regulating seed fill can negatively impact seed vigor expression when seeds are grown for propagation. High temperatures, for example, during seed development decreased seed weight, caused shriveling, and decreased seed quality of soybean [36] and reduced soybean seed vigor in the absence of mechanical injury and seedborne diseases [37]. Drought stress on the parent soybean plant had little effect on seed quality although yields were reduced [38,39]. To mitigate maternal environmental effects on quality, seeds of all six genotypes used in this study were produced in the same season and location.

Seed vigor, another important determinant of emergence particularly under stressful field conditions, is affected by a number of factors such as: seed maturity at harvest, physical seed damage during harvest and transport, and improper storage. McDonald (1985) reviewed losses in seed vigor from maturation to planting in soybean as well as identifying seed quality tests that detected physical seed damage [40]. Although the seeds tested in the current study were grown at the same location to minimize differences in seed vigor, tailoring the time of harvest for highest seed vigor was not a focus. Delayed harvest may reduce soybean seed vigor [41]. Maximum seed quality and vigor often correlates with maximum dry weight accumulation [42]. However, physiological maturity can be better detected morphologically in some seeds. For example, maximum seed dry weight was not the best indicator of physiological maturity in common bean as pod color change [43]. Bean seeds with low quality produced fewer nodules, less nodule weight, and less nitrogen fixation that resulted in less plant growth and yield [44]. Vigor tests are more sensitive measures of seed quality than the standard germination tests or field stand counts, which are often used to assess germination of low phytate genotypes. Vigor tests in future studies could yield additional valuable information about the poor emergence sometimes observed in LPA soybean genotypes.

Improper post-harvest handling compromises seed quality. Open storage in combination with high relative humidity and high temperature can quickly result in a loss of seed vigor. All genotypes were grown and stored under identical conditions, so differences were most likely due to seed genotypes and not environment. Chauhan (1985) found the growing points of the embryonic axis in soybean were most prone to aging than other seed tissues [45]. This illustrates that seed tissues do not age simultaneously, and cotyledons may be healthy even after embryonic axis is damaged resulting in poor emergence. In this study all seeds were adjusted to the same moisture content after harvest and stored in paper bags at a room temperature. Seeds may have aged under these conditions, but all genotypes were exposed to the same aging conditions.

Hoy and Gamble (1985, 1987) found that soybean seed size had no effect on specific growth rate or seedling weight from planted seeds possessing no mechanical injury [46,47]. No improvement in speed of field emergence or final yield was detected when soybean seeds were separated into varying seed density classes [48]. Thus, in this study, seeds were not sized before field planting due to the poor correlation between seed size and seed vigor.

Seed treatments may benefit field emergence and were investigated as a strategy for improving establishment of LPA soybean genotypes. While there is no consensus about the exact reason for low emergence by LPA soybean genotypes, there are likely causes. Because some fungicide treatments improved LPA emergence, disease pressure before emergence is likely higher for LPA than NPA genotypes. Soil-borne pathogens are possibly the main cause of poor emergence in some seed lots planted in wet soils. Cellular leakage occurs in all seeds during imbibition because of cell membrane damage that occurs during desiccation that must be repaired. Cells repair membrane damage during hydration, and the duration of this process depends on seed quality. Electrolyte leakage is widely used vigor test to assess soybean seed quality [49]. Aged seeds leak more solutes and electrolytes than newly harvested undamaged high vigor seeds. Evidence suggests that some LPA genotypes naturally leak more compounds that attract seed/seedling pathogens because 75% of their phosphorus is inorganic [1,33]. The loss of inorganic P from the cytoplasm of LPA varieties due to imbibitional leakage could increase disease since leaked P can attract soil-borne pathogens to the emerging seedling explaining the benefits of fungicide treatments [50]. In addition, Douglass, et al. (1993) found a negative correlation between seed sugar content of differing sweet corn genotypes and emergence in cold soils [51] while this correlation is still unclear for soybean.

The lower emergence in irrigated plots supports the hypothesis that LPA are more prone to fungal attack since moist soils would create favorable conditions for disease development possibly leading to greater seed/seedling mortality. Fungicide was the most effective seed treatment in this study. Both fungicides significantly increased the field emergence of LPA genotype MD 03-5453, supported the hypothesis higher pre-emergence disease pressure could be a major cause of the low field emergence of LPA soybeans.

Osmotic priming is a common preplant controlled hydration treatment often applied to high value flower and vegetable seeds. Benefits of priming include faster germination, advancing seed maturity, leaching of inhibitors, and removal of dormancy. However, priming also reduces the storage life of seeds [18]. Priming treatments are less often applied to lower value agronomic seeds because the cost of application may outweigh benefits. Osmotic priming was used to increase germination rate so that seedlings would establish before diseases could infect vulnerable young plants [52]. Surprisingly, osmotic priming did not improve establishment and reduced field emergence similar to hydroprimed soybeans [53]. In the current study, seeds were primed in potassium phosphate solution which was not removed by washing at the end of treatment. These salts combined with the leakage of electrolytes that occurred during the controlled hydration priming treatment, described above, likely increased susceptibility to pathogenic attack as nutrients surrounded seeds and aided the proliferation of plant pathogens. Similarly, MicroCel-E, a calcium silicate processed from diatomaceous earth with a low salt index that contains small amounts of plant nutrients including phosphate, was applied as a seed coating. Ideally the nutrients would stimulate early seedling growth and the antipathogenic properties of diatomaceous earth may provide protection from insect and fungal predation. However, MicroCel-E consistently failed to improve emergence unless it was combined with a fungicide. The antifungal properties of diatomaceous earth were likely ineffective against seedling pathogens and insect predation. The fertilizer may have attracted and stimulated microbial growth unless fungicides Rancona Summit or ApronMaxx were present.

Emergence results were variable in this study, making it difficult to draw simple conclusions about treatments or genotypes. This is because of the complexity of factors interacting to affect field emergence. Edaphic stressors in the field commonly reduce emergence compared to results obtained from standardized laboratory germination tests conducted under near ideal conditions. In some plots, LPA seeds with *lpa1/lpa2* and *mips1* alleles had satisfactory field emergence compared to NPA. In other trials emergence of LPA genotypes was less than NPA likely because of conditions favoring seedling disease due to greater metabolite leakage from LPA seeds because of the altered phosphate and sugar metabolism which increased mortality. Osmotic priming and diatomaceous earth coating were ineffective. In some plots, seed fungicide treatments improved emergence of certain genotypes likely by protecting seeds/seedlings from pathogens that reduce emergence.

Author Contributions: Conceptualization, B.J.A., G.E.W. and B.Z.; methodology, G.E.W. and B.Z.; software, B.J.A., E.P. and J.Q.; formal analysis, B.J.A., E.P.; investigation, B.J.A.; resources, B.Z.; data curation, B.J.A.; writing—original draft preparation, B.J.A.; writing—review and editing, G.E.W., X.L., E.P. and J.Q., and B.Z.; visualization, B.J.A.; supervision, B.Z.; project administration, G.E.W. and B.Z.; funding acquisition, B.Z. All authors have read and agreed to the published version of the manuscript.

Funding: This research was funded by United Soybean Board.

Acknowledgments: This work was conducted under US multi-state project, W-468. Thanks to Luciana Rosso, Tom Pridgen, Steve Gulick, and Andy Jensen for technical support. Thanks to Hwasoo Shin for helping to make the graph.

Conflicts of Interest: The authors declare no conflict of interest.

References

1. Wilcox, J.R.; Premachandra, G.S.; Young, K.A.; Raboy, V. Isolation of high seed inorganic P, low-phytate soybean mutants. *Crop Sci.* **2000**, *40*, 1601–1605. [CrossRef]
2. Boehm, J.D.; Walker, F.R.; Bhandari, H.S.; Kopsell, D.; Pantalone, V.R. Seed inorganic phosphorus stability and agronomic performance of two low-phytate soybean lines evaluated across six southeastern US environments. *Crop Sci.* **2017**, *57*, 2555–2563. [CrossRef]

3. Schindler, D.W.; Hecky, R.E.; Findlay, D.L.; Stainton, M.P.; Parker, B.R.; Paterson, M.J.; Beaty, K.G.; Lyng, M.; Kasian, S.E.M. Eutrophication of lakes cannot be controlled by reducing nitrogen input: Results of a 37-year whole-ecosystem experiment. *Proc. Natl. Acad. Sci. USA* **2008**, *105*, 11254–11258. [CrossRef]
4. Sinkko, H.; Lukkari, K.; Sihvonen, L.M.; Sivonen, K.; Leivuori, M.; Rantanen, M.; Paulin, L.; Lyra, C. Bacteria contribute to sediment nutrient release and reflect progressed eutrophication-driven hypoxia in an organic-rich continental sea. *PLoS ONE* **2013**, *8*, e67061. [CrossRef]
5. Averitt, B.; Shang, C.; Rosso, L.; Qin, J.; Zhang, M.; Rainy, K.M.; Zhang, B. Impact of *mips1*, *lpa1*, and *lpa2* alleles for low phytic acid content on agronomic, seed quality, and seed composition traits of soybean. *Crop Sci.* **2017**, *57*, 2490–2499. [CrossRef]
6. Gillman, J.D.; Baxte, I.; Bilyeu, K. Phosphorus partitioning of soybean lines containing different mutant alleles of two soybean seed-specific adenosine triphosphate-binding cassette phytic acid transporter paralogs. *Plant Genome* **2013**, *6*, 1–10. [CrossRef]
7. Hegeman, C.E.; Good, L.L.; Grabau, E.A. Expression of D-myo-inositol-3-phosphate synthase in soybean. Implications for phytic acid biosynthesis. *Plant Physiol.* **2001**, *125*, 1941–1948. [CrossRef]
8. Redekar, N.; Pilot, G.; Raboy, V.; Li, S.; Saghai Maroof, M.A. Inference of transcription regulatory network in low phytic acid soybean seeds. *Front. Plant Sci.* **2017**, *8*, 2029. [CrossRef]
9. Yuan, F.; Yu, X.; Dong, D.; Yang, Q.; Fu, X.; Zhu, S.; Zhu, D. Whole genome-wide transcript profiling to identify differentially expressed genes associated with seed field emergence in two soybean low phytate mutants. *BMC Plant Biol.* **2017**, *17*, 1–17. [CrossRef]
10. Yu, X.; Jin, H.; Fu, X.; Yang, Q.; Yuan, F. Quantitative proteomic analyses of two soybean low phytic acid mutants to identify the genes associated with seed field emergence. *BMC Plant Biol.* **2019**, *19*. [CrossRef]
11. Anderson, B.P.; Fehr, W.R. Seed source affects field emergence of low-phytate soybean lines. *Crop Sci.* **2008**, *48*, 929–932. [CrossRef]
12. Maupin, L.M.; Rainey, K.M. Improving emergence of modified phosphorus composition soybeans: Genotypes, germplasm, environments, and selection. *Crop Sci.* **2011**, *51*, 1946–1955. [CrossRef]
13. Meis, S.J.; Fehr, W.R.; Schnebly, S.R. Seed source effect on field emergence of soybean lines with reduced phytate and raffinose saccharides. *Crop Sci.* **2003**, *43*, 1336–1339. [CrossRef]
14. Punjabi, M.; Bharadvaja, N.; Jolly, M.; Dahuja, A.; Sachdev, A. Development and evaluation of low phytic acid soybean by siRNA triggered seed specific silencing of inositol polyphosphate 6-/3-/5-kinase gene. *Front. Plant Sci.* **2018**, *9*, 804. [CrossRef] [PubMed]
15. Akamatsu, H.; Kato, M.; Ochi, S.; Mimuro, G.; Matsuoka, J.I.; Takahashi, M. Variation in the resistance of japanese soybean cultivars to phytophthora root and stem rot during the early plant growth stages and the effects of a fungicide seed treatment. *Plant Pathol. J.* **2019**, *35*, 219–233. [CrossRef]
16. Costa, E.M.; Nunes, B.M.; Ventura, M.V.A.; Mortate, R.K.; Vilarinho, M.S.; Da Silva, R.M.; Chagas, J.F.R.; Nogueira, L.C.A.; Arantes, B.H.T.; Lima, A.P.A.; et al. Physiological effects of insecticides and fungicide, applied in the treatment of seeds, on the germination and vigor of soybean seeds. *J. Agric. Sci.* **2019**, *11*. [CrossRef]
17. Rouhi, H.R.; Surki, A.A.; Sharif-Zadeh, F.; Afshari, R.T.; Aboutalebian, M.A.; Ahmadvand, G. Study of different priming treatments on germination traits of soybean seed lots. *Not. Sci. Biol.* **2011**, *3*, 101–108. [CrossRef]
18. Welbaum, G.E.; Shen, Z.; Oluoch, M.O.; Jett, L.W. The evolution and effects of priming vegetable seeds. *Seed Technol.* **1998**, *20*, 209–235. Available online: http://www.jstor.org/stable/23433024 (accessed on 1 October 2020).
19. Farooq, M.; Wahid, A.; Siddique, K.H.M. Micro-nutrient application through seed treatments—A review. *J. Soil Sci. Plant Nutr.* **2012**, *12*, 125–142. [CrossRef]
20. Sharma, K.K.; Singh, U.S.; Sharma, P.; Kumar, A.; Sharma, L. Seed treatments for sustainable agriculture—A review. *J. Appl. Nat. Sci.* **2015**, *7*, 521–539. [CrossRef]
21. Xue, A.G.; Cober, E.; Morrison, M.J.; Voldeng, H.D.; Ma, B.L. Effect of seed treatments on emergence, yield, and root rot severity of soybean under *Rhizoctonia solani* inoculated field conditions in Ontario. *Can. J. Plant Sci.* **2007**, *87*, 167–173. [CrossRef]
22. Nyandoro, R.; Chang, K.F.; Hwang, S.F.; Ahmed, H.U.; Turnbull, G.D.; Strelkov, S.E. Management of root rot of soybean in Alberta with fungicide seed treatments and genetic resistance. *Can. J. Plant Sci.* **2019**, *99*, 499–509. [CrossRef]

23. Pantalone, V.R.; Allen, F.; Landau-Ellis, D. Registration of '5002T' soybean. *Crop Sci.* **2004**, *44*, 1483–1484. [CrossRef]
24. Hayter Series. Available online: https://soilseries.sc.egov.usda.gov/OSD_Docs/H/HAYTER.html (accessed on 1 October 2020).
25. Fehr, W.R.; Caviness, C.E. *Stages of Soybean Development*; Special Report 80; Iowa State University: Ames, IA, USA, 1977.
26. Burleson, S.A.; Shang, C.; Luciana Rosso, M.; Maupin, L.M.; Rainey, K.M. A modified colorimetric method for selection of soybean phytate concentration. *Crop Sci.* **2012**, *52*, 122–127. [CrossRef]
27. Wei, T.; Simko, V. R Package "corrplot": Visualization of a Correlation Matrix (Version 0.84). 2017. Available online: https://github.com/taiyun/corrplot (accessed on 1 June 2019).
28. Sarkar, D. *Lattice: Multivariate Data Visualization with R*; Springer: New York, NY, USA, 2008.
29. Fox, J.; Weisberg, S. *An {R} Companion to Applied Regression*, 2nd ed.; Sage Publishing: Thousand Oaks, CA, USA, 2011; Available online: http://socserv.socsci.mcmaster.ca/jfox/Books/Companion (accessed on 1 June 2019).
30. de Mendiburu, F. Agricolae: Statistical Procedures for Agricultural Research. R package Version 1.3-0. 2019. Available online: https://CRAN.R-project.org/package=agricolae (accessed on 1 June 2019).
31. Raboy, V. Seeds for a better future:'low phytate'grains help to overcome malnutrition and reduce pollution. *Trends Plant Sci.* **2001**, *6*, 458–462. [CrossRef]
32. Bregitzer, P.; Raboy, V. Effects of four independent low-phytate mutations on barley agronomic performance. *Crop Sci.* **2006**, *46*, 1318–1322. [CrossRef]
33. Bilyeu, K.D.; Zeng, P.; Coello, P.; Zhang, Z.J.; Krishnan, H.B.; Bailey, A.; Beuselinck, P.R.; Polacco, J.C.; Department, B. Quantitative conversion of phytate to inorganic phosphorus in soybean seeds expressing a bacterial phytase. *Am. Soc. Plant Biol.* **2008**, *146*, 468–477. [CrossRef]
34. Shi, J.; Wang, H.; Schellin, K.; Li, B.; Faller, M.; Stoop, J.M.; Meeley, R.B.; Ertl, D.S.; Ranch, J.P.; Glassman, K. Embryo-specific silencing of a transporter reduces phytic acid content of maize and soybean seeds. *Nat. Biotechnol.* **2007**, *25*, 930–937. [CrossRef]
35. Lueschen, W.E.; Hicks, D.R. Influence of plant population on field performance of three soybean cultivars. *Agron. J.* **1977**, *69*, 390–393. [CrossRef]
36. Franca Neto, J.B.; Krzyzanowski, F.C.; Henning, A.A.; West, S.H.; Miranda, L.C. Soybean seed quality as aff ected by shriveling due to heat and drought stress during seed filling. *Seed Sci. Technol.* **1993**, *21*, 107–116.
37. Spears, J.; TeKrony, D.; Egli, D.B. Temperature during seed filling and soybean seed germination and vigor. *Seed Sci. Technol.* **1997**, *25*, 233–244.
38. Vieira, R.D.; TeKrony, D.M.; Egli, D.B. Effect of drought and defoliation stress in the field on soybean seed germination and vigor. *Crop Sci.* **1992**, *32*, 471–475. [CrossRef]
39. Vieira, R.; TeKrony, D.; Egli, D.B. Effect of drought stress on soybean seed germination and vigor. *Seed Sci. Technol.* **1991**, *15*, 12–21.
40. McDonald, M.B. Physical seed quality of soybean. *Seed Sci. Technol.* **1985**, *13*, 601–628.
41. Mugnisjah, W.; Nakamura, S. Vigour of soybean seed produced from different harvest date and phosphorus fertiliser application. *Seed Sci. Technol.* **1984**, *12*, 483–491.
42. Miles, D.F.; TeKrony, D.M.; Egli, D.B. Changes in viability, germination, and respiration of freshly harvested soybean seed during development. *Crop Sci.* **1988**, *28*, 700–704. [CrossRef]
43. Chamma, H.; Filho, J.M.; Crocomo, O.J. Maturation of seeds of "Aroana" beans (*Phaseolus vulgaris* L.) and its influence on the storage potential. *Seed Sci. Technol.* **1990**, *18*, 371–382.
44. Rodriguez, A.; McDonald, M.B. Seed quality influence on plant growth and dinitrogen fixation of red field bean. *Crop Sci.* **1989**, *29*, 1309–1314. [CrossRef]
45. Chauhan, K. The incidence of deterioration and its localisation in aged seeds of soybean and barley. *Seed Sci. Technol.* **1985**, *13*, 769–773.
46. Hoy, D.J.; Gamble, E.E. Field performance in soybean with seeds of differing size and density. *Crop Sci.* **1987**, *27*, 121–126. [CrossRef]
47. Hoy, D.J.; Gamble, E.E. The effects of seed size and seed density on germination and vigor in soybean (*Glycine max* (L.) Merr.). *Can. J. Plant Sci.* **1985**, *65*, 1–8. [CrossRef]
48. Egli, D.B.; TeKrony, D.M.; Wiralaga, R.A. Effect of soybean seed vigor and seed size on soybean seedling growth. *Seed Sci. Technol.* **1990**, *14*, 1–12.

49. Elias, S.G.; Copeland, L.O.; McDonald, M.B.; Baalbaki, R.Z. *Seed Testing: Principles and Practices*; Michigan State University Press: East Lansing, MI, USA, 2012.
50. Veresoglou, S.D.; Barto, E.K.; Menexes, G.; Rillig, M.C. Fertilization affects severity of disease caused by fungal plant pathogens. *Plant Pathol.* **2013**, *62*, 961–969. [CrossRef]
51. Douglass, S.K.; Juvik, J.A.; Splittstoesser, W.E. Sweet corn seedling emergence and variation in kernel carbohydrate reserves. *Seed Sci. Technol.* **1993**, *21*, 433–445.
52. Mushtaq, S.; Hafiz, A.I.; Hasan, S.Z.; Arif, M.; Shehzad, M.A.; Rafique, R.; Rasheed, M.; Ali, M.; Iqbal, M.S. Evaluation of seed priming on germination of *Gladiolus alatus*. *Afr. J. Biotechnol.* **2012**, *11*, 11520–11523. [CrossRef]
53. Kering, M.; Zhang, B. Effect of priming and seed size on germination and emergence of six food-type soybean varieties. *Int. J. Agron.* **2015**, *1*, 1–6. [CrossRef]

 agriculture

MDPI

Article

Tomato Seed Coat Permeability: Optimal Seed Treatment Chemical Properties for Targeting the Embryo with Implications for Internal Seed-Borne Pathogen Control

Hilary Mayton [1,†], Masoume Amirkhani [1,†], Daibin Yang [2], Stephen Donovan [3] and Alan G. Taylor [1,*]

1. Cornell AgriTech, School of Integrative Plant Science, Horticulture Section, Cornell University, New York, NY 14456, USA; hsm1@cornell.edu (H.M.); ma862@cornell.edu (M.A.)
2. Institute of Plant Protection, Chinese Academy of Agricultural Sciences, Beijing 100193, China; yangdaibin@caas.cn
3. Retired Agricultural Chemist, Currently Fellow at The Center for Forensic Science Research & Education, Willow Grove, PA 19090, USA; sdonovan@ptd.net

* Correspondence: agt1@cornell.edu; Tel.: +1-315-787-2243

† These authors contributed equally in this study.

Abstract: Seed treatments are frequently applied for the management of early-season pests, including seed-borne pathogens. However, to be effective against internal pathogens, the active ingredient must be able to penetrate the seed coat. Tomato seeds were the focus of this study, and the objectives were to (1) evaluate three coumarin fluorescent tracers in terms of uptake and (2) quantify seed coat permeability in relation to lipophilicity to better elucidate chemical movement in seed tissue. Uptake in seeds treated with coumarin 1, 120, and 151 was assessed by fluorescence microscopy. For quantitative studies, a series of 11 *n*-alkyl piperonyl amides with log K_{ow} in the range of 0.02–5.66 were applied, and two portions, namely, the embryo, and the endosperm + seed coat, were analyzed by high-performance liquid chromatography (HPLC). Coumarin 120 with the lowest log K_{ow} of 1.3 displayed greater seed uptake than coumarin 1 with a log K_{ow} of 2.9. In contrast, the optimal log K_{ow} for embryo uptake ranged from 2.9 to 3.3 derived from the amide series. Therefore, heterogeneous coumarin tracers were not suitable to determine optimal log K_{ow} for uptake. Three tomato varieties were investigated with the amide series, and the maximum percent recovered in the embryonic tissue ranged from only 1.2% to 5%. These data suggest that the application of active ingredients as seed treatments could result in suboptimal concentrations in the embryo being efficacious.

Keywords: tissue lipophilicity; systemic uptake; coumarin; piperonyl amides

check for updates

Citation: Mayton, H.; Amirkhani, M.; Yang, D.; Donovan, S.; Taylor, A.G. Tomato Seed Coat Permeability: Optimal Seed Treatment Chemical Properties for Targeting the Embryo with Implications for Internal Seed-Borne Pathogen Control. *Agriculture* **2021**, *11*, 199. https://doi.org/10.3390/agriculture11030199

Academic Editor: Rentao Song

Received: 13 February 2021
Accepted: 23 February 2021
Published: 28 February 2021

1. Introduction

Seed-borne pathogens are responsible for the initiation of numerous plant diseases and are one of the primary mechanisms for the global spread of plant pathogens [1–4]. Internal infection of seeds and colonization of the embryo and endosperm are most often associated with infection of the mother plant via the xylem, stigma, or non-vascular tissue [4–6]. Seed-borne pathogens have been observed in the seed embryo, storage tissue (endosperm and perisperm), and seed coat or testa [4,7]. Disinfection techniques can be used to remove and clean contaminants from the seed surface; however, plant pathogenic organisms located within the seed endosperm and embryo are much more difficult to control. Tomatoes are an important high-value vegetable crop and are susceptible to multiple pathogens. Tomato seeds can harbor fungal, bacterial, and viral pathogens [8–10]. Several systemic conventional pesticide seed treatments are available for fungal pathogens of tomato, but options are more limited for organic production and control of bacterial pathogens [3,11,12].

Seed treatments are applied worldwide for crop protection against pests and plant pathogens [13,14]; however, the systemic uptake and distribution of active ingredients of pesticide seed treatments in seed tissue have not been as well defined as root and leaf transport [15,16]. The long-term efficacy of seed treatments and control of seed-borne pathogens are dependent on seed coat permeability, as the active ingredients must be able to penetrate the seed coat and diffuse to the embryo. There have been several studies focused on the physiochemical barriers that prevent or allow a chemical to permeate the seed coats of several plant species [17–19]. Taylor and Salanenka (2012) developed a system to classify seed coat permeability based on the passage of ionic and non-ionic compounds through the seed coat of ten plant species from seven plant families [18]. Tomato seeds have selective permeability defined as only non-ionic compounds diffusing through the seed coat, while ionic compounds are blocked [17,18].

Seed uptake research on potential chemical pesticides applied as seed treatments is problematic due to the potential human risk of exposure to agrochemicals and/or radioactively labeled compounds. Fluorescent tracers provide an alternative approach, and coumarins are one group that includes several fluorescent, non-ionic tracers differing in chemical properties and that allows for a more comprehensive analysis of seed coat permeability characteristics [16]. Therefore, coumarin compounds were used both for qualitative uptake [16,17,19] and, using a single coumarin compound, for quantitative uptake research [20]. One objective of this research is to use three coumarin compounds with different chemical properties for tomato seed uptake to assess optimum log K_{ow}.

A key chemical property that affects the uptake of an organic compound in a seed is the log K_{ow}, also known as the log P [20–23]. A compound's lipophilicity is measured as the log K_{ow} and is the ratio of its chemical concentration in octanol (o) to its concentration in the aqueous (w) phase expressed on a log10 scale [24]. A series of fluorescent piperonyl amides were synthesized, and a novel combinatorial pharmacokinetic technique was developed to provide a range of compounds with log K_{ow} from 0.2 to 5.8. This series of fluorescent piperonyl amides was used to explore seed coat permeability and systemic uptake in soybean and corn seeds [23]. This same approach was adopted for tomato seed in this study.

Understanding the chemical/physical properties associated with the uptake of active ingredients in tomato seed tissue will aid in the development of new products for the control of internal seed-borne pathogens. The key objectives of this study were to evaluate the movement of selected coumarin compounds in uptake by fluorescence imaging and assess the role of log K_{ow} in seed tissue permeability using a homologous series of 11 fluorescent piperonyl amides quantified by high-performance liquid chromatography (HPLC).

2. Materials and Methods

2.1. Fluorescence Microscopy of Coumarin 1, 120, and 151 in Tomato Seeds

The first study was on the uptake of selected coumarin tracers in tomato seeds imaged by fluorescence microscopy. Tomato seeds of the variety "Hypeel 696" were provided by Seminis, Oxnard, CA, and coumarin 1, 120, and 151 were purchased from TCI America, Portland, OR. The chemical and other properties of these three coumarin compounds are shown in Table 1. Tomato seeds were treated with 3 μmoles of each coumarin per gram of seed, which was 0.833, 0.631, and 0.825 mg coumarin 1, 120, or 151, respectively, per gram of seed. Each coumarin compound was mixed with 3.8 mg L650 seed treatment binder (Incotec, Salinas, Canada), 250 μL deionized water, and mixed in a 50 mL centrifuge tube using a vortex mixer (Scientific Industries, Inc., Model 2-Genie No. G560, New York, NY, USA). Ten non-treated and treated tomato seeds of each tracer were sown in 20% moisture content silica sand (#1 Q-ROK, 0.15–0.84 mm, New England Silica, Inc., South Windsor, CT, USA) and maintained in a germinator at 20 °C for 40 h in the dark. Imbibed seeds were then removed and washed with deionized water, and then the seeds were dissected with scalpel blades and imaged under an Olympus microscope (SZX12, Tokyo, Japan), imaging camera (Infinity 3- 3URC, Lumenera Corp., Ottawa, ON, Canada), and Infinity Analyze (Revision 6.5.2, Teledyne Lumenera, Ottawa, ON, Canada). Seed tissue was illuminated with long UV light, UV lamp (Model 9-circular illuminator, Stocker & Yale, Salem, NH, USA). Non-treated seeds were used as the control.

Table 1. Physical/chemical properties of coumarin 120, 151, and 1.

Coumarin Compound	CAS Number	MW, g/mol	* Log K_{ow}	* Water Solubility, Log S	Excitation/Emission Max, nm	Molar Abs Coefficient, cm^{-1}	Quantum Yield
120	26093-31-2	175.2	1.25	1.25	342/409	3.50×10^8	0.63
151	53518-15-3	229.2	1.62	−3.56	364/460	4.58×10^8	0.53
1	91-44-1	231.3	2.90	−3.69	369/431	4.63×10^8	0.73

* Log K_{ow} and water solubility data obtained from Chemicalize, ChemAxon's cheminformatic tool. Excitation/emission, molar absorbance coefficient, and quantum yield information were obtained from Aazam (2010) [25] and Taniguchi and Lindsey (2018) [26].

2.2. Chemicals and Synthesis of N-alkyl Piperonyl Amides

An experimental series of *n*-alkyl piperonyl amides developed by S. Donovan and B. Black [23] was used in this study. There were 11 custom synthesized homologous piperonyl amides with carbons ranging from 1 to 14, with molecular weights of 189.2 to 361.5 g/mol. The methods and materials are described in Yang et al. (2018B) [23]. Briefly, 3.0 g of piperonylic acid was added to 5 mL of thionyl chloride. The solution was then refluxed for 30 min after which 5 mL pyridine, 25 mL toluene, and 18.1 mM amines were added and the solution was refluxed for 1 h. After cooling to ambient temperature, ethyl acetate (50 mL) was added and the solution was washed with saturated NaCl, 5% NaOH, and 5% HCl. The solution was then dried (using anhydrous sodium sulfate), filtered, and concentrated using a rotary evaporator. Recrystallization was achieved by refluxing 50 mL of methylcyclohexane until a solution was attained. The C1, C2, C3, and C4 *n*-alkyl piperonyl amides were made by adding a small amount of methylene chloride until the desired solution was completed. Lastly, all solutions were cooled and vacuum filtered before use. Each piperonyl amide (0.56 mM) solution consisted of 70% acetone + 30% water.

A short octadecyl-poly (vinyl alcohol) column was used to determine the log of the octanol-water partition coefficient for each compound by HPLC [27,28]. The HPLC-log K_{ow} of the *n*-alkyl piperonyl amides series is shown in relation to the number of carbon groups and water solubility determined by Chemicalize, ChemAxon's cheminformatics tool (Figure 1).

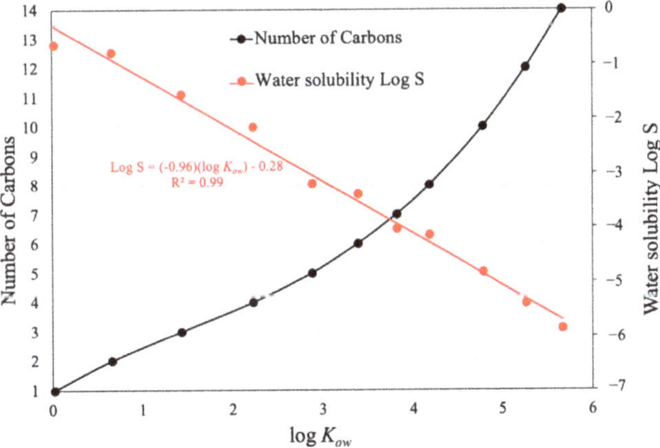

Figure 1. The log K_{ow} of the piperonyl amide fluorescent tracers with corresponding number of carbon atoms and water solubility, log S.

2.3. Sample Preparation for High-Performance Liquid Chromatography (HPLC) Analysis

2.3.1. Coating Tomato Seeds with Amides

Tomato seeds "Florida 47" and "Hypeel 696" were donated by Seminis, Oxnard, CA, and "OH88119" was provided by The Ohio State University, Columbus, OH. A seed coating formulation was developed to apply high loading rates of the fluorescent tracer series as a single seed treatment. A thin adsorbent seed coating was first applied to single seeds to facilitate the high loading rates of the fluorescent tracer series in a single seed treatment. General methods and materials are described in Yang et al. (2018B) [23]. Twenty grams of diatomaceous earth (DE) was dispersed in 80 g of 4% polyvinyl alcohol (PVA) aqueous solution to prepare a 20% DE suspension concentrate. One gram of tomato seeds and 1.5 g of 20% DE suspension concentrate were stirred until a layer of dry DE was coated on the surface of each seed. The coated seeds were allowed to dry in a gentle air stream. A 1.2 mL solution of amides (approximately 6 μL for each seed) was loaded gradually onto 200 tomato seeds with a micropipette. The resulting dosage was 1 μmole of each amide per gram of seed—applied to each tomato variety. Now considering the molecular weights of the 11 amides, which ranged from 179.2 to 361.5 [23], the seed treatment dosage ranged from 0.179 to 0.361 mg per gram of seed. The seeds were again dried with a gentle air stream.

2.3.2. Incubation and Harvest of Treated Seeds in Growth Chamber

Seeds treated with the piperonyl amide series were imbibed as described in Section 2.1. Seed tissue was separated after imbibition, just prior to visible germination. Seeds were removed and washed (to remove seed treatment) with sterile distilled water, cut with a razor blade, and the embryo was removed. Embryos of 50 seeds were pooled to comprise one replicate. The endosperm and seed coat of 50 seeds were also pooled together as one sample. Three replicates were evaluated for each treatment. Ten seeds were pooled together as one sample or replicate. The covering layers consisted of the endosperm and seed coat, while the internal tissues were comprised of the embryo.

2.3.3. Harvesting Tomato Seed Tissue for HPLC Analysis

For each embryo sample, 1.5 mL of acetonitrile (MeCN) was added and the embryos were homogenized with a glass rod. For each sample containing the endosperm and seed coat, the samples were frozen with liquid nitrogen and then homogenized in a mortar after which 1.5 mL of MeCN was added [23]. The homogenized samples were vortexed for 2 min. The extract was transferred into a tube containing 20 mg of PSA, 5 mg of GCB, and 50 mg of MgSO4, then shaken for 1 min, and was then passed through a 0.22 μm syringe filter. The recovery is shown in Table S1 in the supporting information of Yang et al. (2018B) [23].

The tomato embryo and internal tissue samples were extracted by the QuEChERS (Quick, Easy, Cheap, Effective, Rugged, and Safe) method as described for soybean and corn seeds [23]. Ten tomato embryos or ten tomato endosperm + seed coats were placed into a frozen mortar and frozen with liquid nitrogen, and ground into a fine powder. The powder was transferred into a 50 mL centrifuge tube with a screw cap, and 8 mL of MeCN was added and the mixture was shaken for 2 min using a Vortex mixer at room temperature. Following this, a mixture of 2.5 g of $MgSO_4$ and 1.0 g of NaCl was added. The tube was immediately shaken vigorously for 1 min to prevent the formation of $MgSO_4$ agglomerates and centrifuged at 3500 rpm for 5 min. Then, 3.0 mL of the supernatant was subjected to dispersive solid-phase extraction (SPE) using a mixture of 8 mg GCB, 50 mg PSA, and 100 mg $MgSO_4$. The mixture was shaken vigorously for 1 min using a Vortex mixer. Finally, the extract was filtered through a 0.22 μm syringe filter. In developing the HPLC method, the percent recoveries were determined for the eleven amides from soybean embryo + testa (seed coat), and corn endosperm + embryo, and pericarp + testa. The recovery at $\leq$3.82 log K_{ow} for both seed tissues was > 82% for soybean and > 85% for corn [23].

2.3.4. HPLC Analysis of Tomato Seed Tissue

The amides content was determined using an Agilent 1100 HPLC equipped with a 1200 fluorescence detector (FLD) using an ODS-3 column (GL Sciences Inc., 5 μm, 4.6 mm × 75 mm column). The wavelengths of FLD were set at 292 nm (excitation) and 340 nm (emission). The mobile phase used was 0 min 30% MeCN + 70% water, 22 min 40% MeCN + 60% water, 25 min 80% MeCN + 20% water, 40 min 90% MeCN + 10% water. The temperature of the column was 30 °C. The injected volume was 20 μL. The retention time in minutes for each amide derivative was 3.63 (C1), 5.94 (C2), 10.38 (C3), 15.43 (C4), 18.59 (C5), 21.04 (C6), 23.15 (C7), 24.96 (C8), 27.70 (C10), 31.08 (C12), and 35.58 (C14).

2.3.5. Tomato Seed Coat Permeability Data Calculation

Percent uptake in relation to the maximal log K_{ow} (relative amount)

$$= \frac{\text{Concentration of each amide in tissue}}{\text{Concentration of the amide at the maximal log } Kow \text{ in the same tissue}} \times 100\%$$

Percent uptake based on amount applied (uptake efficiency)

$$= \frac{\text{Amount of each amide absorbed by a seed}}{\text{Applied amount of each amide}} \times 100\%$$

Percent in embryo of total seed uptake

$$= \frac{\text{Amount of each amide in the embryo}}{\text{Sum amount of each amide in the covering + internal tissues}} \times 100\%$$

3. Results

3.1. Fluorescence Microscopy of Coumarin 1, 120, and 151 in Tomato Seeds

Assessment of coumarin 1, 120, and 151 uptake was conducted by visual fluorescence observation of the applied seed treatment tracers in tomato seed tissue. Coumarin tracers evaluated in this study were all non-ionic and therefore were expected to permeate the seed coat and move to the embryo [17]. Results showed that coumarin 120, with the greatest water solubility and the lowest log K_{ow} (Table 1), was readily taken up in the embryonic tissue, whereas coumarin 1 and coumarin 151 were only partially taken up in the embryo (Figure 2). The low level of fluorescence in the non-treated control was attributed to autofluorescence in the embryonic tissues.

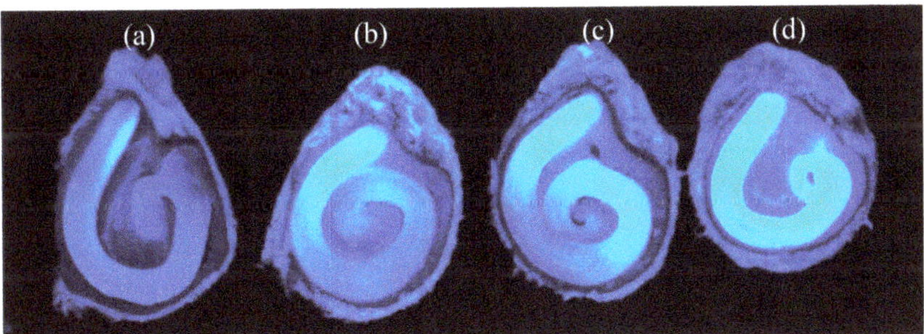

Figure 2. Tomato "Hypeel 696" seed coat permeability of three different coumarin tracers: (**a**) non-treated, (**b**) coumarin 1, (**c**) coumarin 151, and (**d**) coumarin 120.

3.2. Tomato Seed Coat Permeability

3.2.1. Maximal Uptake of Piperonyl Amides in Relation to log K_{ow}

The maximum (100%) relative amount of *n*-alkyl piperonyl amide recovered in tomato seed tissue was in the range of 2.88–3.39 log K_{ow} in embryonic seed tissue and 3.39–4.18 log K_{ow} in endosperm + seed coat tissue (Figures 3 and 4). Amide diffusion to the embryo was limited when log K_{ow} exceeded 4.18 (Figure 3). The maximal uptake in relation to log K_{ow} was achieved at lower log K_{ow} values for Hypeel 696 in both types of seed tissue compared with OH88119 and Florida 47, which had very similar uptake profiles (Figures 3 and 4).

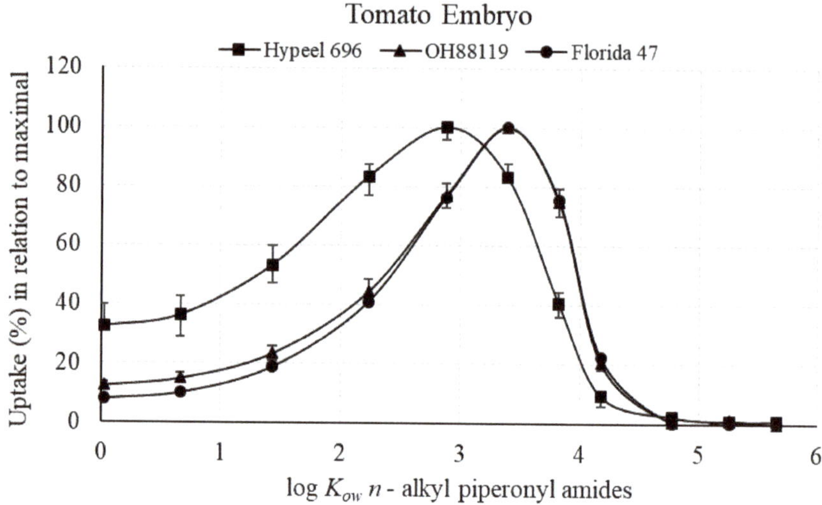

Figure 3. *N*-alkyl piperonyl amide uptake in tomato embryo in relation to maximal log K_{ow} of 100%. Means with standard error bars are shown.

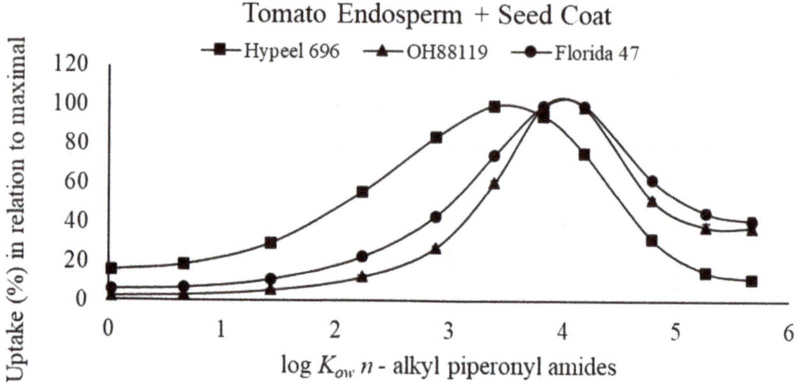

Figure 4. *N*-alkyl piperonyl amide uptake of tomato seed coat + endosperm in relation to maximal log K_{ow} of 100%. Means with standard error bars are shown.

3.2.2. Uptake Efficiency (%) of Piperonyl Amides in Seed Tissue in Relation to Amount Applied

The uptake efficiency, based on the total amount recovered in the embryo compared with the amount applied, showed that maximum uptake associated with log K_{ow} for the embryo occurred at 2.88 for Hypeel 696 and 3.39 for OH99119 and Florida 47 (Figure 5). However, even at the maximal log K_{ow}, the percent uptake was only 5.0% for Florida 47, 4.3% for Hypeel 696, and 1.2% for OH99119. In contrast, uptake efficiency for the entire seed was much greater than the embryo, and ranged from 27% to 36% for the three varieties (Figure 6).

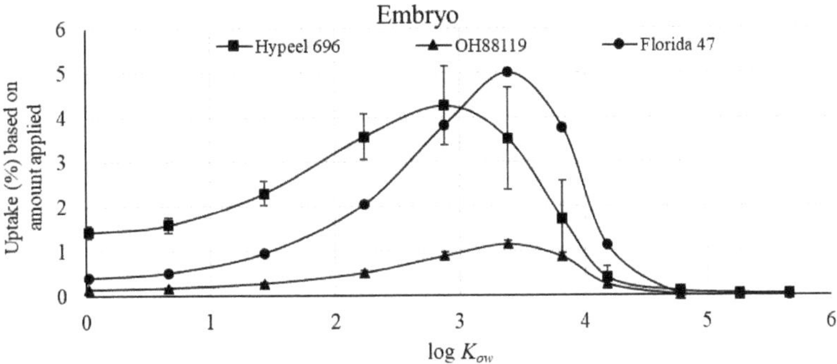

Figure 5. Uptake efficiency of piperonyl amides in the embryo, measured as percent compound applied of *n*-alkyl piperonyl amides. Means with standard error bars are shown.

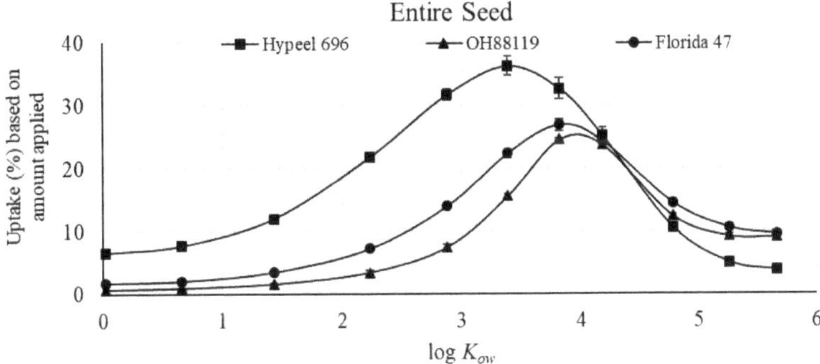

Figure 6. Uptake efficiency of piperonyl amides in the seed, measured as percent compound applied of *n*-alkyl piperonyl amides. Means with standard error bars are shown.

3.2.3. Percent of Piperonyl Amides in the Embryo Compared with the Entire Seed

The percent of the lipophilic amide series in the tomato embryo declined with log K_{ow} from 0.02 to 4.18 for Hypeel 696 and OH8819 (Figure 7). In contrast, Florida 47 revealed a slight increase in the percent embryo distribution from log K_{ow} 0.02 to 2.88–3.18, followed by a decrease to 4.78.

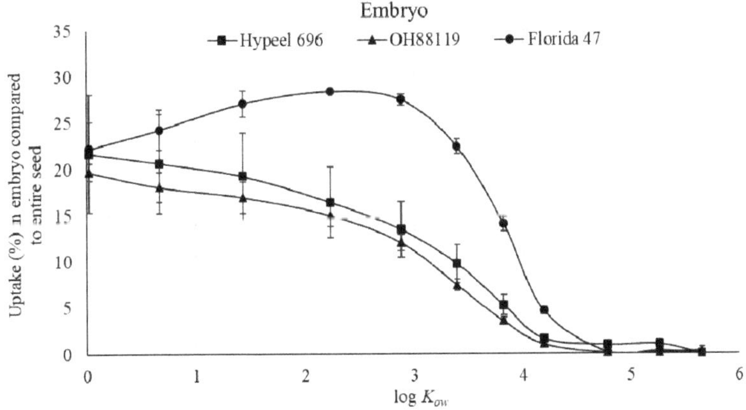

Figure 7. Percent of the absorbed *n*-alkyl piperonyl amides in seed embryo compared with the entire seed uptake of three tomato varieties. Means with standard error bars are shown.

4. Discussion

Fluorescent tracers were used in many previous studies in our lab to examine seed coat permeability in vegetable and field crop seed species. Application of single tracer compounds was used in these qualitative studies to determine seed coat permeability characteristics, resulting in three categories: (1) permeable, (2) selectively permeable, and (3) non-permeable [17]. A dual fluorescent tracer method was later developed to investigate corn pericarp/testa permeability of 27 maize lines [19]. This method could be readily adopted to determine the seed coat permeability category of other seed species. Collectively, both tomato [17] and corn [19] have selective permeability as only non-ionic compounds diffused through the seed coat, while ionic compounds were blocked. A single coumarin compound, coumarin 120, was used in quantitative uptake studies, and a linear increase in seed uptake was measured for corn seed treatment dosage in the range of 0.01 to 1.0 mg coumarin 120 applied per gram of seed [20]. In this study, coumarin 120 and each amide in the series were applied at 3 and 1 μmole per gram of seed, respectively. These seed treatment dosages convert to a range of 0.83 to 0.17 mg per gram of seed, which was in the linear uptake range of corn [20].

The objective of the first investigation in this study was the evaluation of the uptake of three coumarin fluorescent tracers in an attempt to develop a simple method to assess the optimum log K_{ow} for the penetration of neutral compounds through the tomato seed coats. The major advantage was the use of readily available chemical compounds, and these tracers were previously documented with systemic uptake in seeds and seedlings [16]. In addition, fluorescence microscopy could be used for rapid assessment for comparisons without the need for chemical extraction and chemical analyses. Unfortunately, fluorescence intensity was more related to water solubility than log K_{ow} (Table 1 and Figure 2). Moreover, limited conclusions can be drawn using only three tracer compounds. These inconclusive results were attributed to the use of heterogeneous compounds with different physical/chemical properties (Table 1). In addition, there are other properties unique to each coumarin compound including polarizability, topological polar surface area (TPSA), polar surface area (PSA), calculated molar refractivity (CMR), the number of hydrogen bond donors and acceptors, and pK_a (Chemicalize, ChemAxon's cheminformatics tool), and these properties may play a role in seed uptake. Moreover, fluorescence microscopy images may produce false-positive images with the confounding effect of auto-fluorescence from internal seed structure constituents.

There is great value in understanding the chemical/physical properties of active ingredients, and this information can guide a directed chemical synthesis program giving optimal uptake. Alternatively, potential uptake of an existing active ingredient with known or predicted log K_{ow} can be assessed through knowledge of the optimal lipophilicity for seed coat permeability. For this second objective, a combinatorial pharmacodynamic technique was employed using a homogeneous series of 11 *n*-alkyl piperonyl amides that varied in log K_{ow} from 0.02 to 5.66. The mixture of amides was applied as a seed treatment, and tomato seeds were imbibed and dissected into two portions, the embryo, and the endosperm + seed coat. The relative amounts of the amides in these two fractions were quantified by HPLC and plotted as a function of log K_{ow}. This allowed a clear understanding of the role of lipophilicity as it relates to uptake through the tomato seed coat and endosperm and the resulting transport into the embryo by neutral compounds. This knowledge of the optimal physical properties is an invaluable guide in the targeted control of internal seed-borne pathogens.

The overall uptake profile for the tomato embryo and endosperm + seed coat of all three varieties revealed a Gaussian distribution (Figure 3) that is similar to root uptake in plants [21–23]. This similar Gaussian distribution for both tomato roots and seeds was expected based on the composition of the barrier layers. Suberin is found in the endodermis and exodermis of tomato roots [29] and also the inner layer of the tomato seed coat [30]. In contrast, the Gaussian distribution pattern in root uptake in corn and soybean was not revealed for corn or soybean seed uptake [23].

The maximal uptake for the tomato embryo tissue of the three varieties ranged from 2.88 to 3.39 log K_{ow} (Figure 3), while the maximal uptake ranged from 3.39 to 3.88 log K_{ow} for the endosperm + seed coat (Figure 4). Thus, a slight shift to lower log K_{ow} for the embryo tissue in comparison with the other seed tissues was revealed. In the case of corn, the maximal uptake was 3.39 log K_{ow} for both the endosperm + embryo and pericarp/testa using a similar method with the 11 *n*-alkyl piperonyl amides [20]. Therefore, both tomato and corn have similar maximal uptake profiles. Unfortunately, the tomato embryo readily detached from the endosperm during dissection of the fully imbibed seed, which did not allow the measurement of the sum of endosperm with embryo, so a comparison of the effect of the endosperm on shifting the maximal log K_{ow} could not be directly made between corn and tomato.

The uptake efficiency was calculated as the percent of each amide taken up in relation to the amount applied. The maximum uptake efficiency of the entire seed of the three varieties ranged from 27% to 36% (Figure 6), while the maximum uptake efficiency of corn was 43% [20]. Therefore, tomato had lower seed coat permeability than corn, which may be attributed to seed coat composition. The inner layer of the tomato seed coat is known as the semipermeable layer [31] and was shown to be composed of suberin [30], while corn has a semipermeable cutinized or suberized membrane that is located below the inner integument [32].

The maximum uptake efficiency of the embryo in relation to the amount applied of the three varieties ranged from 1.2% to 5.0% (Figure 5). Another calculation based only on the absorbed *n*-alkyl piperonyl amides uptake revealed that less than 30% of an amide was measured in the embryo compared with the entire seed (Figure 7). These data demonstrate that most of the piperonyl compounds were unable to reach the embryo. However, fluorescence images revealed the greatest fluorescence intensity in the embryo compared with the endosperm or seed coat (Figure 2). Therefore, fluorescence imaging that provides excellent qualitative data on the presence or absence of a tracer in the seed tissue was not related to quantitative results from our analytical method.

The pathway by which the applied seed treatment moved to the embryo was not investigated in this study. In the dicot seed *Sedum acre*, movement between the seed compartments was attributed to symplastic movement with cell-to-cell movement through the plasmodesmata [33]. We assume that movement from the endosperm to embryo in tomato seed is by the same symplastic pathway.

Three tomato varieties were investigated with the 11 *n*-alkyl piperonyl amides. The maximal log K_{ow} for both the embryo and endosperm + seed was shifted to a slightly lower value for Hypeel 696 compared with the two other varieties (Figures 3 and 4). Florida 47 had the greatest accumulation in the embryo with 5.0% (Figure 5), while Hypeel 696 had the greatest accumulation in the entire seed with 36% (Figure 6). After an amide was absorbed, Florida 47 had a greater distribution in the embryo than the other two varieties (Figure 7). These varietal differences may be attributed to differences in seed coat composition and/or structural properties. Varietal differences in tomato seed coat permeability were related to the efficacy of jasmonic acid seed treatments used as an elicitor of defense against western flower thrips [34]. The thickness and compactness of the inner tomato seed coat layer composed of suberin [28] may be responsible for varietal differences. Further, the composition of the seed coat and embryo may differ and thus can affect both permeability and affinity for a compound. Further study is needed to investigate varietal differences in seed uptake, as retention of an active ingredient in the seed coat could result in a suboptimal concentration in the embryo being efficacious.

5. Conclusions

This study quantitively described the relationship between the log K_{ow} and the permeation capacity of a chemical through the seed coat to the embryo of tomato seeds. The relatively hard and thick tomato testa attenuated the movement of the seed treatment to the embryo tissue. Less than 5% of the applied compound was measured in the embryo, while most resided in the seed coat + endosperm. For the control of internal seed-borne pathogens, seed treatment with log K_{ow} in the range of 2.9 to 3.8 log K_{ow} is suggested as these chemicals were found to most effectively reach the tomato embryo tissues.

The piperonyl amide method uses a combinatorial pharmacodynamic technique to probe the uptake and transport of xenobiotic compounds in seeds. This is in contrast with the use of heterogeneous compounds that differ in a multitude of physical properties, isolation efficiencies, and detection sensitivities. When using heterogeneous compounds, often the experimental method involves a separate experiment for each compound to generate an uptake and/or transport parameter. Combining the ensemble of data from the piperonyl amide method resulted in a trend that identified the optimum chemical properties for uptake and accumulation in specific seed tissues. Moreover, the combinatorial pharmacodynamic method used eleven piperonyl amides combined into a single experiment, with significant benefits with regard to time, cost, and many experimental variables being eliminated. Further, very subtle absorption and transport trends were quantified in different crop seeds [23], and also in plants and insects [Donovan and Black, unpublished]. Thus, the method can be broadly adapted for agricultural research, and provides detailed physical property space information at a level of precision that is not available using other techniques.

Author Contributions: Conceptualization, A.G.T. and S.D.; methodology, H.M., M.A. and D.Y.; software, H.M. and M.A.; validation, A.G.T., H.M. and M.A.; formal analysis, H.M. and M.A.; investigation, H.M., M.A. and D.Y.; resources, A.G.T.; data curation, A.G.T., S.D., H.M., M.A. and D.Y.; writing—original draft preparation, H.M. and M.A.; writing—review and editing, A.G.T., S.D., H.M., M.A. and D.Y.; visualization, A.G.T. and M.A.; supervision, A.G.T.; project administration, A.G.T.; funding acquisition, A.G.T. All authors have read and agreed to the published version of the manuscript.

Funding: This work was supported by the Specialty Crop Research Initiative (grant no. 2015-51181-24312) from the USDA National Institute of Food and Agriculture. D. Yang was partially supported by the China Scholarship Council (grant No. 201503250009). This material is based upon work that is supported by the National Institute of Food and Agriculture, US Department of Agriculture, Multi-state Project W-4168, under accession #1007938.

Institutional Review Board Statement: Not applicable.

Informed Consent Statement: Not applicable.

Data Availability Statement: Not applicable.

Acknowledgments: Authors would like to thank Lailiang Cheng, Cornell University, for providing access to his HPLC for this project.

Conflicts of Interest: The authors declare no conflict of interest.

References

1. Maude, R.B. *Seedborne Diseases and Their Control: Principles and Practice*; CAB International: Wallingford, Oxon, UK, 1996.
2. Elmer, W.H. Seeds as vehicles for pathogen importation. *Biol. Invasions.* **2001**, 3, 263–271. [CrossRef]
3. Gitaitis, R.; Walcott, R. The epidemiology and management of seedborne bacterial diseases. *Ann. Rev. Phytopathol.* **2007**, *45*, 371–397. [CrossRef]
4. Shade, A.; Jacques, M.A.; Barret, M. Ecological patterns of seed microbiome diversity, transmission, and assembly. *Curr. Opin. Microbiol.* **2017**, *37*, 15–22. [CrossRef]
5. Nelson, E.B.; Simoneau, P.; Barret, M.; Mitter, B.; Compant, S. Editorial special issue: The soil, the seed, the microbes and the plant. *Plant. Soil.* **2018**, *422*, 1–5. [CrossRef]
6. Barret, M.; Guimbaud, J.F.; Darrasse, A.; Jacques, M.A. Plant microbiota affects seed transmission of phytopathogenic microorganisms. *Mol. Plant. Pathol.* **2016**, *17*, 791. [CrossRef]
7. Rodríguez, C.E.; Mitter, B.; Barret, M.; Sessitsch, A.; Compant, S. Commentary: Seed bacterial inhabitants and their routes of colonization. *Plant. Soil.* **2018**, *422*, 129–134. [CrossRef]
8. Dombrovsky, A.; Smith, E. Seed transmission of Tobamoviruses: Aspects of global disease distribution. *Adv. Seed Biol.* **2017**, 233–260. [CrossRef]

9. Nandi, M.; Macdonald, J.; Liu, P.; Weselowski, B.; Yuan, Z.C. Clavibacter michiganensis ssp. michiganensis: Bacterial canker of tomato, molecular interactions and disease management. *Mol. Plant Patho.* **2018**, *19*, 2036–2050. [CrossRef] [PubMed]
10. Yoon, T.; Ha Lee, B. Identification of fungus-infected tomato seeds based on full-field optical coherence tomography. *Curr. Opt. Photonics* **2019**, *3*, 571–576.
11. Mink, G.I. Pollen and seed-transmitted viruses and viroids. *Ann. Rev. Phytopathol.* **1993**, *31*, 375–402. [CrossRef]
12. Maude, R.B.; Kyle, A.M. Seed treatments with benomyl and other fungicides for the control of *Ascochyta pisi* on peas. *Ann. Appl. Biol.* **1970**, *66*, 37–41. [CrossRef]
13. Taylor, A.G. Seed treatments. In *Encyclopedia of Applied Plant Sciences*; Thomas, B., Murphy, D.J., Murray, B.G., Eds.; Elsevier Academic Press: Amsterdam, The Netherlands, 2003; pp. 1291–1298, ISBN 9780122270505.
14. Afzal, I.; Javed, T.; Amirkhani, M.; Taylor, A.G. Modern seed technology: Seed coating delivery systems for enhancing seed and Crop performance. *Agriculture* **2020**, *10*, 526. [CrossRef]
15. Su, W.H.; Fennimore, S.A.; Slaughter, D.C. Development of a systemic crop signaling system for automated real-time plant care in vegetable crops. *Biosyst. Eng.* **2020**, *193*, 62–74. [CrossRef]
16. Wang, Z.; Amirkhani, M.; Avelar, S.A.; Yang, D.; Taylor, A.G. Systemic uptake of fluorescent tracers by soybean (*Glycine max* (L.) Merr.) seed and seedlings. *Agriculture* **2020**, *10*, 248. [CrossRef]
17. Salanenka, Y.A.; Taylor, A.G. Seedcoat permeability: Uptake and post-germination transport of applied model tracer compounds. *HortScience* **2011**, *46*, 622–626. [CrossRef]
18. Taylor, A.G.; Salanenka, Y.A. Seed treatments: Phytotoxicity amelioration and tracer uptake. *Seed Sci. Res.* **2012**, *22*, S86–S90. [CrossRef]
19. Dias, M.A.N.; Taylor, A.G.; Cicero, S.M. Uptake of systemic treatments by maize seeds evaluated with fluorescent tracers. *Seed Sci. Technol.* **2014**, *42*, 101–107.
20. Yang, D.; Avelar, S.A.; Taylor, A.G. Systemic seed treatment uptake during imbibition by corn and soybean. *Crop. Sci.* **2018**, *58*, 2063–2070. [CrossRef]
21. Briggs, G.G.; Bromilow, R.H.; Evans, A.A. Relationships between lipophilicity and root uptake and translocation of non-ionised chemicals by barley. *Pesticide Sci.* **1982**, *13*, 495–504. [CrossRef]
22. Clarke, E.D.; Delaney, J.S. Physical and molecular properties of agrochemicals: An analysis of screen inputs, hits, leads, and products. *CHIMIA Int. J. Chem.* **2003**, *57*, 731–734. [CrossRef]
23. Yang, D.; Donovan, S.; Black, B.C.; Cheng, L.; Taylor, A.G. Relationships between compound lipophilicity on seed coat permeability and embryo uptake by soybean and corn. *Seed Sci. Res.* **2018**, *28*, 229–235. [CrossRef]
24. Lipinski, C.A.; Lombardo, F.; Dominy, B.W.; Feeney, P.J. Experimental and computational approaches to estimate solubility and permeability in drug discovery and development settings. *Adv. Drug Deliv. Rev.* **2001**, *23*, 3–26. [CrossRef]
25. Aazam, E.S. Synthesis and characterization of mononuclear and binuclear metal complexes of a new fluorescent dye derived from 2-hydroxy-1-naphthaldehyde and 7-amino-4-methylcoumarin. *J. King Abdulaziz Univ. Sci.* **2010**, *148*, 1–32. [CrossRef]
26. Taniguchi, M.; Lindsey, J.S. Database of absorption and fluorescence spectra of >300 common compounds for use in photochemCAD. *Photochem. Photobiol.* **2018**, *94*, 290–327. [CrossRef] [PubMed]
27. Donovan, S.F.; Pescatore, M.C. Method for measuring the logarithm of the octanol–water partition coefficient by using short octadecyl–poly (vinyl alcohol) high-performance liquid chromatography columns. *J. Chromatogr. A* **2002**, *952*, 47–61. [CrossRef]
28. Anastassiades, M.; Lehotay, S.J.; Štajnbaher, D.; Schenck, F.J. Fast and easy multiresidue method employing acetonitrile extraction/partitioning and "dispersive solid-phase extraction" for the determination of pesticide residues in produce. *J. AOAC. Int.* **2003**, *86*, 412–431. [CrossRef] [PubMed]
29. Quiroga, M.; Guerrero, C.; Botella, M.A.; Barceló, A.; Amaya, I.; Medina, M.I.; Alonso, F.J.; de Forchetti, S.M.; Tigier, H.; Valpuesta, V. A tomato peroxidase involved in the synthesis of lignin and suberin. *Plant. Physiol.* **2000**, *122*, 1119–1127. [CrossRef]
30. Beresniewicz, M.B.; Taylor, A.G.; Goffinet, M.C.; Koeller, W.D. Chemical nature of a semipermeable layer in seed coats of leek, onion (Liliaceae), tomato and pepper (Solanaceae). *Seed Sci. Technol.* **1995**, *23*, 135–145.
31. Beresniewicz, M.B.; Taylor, A.G.; Goffinet, M.C.; Terhune, B.T. Characterization and location of a semipermeable layer in seed coats of leek, onion, tomato and pepper. *Seed Sci. Technol.* **1995**, *23*, 123–134.
32. Kiesselbach, T.A.; Walker, E.R. Structure of certain specialized tissues in the kernel of corn. *Am. J. Bot.* **1952**, *39*, 561–569. [CrossRef]
33. Wróbel-Marek, J.; Kurczyńska, E.; Płachno, B.J.; Kozieradzka-Kiszkurno, M. Identification of symplasmic domains in the embryo and seed of *Sedum acre* L. (Crassulaceae). *Planta* **2017**, *245*, 491–505. [CrossRef] [PubMed]
34. Mouden, S.; Kappers, I.F.; Klinkhamer, P.G.L.; Leiss, K.A. Cultivar variation in tomato seed coat permeability is an important determinant of jasmonic acid elicited defenses against western flower thrips. *Front. Plant. Sci.* **2020**, *11*, 576505. [CrossRef] [PubMed]

Article

Systemic Uptake of Fluorescent Tracers by Soybean (*Glycine max* (L.) Merr.) Seed and Seedlings

Zhen Wang [1,2,†], Masoume Amirkhani [1,*,†], Suemar A.G. Avelar [3], Daibin Yang [4] and Alan G. Taylor [1,*]

1 Horticulture Section, School of Integrative plant Science, Cornell AgriTech, Cornell University, Geneva, New York, NY 14456, USA; cau1022@imau.edu.cn

2 Inner Mongolia Agricultural University, College of Horticulture and Plant Protection, Hohhot 010018, China

3 Seed Analysis Laboratory of APROSMAT–Mato Grosso Seed Grower Association–APROSMAT Rua dos Andradas, Rua dos Andradas, Rondonópolis 688, Brazil; suemaralexandre@yahoo.com.br

4 Institute of Plant Protection, Chinese Academy of Agricultural Science, Beijing 100193, China; yangdaibin@caas.cn

* Correspondence: ma862@cornell.edu (M.A.); agt1@cornell.edu (A.G.T.)

† These authors contributed equally to this work.

Received: 3 June 2020; Accepted: 23 June 2020; Published: 26 June 2020

Abstract: Systemic seed treatment uptake was investigated in seeds and seedlings using fluorescent tracers to mimic systemic agrochemicals. Soybean was used as the model as soybean has the permeable seed coat characteristic to both charged and noncharged molecules. The purpose of the paper is to (1) screen 32 fluorescent tracers and then use optimal tracers for seed and seedling uptake, (2) investigate varietal differences in seed uptake, (3) examine the distribution of tracer uptake into 14-day-old seedlings, and (4) study the relationship between seed treatment lipophilicity, measured as log *P* on seed and root uptake. The major chemical families that displayed both seed and seedling uptake were coumarins and xanthenes. Seed uptake of coumarin 120 ranged from 1.1% to 4.8% of the applied seed treatment tracer from 15 yellow-seeded varieties. Rhodamine B, a xanthene compound uptake in seedlings, showed translocation from the applied seed treatment to all seedling tissues. Most of the tracer was measured in the hypocotyl and root, with lesser amounts in the epicotyl and true leaves. Log *P* is well documented in the literature to model systemic uptake by roots, but log *P* of the tracers were not related to seed uptake.

Keywords: fluorescent tracer; systemic uptake; soybean; in vivo imaging system (IVIS)

1. Introduction

Crop seeds are treated commercially by the seed industry or on the farm to protect seeds and seedlings from attack by insect pests and pathogens that cause plant diseases [1]. Active seed treatment ingredients may have contact activity or be systemic in nature. Compounds with contact activity are restricted to control of pests in the immediate vicinity of the sown treated seed. In contrast, a compound with systemic activity protects in the immediate vicinity and is also translocated within the plant [2]. Thus, systemic movement allows the protection of plants during the early stages of seedling growth after emergence. The agronomic benefit of using systemic seed treatments is to reduce the need for foliar applications for early season pest management, thus avoiding a field operation that may not be possible or be delayed due to weather-related events and/or wet soils. Therefore, seed treatments are used globally for early season pest management and have less potential environmental impact than foliar applications due to lower pesticide usage per hectare [3].

The physiochemical properties required for a systemic seed treatment to permeate through the seed coat to the embryo and ultimately be taken up into the seedling and transpiring leaves are still not well understood. One pathway for systemic uptake in seeds is for a compound to diffuse through the seed coat to the embryo during imbibition. For this to occur, the seed coat must be permeable to the specific compound. Seed coat permeability has been previously investigated on several crops using fluorescent tracers [4–6]. Taylor et al. [5] reported that the passage of organic compounds applied as seed treatments to the embryo during imbibition is dependent on the chemical properties of the treatment and crop species. As a result of this research, crop species were grouped into three categories based on seed coat permeability, those with permeable, selectively permeable and nonpermeable seed coats [4,5]. Crops with a permeable seed-coats, including soybean (*Glycine max* (L.) Merr.), snap bean (*Phaseolus vulgaris* L.), and pea (*Pisum sativum* L.), were reported to be permeable to both nonionic and ionic compounds, while selectively permeable crop seeds, including corn (*Zea mays* L.), onion (*Allium* cepa L.), tomato (*Solanum lycopersicum* L.), and pepper (*Capsicum annuum* L.) were only permeable to nonionic compounds. Cucumber (*Cucumis sativus* L.) and lettuce (*Lactuca sativa* L.) seeds were categorized as having nonpermeable seed coats as they were not permeable to either nonionic or ionic charged compounds. In these experiments, systemic tracer translocation was observed under long-UV light, eliminating the use of pesticides and radioactively labeled compounds. The nine fluorescent tracers used in these experiments varied by ionic charge (nonionic, cationic, or anionic) [4,6]. Investigation of seed uptake and systemic activity using fluorescent tracers with a range of physio/chemical properties will help to expand our knowledge and utility of these tracers in seed technology research. Moreover, an expanded set of tracers could aid in understanding the relationship between seed uptake and root uptake.

Both nonionic (coumarin 151) and ionic (rhodamine B) tracers diffused from treated soybean seed, with a permeable seed coat, to the embryo after sowing in a moist medium [6]. In a separate experiment with a nonionic fluorescent tracer, coumarin 120, the maximum or saturated seed uptake of two yellow-seeded varieties was more than 50% greater than a black-seeded variety, illustrating varietal differences between genotypes [5]. Due to these varietal differences, additional research is needed to understand agrochemical uptake in seeds more fully.

The fluorescent tracer coumarin 151, used to investigate uptake into seedlings in snap bean and cucumber, was found in the xylem vessels of all seedling structures, including the roots, hypocotyl, cotyledons, petiole, and true leaves [7]. These observations indicate that systemic uptake of coumarin 151 is by apoplastic or acropetal movement. A more recent report showed uptake of rhodamine B uptake in snap bean seedlings [8]. Collectively, fluorescent tracers have great utility to assess seed coat permeability and systemic uptake into plants.

As previous research has demonstrated, fluorescent tracers are convenient tools to visualize the movement of compounds in animals, plants, and seeds for qualitative and quantitative measurements. Fluorescence imaging systems are also useful tools that can quantify excitation and emission signals over a wide range of wavelengths. The in vivo imaging system (IVIS) was developed for in vivo fluorescence and bioluminescence imaging. IVIS consists of a stationary charge-coupled device (CCD) imaging camera with illumination and a set of excitation filters from 415–760 nm in 30 nm bandwidths and a set of emission filters from 490–850 nm in 20 nm bandwidths [9]. IVIS is used in small animal imaging for nondestructive, noninvasive, internal imaging of fluorophores [10]. In addition, the IVIS spectrum system is also used in plant science research [11–13], and has potential for fluorescence imaging of seeds of large-seeded crops.

Several physical/chemical properties of an organic compound contribute to systemic activity and plant root uptake. Much of our understanding and rules that govern the ability for uptake was first illustrated for oral pharmaceuticals and described as Lipinski's rule of five, or simply, the rule of five (RO5) [14]. A compound having chemical properties satisfying the RO5 has potential pharmacological or biological activity as an orally active drug in humans [14]. The RO5 approach used in pharmacology was quickly adopted with some modifications to profile agrochemical uptake, and several new "rules" were established by Briggs, Carr, Tice, and Hao, cited by Jampilek (2016) [15]. Of particular interest to seed science were the properties described by Clarke, known to influence the absorption and distribution of agrochemicals in crop plants, termed the rule of two. The parameters or criteria important for the rule of two are molecular mass from 200–400, log $P \leq 4$, and hydrogen-bond donors ≤2 [15,16].

The purpose of this study is to advance the understanding of systemic seed uptake through the evaluation of fluorescent tracers representing a range of log *P* and electrical charge and the use of fluorescence imaging of seeds and seedlings, using soybean as the model. The specific objectives of the paper are to (1) screen a wide range of fluorescent tracers and then use optimal tracers for seed and seedling uptake, (2) investigate varietal differences in seed uptake, (3) examine the distribution of tracer uptake into 14-day-old seedlings, and (4) contrast the effect of log *P* on seed and root uptake.

2. Materials and Methods

2.1. Fluorescent Tracers, Seed Varieties, and Qualitative Evaluation of Fluorescent Tracer Uptake by Soybean Seed and Seedlings

Thirty-two fluorescent tracers belonging to 10 chemical families or classes were used to investigate soybean seed and seedling uptake (Table 1). Some of these tracers were both fluorescent and colored chemicals and were either purchased from Sigma-Aldrich, St. Louis, MO, and Exciton Dye Technologies, Lockbourne, OH or provided by Day-Glo Color Corp, Cleveland, OH, AaKash Chemicals, Glendale, IL, and Milliken Chemical (formerly Keystone), Spartanburg, SC. The Chemical Abstracts Service Registry Number (CAS number), log *P* and log D at pH 6.5, molecular weight, and the electrical charge was obtained from Chemicalize, ChemAxon's cheminformatic tool, ([17]; Table 1). Log *P* and log *D* are the partition coefficient and distribution coefficient, respectively, and are the ratio of a compound in a mixture of water and 1-octanol at equilibrium [18]. The partition coefficient is the concentration ratio of un-ionized species of a compound, whereas the distribution coefficient refers to the concentration ratio of all species of the compound (ionized plus un-ionized). Collectively, log *P* and log *D* values are a measure of a compound's lipophilic characteristic. Clarke's rule of two was assessed as "yes" if in agreement with or "no" if in violation of the following criteria: molecular mass from 200–400, log $P \leq 4$, and hydrogen-bond donors ≤2 [16]. Fifteen yellow seed coat soybean varieties and two black seed coat varieties were obtained from multiple sources listed in Table 2. Seeds were stored at 5 °C until used for experiments.

Table 1. List of 32 fluorescent and/or colored tracers grouped by chemical family/class, chemical name, Chemical Abstracts Service Registry Number (CAS number), log *P* or log *D* at pH 6.5, molecular weight (MW), compliance with the rule of two, and electrical charge at pH 6.5. Values obtained from the Chemicalize database [17], in compliance with the rule of two with the following parameters or criteria: molecular mass from 200–400, log P ≤ 4, and hydrogen-bond donors ≤2 [15].

Chemical Family/Class	Chemical Name	CAS Number	Log *P*/ Log *D*	MW (g/mol)	Rule of Two	Electrical Charge
Acridine	9-Aminoacridine hydrochloride hydrate	52417-22-8	2.68/1.01	248.7	Yes	Cationic
Arylmethane Dye	Auramine O	2465-27-2	3.66/0.34	303.8	Yes	Cationic
	Crystal Violet	548-62-9	1.39	408.0	No	Cationic
Azine Dye	Neutral Red	553-24-2	2.85	288.8	Yes	Nonionic
Benzotriazole	1,2,3-Benzotriazole	95-14-7	1.30	119.1	No	Nonionic
Benzoxathiole	Phenol Red	143-74-8	4.11	354.4	No	Nonionic

Table 1. *Cont.*

Chemical Family/Class	Chemical Name	CAS Number	Log *P*/ Log *D*	MW (g/mol)	Rule of Two	Electrical Charge
Coumarin	7-Amino-4-Methyl-3-coumarinylacetic Acid (AMCA)	106562-32-7	0.25/−1.76	233.2	Yes	Anionic
	3-(Benzoxazolyl-2′)-7-diethylaminocoumarin	35773-42-3	4.00	334.4	Yes	Nonionic
	Coumarin	91-64-5	1.78	146.1	No	Nonionic
	Coumarin 1	91-44-1	2.90	231.3	Yes	Nonionic
	Coumarin 120	26093-31-2	1.25	175.2	No	Nonionic
	Coumarin 151	53518-15-3	1.62	229.2	Yes	Nonionic
	Coumarin 152	53518-14-2	2.56	257.2	Yes	Nonionic
	Coumarin 314	55804-66-5	3.18	313.3	Yes	Nonionic
	o-Coumaric Acid	614-60-8	1.83/−0.77	164.2	No	Anionic
	m-Coumaric Acid	588-30-7	1.83/−0.79	164.2	No	Anionic
	6,7-Dihydroxy coumarin	305-01-1	1.18/1.16	178.1	No	Nonionic
Naphtalimide	Fluorescent Brightener 162	3271-05-4	1.75	241.2	Yes	Nonionic
	Solvent Yellow 131	52821-24-6	0.12	328.4	No	Nonionic
	Solvent Yellow 43	19125-99-6	4.03	324.4	No	Nonionic
	Solvent Yellow 44	2478-20-8	3.76	316.4	Yes	Nonionic
Pyrimidine Anthrone	Solvent Red 149	21295-57-8	4.30	358.4	No	Nonionic
Thiazine	Methylene Blue	122965-43-9	−0.62	337.9	Yes	Cationic
Xanthene	5(6)-Carboxy-fluorescein	72088-94-9	3.54/0.92	376.3	No	Anionic
	2′.7′-Dichloro-fluorescein	76-54-0	5.09/4.99	401.2	No	Nonionic
	Fluorescein	2321-07-5	3.88	332.3	Yes	Nonionic
	Fluorescein sodium salt	518-47-8	3.01/−0.11	376.3	Yes	Anionic
	Pyronin Y	92-32-0	4.23	302.8	No	Cationic
	Rhodamine 800	101027-54-7	6.03	496.0	No	Cationic
	Rhodamine B	81-88-9	1.78/2.34	479.0	No	Zwitterion
	Rhodamine B base	509-34-2	6.13	442.6	No	Nonionic
	Uranine K	6417-85-2	3.01/−0.11	408.5	No	Nonionic

Table 2. Soybean genotypes, the origin of production and seed source maturity group (MG), thousand-seed weight in grams (TSW) used to evaluate seed coat permeability to fluorescent tracers, and seed coat color (SCC).

Genotype	Origin	Breeder/Source	MG	TSW	SCC
92Y51	United States	Pioneer Hi-Bred International,	2.5	174.5	Yellow
AG1901	United States	AsGrow Seed Company	1.9	165.2	
Anta 82	Brazil	TMG, Fundação MT, UNISOJA	7.4	109.7	
GB 874RR	Brazil	MONSOY LTDA	8.7	143.3	
IAR 1902 SCN	United States	Iowa State University Research Foundation	1.9	146.7	
IAR 2601 SCN	United States	Iowa State University Research Foundation	2.6	168.0	
M6972 IPRO	Brazil	MONSOY LTDA	6.9	158.9	
M7739 IPRO	Brazil	MONSOY LTDA	7.7	150.0	
TMG 1174RR	Brazil	TMG, Fundação MT, UNISOJA	8.2	118.2	
TMG 1175RR	Brazil	TMG, Fundação MT, UNISOJA	7.4	128.8	
TMG 1176RR	Brazil	UNISOJA S/A	7.5	111.1	
TMG 1179RR	Brazil	TMG, Fundação MT, UNISOJA	7.6	129.3	
TMG 132RR	Brazil	TMG, Fundação MT, UNISOJA	7.9	144.4	
TMG 4182	Brazil	TMG, Fundação MT, UNISOJA	8.5	148.9	
TMG 4185	Brazil	TMG, Fundação MT, UNISOJA	8.5	150.4	
Black Jet	United States	Johnny's Selected Seed	N/A	356.0	Black
V12-1223	United States	Virginia Tech	N/A	169.5	

Seed and seedling uptake of the 32 fluorescent tracers (Table 1) were investigated with the yellow-seeded soybean variety AG1901. Each tracer was applied at a dosage of 5 mg of tracer/g seed, and 10 g seeds were treated in a 250-mL glass Erlenmeyer flask; 4 drops of water were added and mixed in the flask for 30 seconds, resulting in uniform coverage. Treated seeds were sown in silica sand (#2 Q-ROK, 0.3–1.19 mm, U.S. Silica Company, New Philadelphia, OH, USA), and water added at 20% of dry weight and maintained in a germinator at 25 °C for 12 hours in the dark. Imbibed seeds were removed, washed with deionized water, hand-dissected, and observed under long-UV (365 nm).

A subset of 9 treated seeds was kept in the germinator in the dark for 4 days to evaluate fluorescence in seedlings. Nontreated seeds were sown as controls.

2.2. Fluorescence Microscopy and Quantification of Coumarin 120 Uptake by Soybean Seeds

Fluorescence microscopy was accomplished with seeds of yellow seed coat genotypes TMG 4185 and M7739 IPRO. Seeds were treated with 0.5% coumarin 120, 0.1% L650 seed treatment binder (Incotec, the seed enhancement arm of Croda, Salinas, CA), 63 mg ai/100 g seed of Thiram 42S fungicide (Bayer, RTP, NC, USA), and 4.75% deionized water (on a weight basis) and mixed in a 50-mL centrifuge tube. Six treated soybean seeds were sown in silica sand maintained at 20% moisture content and placed in a growth chamber at 20 °C for 6 hours in the dark. Imbibed seeds were then removed and washed with deionized water. Seed coats were removed by hand, and then the seeds were dissected and imaged under an Olympus microscope (SZX12, Tokyo, Japan; imaging camera Infinity3-3URC, Lumenera Corp., Ottawa, ON, Canada) and Infinity Analyze (Revision 6.5.2, Teledyne Lumenera, Ottawa, ON, Canada). Seed tissue was illuminated with a long-UV light UV lamp (Model 9-circular illuminator, Stocker & Yale, Salem, NH, USA). Seeds treated with the same formulation without coumarin 120 were used as the control.

Six soybean seeds of each variety were selected and treated with a suspension of coumarin 120 (0.065%) in a solution 4% PVA (polyvinyl alcohol) and 0.1% Triton X-100, as described by Yang et al. (2018), to quantify coumarin uptake [19]. Seeds were dried overnight, and then sown in 20% moisture content silica sand (#1 Q-ROK, 0.15–0.84 mm, New England Silica, Inc., South Windsor, CT, USA) and maintained in a germinator at 20 °C for 14 hours in the dark. Then, seeds were removed from the media, washed with deionized water, and the seed coat was removed. There were four replicates per treatment, and one seed from each replicate was ground into a fine powder with liquid nitrogen. The extraction of coumarin 120 was performed according to a procedure developed by Yang et al. (2018) [19]. Solutions of 100, 250, and 500 μg/L of coumarin 120 were prepared in acetonitrile solvent and used to make a standard curve. Fluorescence of coumarin 120 was quantified at 342 nm excitation and 409 nm emission using a luminescence spectrophotometer (LS-50B, Perkin Elmer, Shelton, CT). The extract from the nontreated seeds was used as the zero point.

2.3. Rhodamine B and Rhodamine 800 Uptake in Soybean Seed and Seedlings

Seeds of TMG 4185 (yellow seed coat) were treated with Rhodamine B (RB) and Rhodamine 800 (R800) at 0.05%, 0.1%, or 0.5% by weight, 0.1% L650 seed treatment binder, 63 mg ai/100 g seed of Thiram 42S fungicide and 4.75% deionized water (on a weight basis) and mixed in a 50-mL centrifuge tube. Treated seeds were air-dried under ambient conditions in the laboratory overnight. Seeds treated with the same seed treatment formulation, but no rhodamine compound, were planted as the control. Six treated seeds were sown in silica sand with 20% moisture content and maintained in a growth chamber at 20 °C for 6 hours in the dark. Imbibed seeds were removed and washed with deionized water. Seed coats were removed by hand; then, the seeds were dissected and imaged using IVIS, as described in Section 2.4.

For growth chamber studies, seeds treated with 0.05% RB by weight and nontreated seeds were sown in silica sand (#3 Q-ROK, 0.3–1.68 mm, U.S. Silica Company, Berkeley Springs, WV, USA) in 473 mL plastic cups to prevent leaching of RB from the soil medium. Plants were maintained in a growth chamber at 25/20 °C (14 h/10 h, day/night), with a relative humidity of 60%. A complete nutrient solution was prepared with 2.0 g Peters 5-11-26 Hydroponic Special Fertilizer (Everris NA, Dublin, OH) and 0.65 g calcium nitrate (Sigma-Aldrich chemical company, St. Louis, MO), dissolved in 1 liter of deionized water. The media was maintained at 20% moisture content throughout the growth period and checked daily. Water was initially used to moisten the silica sand to 20%, and Peters fertilizer solution was used after seedling emergence to achieve the desired nutrient level. Fourteen days after sowing, the first true leaves were fully expanded, and the seedlings were used for imaging. There were three replicates of three soybean seeds per treatment, and the experiment was conducted twice.

2.4. IVIS Imaging Rhodamine B (RB) and Rhodamine 800 (R800) Uptake in Soybean Seeds and Seedlings

Dissected control (nontreated) and RB-treated seeds were imaged with the IVIS spectrum image system (PerkinElmer, Waltham, MA, USA) using a 535 nm/620 nm excitation/emission filter and control and R800 treated seeds were imaged with a 640 nm/840 nm excitation/emission filter, with the auto exposure time mode.

The IVIS spectrum image system was also used to image fluorescent tracer uptake in 14-day-old seedlings. Nontreated and RB-treated seedlings were imaged with a 535 nm/620 nm for excitation/emission filter and 0.1 s exposure time. Seedlings were dissected to observe tracer uptake in the adaxial (upper surface of leaf) and abaxial (lower surface of leaf) sides of leaf tissue, along with the epicotyl, hypocotyl, and root tissues. Images of these plant tissues from nontreated and RB-treated seedlings were imaged under the same conditions as that of the whole seedlings. Image analysis: Living Image version 4.7.3 software was used to quantify the intensity of RB and R800 in soybean seeds and different tissues of soybean seedlings. Image adjust in Tool Palette was used to remove noise if needed; subsequently, the region of interest (ROI) in Tool Palette was used to calculate the intensity of the tracer in plant tissue. The fluorescent intensity was quantified as "total radiant efficiency", and the intensity of the controls was subtracted from the intensity of treated samples. The ROI processing tasks in this study were accomplished in accordance with the manual (Living Image software, Caliper Life Sciences, Hopkinton, MA, USA).

2.5. Statistical Analysis

The data obtained from the coumarin 120 content in the soybean seeds study was calculated as the percent of material applied. Prior to the application of ANOVA, data were tested for normality using the goodness-of-fit test and also examined for homogeneity of variance using the Levene test. Data sets were found to normally distributed and passed the variance homogeneity test. Statistical differences were determined using one-way analysis of variance (ANOVA), followed by Tukey HSD test at the 5% level. All statistical analyses were conducted using JMP Pro 14 (SAS Institute, Cary, NC, USA).

3. Results and Discussion

3.1. Qualitative Evaluation of Fluorescent Tracer Uptake by Soybean Seed and Seedlings

Fluorescent tracer uptake by the yellow-seeded soybean variety AG1901 seed and translocation into 4-day-old dark-grown seedlings were investigated by qualitative fluorescence observation in seed and seedling tissue (Table 3). Thirty-two tracers were grouped into 10 chemical families or classes and are listed in Table 1. This list comprises a range of fluorescent tracers of log *P* from –0.62 to 6.03, molecular mass from 119 to 496 (g/mol), and electrical charge characterized as anionic, cationic, nonionic, or zwitterion. When assessed by the rule of two from Clark et al. (2003) [15,16], 18 of 32 the tracers tested were in violation, and therefore half were not predicted to have systemic activity in plants.

Results from our study showed that seed uptake was observed from 18 of the 32 tracers, while 16 were taken up in both seed and seedlings (Table 3). Coumarin derivatives were one of the major chemical families, and 7 of the 16 tracers that displayed both seed and seedling uptake were coumarin compounds, and all were nonionic in nature (Table 1). Most of the coumarin compounds (i.e., coumarin 120, coumarin 151), auramine O, and 9-aminoacridine were observed in the embryonic axis. Xanthene was the other major chemical family, and three rhodamine compounds were taken up in cotyledons, each with unique properties (Table 1): RB is a zwitterion (has both positive and negative charge at pH 6.5), the RB base is nonionic with a high log *P* (6.13) value, and R800 is cationic and has fluorescence in near-infrared (NIR). RB, RB base, and five fluorescein derivatives were observed in the cotyledons, and most showed some degree of phytotoxicity to seedling growth. Consistent with results in this study, fluorescent seed uptake was previously observed from soybean seeds treated with coumarin 1, coumarin 151, AMCA, fluorescein, carboxyfluorescein, uranine, as well as 9-aminoacridine [6].

Table 3. Fluorescent chemical family/class, chemical name, and seed and seedling response from the application of tracers taken up by seeds and seedlings.

Chemical Family/Class	Chemical Name	Seed	Seedlings	Note
Acridine	9-Aminoacridine Hydrochloride Hydrate	blue-embryo axis	blue	
Arylmethane Dye	Auramine O	yellow-embryo axis	-	High phytotoxicity to seeds and seedlings: Delayed germination, shortening and thickening of hypocotyls
Coumarin	7-Amino-4-Methyl-3-Coumarinylacetic Acid (AMCA)	blue-embryo axis	blue	
	Coumarin 1	blue-embryo axis	blue	Seedlings roots showing negative geotropism; hypocotyl and epicotyl shortening
	Coumarin 120	blue-embryo axis	blue	
	Coumarin 151	yellow green-embryo axis	yellow green	
	Coumarin 152	light green-embryo axis	green	
	Coumarin 314	green-cotyledons	green	
	6,7-Dihydroxy coumarin	green-cotyledons	white	
Naphtali-mide	Fluorescent Brightener 162	light green-cotyledons	-	
Xanthene	5(6)-Carboxy-Fluorescein	yellow-cotyledons	yellow	
	2′,7′-Dichloro-Fluorescein	yellow-cotyledons	yellow	Negative geotropism in seedlings
	Fluorescein	yellow-cotyledons	yellow	Seedlings non-uniform height
	Fluorescein sodium salt	yellow-cotyledons	yellow	Seedlings roots showed negative geotropism; seedlings non-uniform height
	Rhodamine 800	NIR	NIR	
	Rhodamine B	red-cotyledons	red	Seedlings roots showed negative geotropism
	Rhodamine B base	red-cotyledons	red	Seedlings roots showed negative geotropism
	Uranine K	yellow-cotyledons	yellow	Seedling non-uniform height

3.2. Fluorescence Microscopy and Quantification of Coumarin 120 Uptake by Soybean Seeds

Two soybean varieties, TMG 4185 and M 7739 IPRO (both yellow-seeded), were selected for observations of coumarin 120 seed uptake by fluorescence microscopy. Fluorescence intensity of coumarin 120 taken up by TMG 4185 was much greater than the fluorescence in M 7739 PRO (Figure 1), thus corroborating our previous investigations that varietal differences exist with respect to uptake in soybean [19] and corn [20].

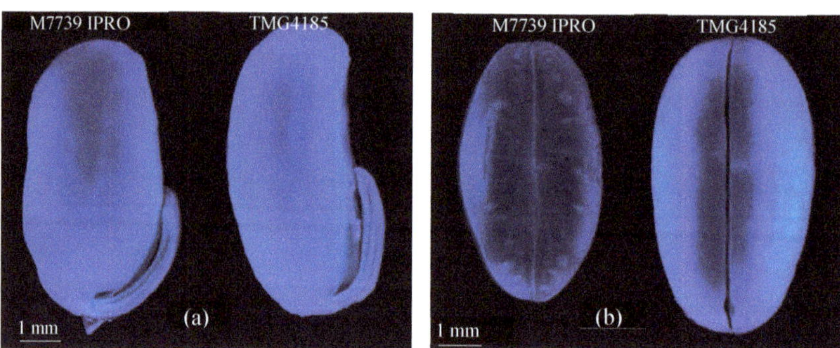

Figure 1. Fluorescence microscopy image of coumarin 120 uptake in low (M7739 IPRO) and high (TMG4185) soybean seed coat permeability varieties: respectively, (**a**) longitudinal section; (**b**) cross-section.

Due to the varietal differences in seed uptake to coumarin 120, an additional 15 soybean varieties were evaluated by fluorescence microscopy and chemical extractions to quantify uptake. Coumarin translocation efficiency was calculated based on the percent of applied material as a seed treatment extracted and recovered from the embryo. The uptake efficiency of yellow-seeded varieties ranged from 1.1% to 4.8%, while the two dark-seeded varieties, Black Jet and V12-1223, had uptake <0.5% (Figure 2). Data from the yellow-seeded varieties showed that TMG 4185 and TMG 1174RR had the highest uptake efficiency, while M 7739 IPRO had the lowest uptake efficiency. The uptake efficiency of coumarin 120 in TMG 4185 was 4.5 times greater than M 7739 IPRO.

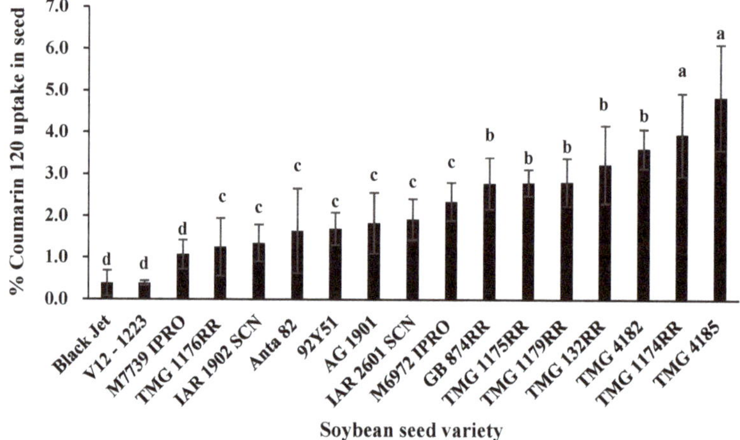

Figure 2. Seed uptake efficiency (percent uptake based on the total amount of coumarin 120 (C120) applied as a seed treatment) in 17 soybean varieties. Means for soybean seed varieties with the same letter are not significantly different from each other (Tukey HSD test, $P < 0.05$). Bars represent standard error of the mean.

3.3. Rhodamine B (RB) and Rhodamine 800 (R800) Uptake in Soybean Seeds

Both RB and R800 were detected and imaged from soybean seeds treated with each tracer (Figure 3). Soybean seeds took up the charged molecules of RB and R800 during imbibition, which is consistent with the classification of large-seeded legumes with the permeable seed coat characteristic [5]. The IVIS images used a false color to quantify fluorescence intensity, and both rhodamine tracers showed increased seed uptake as the applied dosage increased from 0.05% to 0.5% (Figure 3). The fluorescence intensity was quantified with IVIS and expressed as "total radiant efficiency". The total radiant efficiency was consistently greater in RB than in R800-treated seeds by 17 and 40 times at 0.05 and 0.5%, respectively. This is the first reported use of a near-infrared (NIR) fluorescent tracer in seed and seedling uptake studies, which opens the possibility for many other applications of NIR tracers in plant science research.

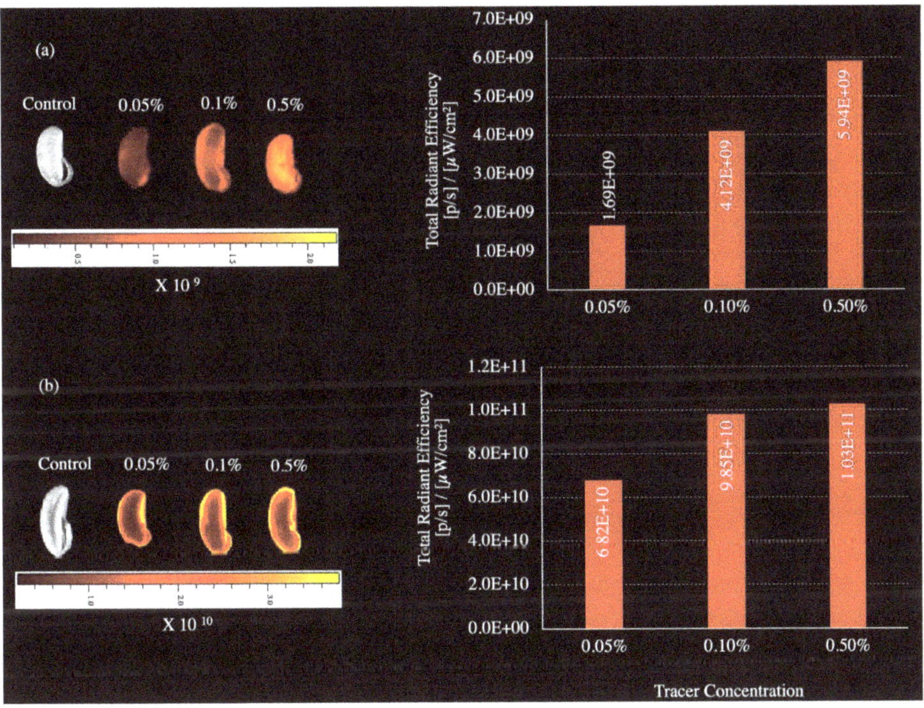

Figure 3. Rhodamine B and rhodamine 800 uptake in yellow soybean seed variety TMG 4185 applied at different concentrations of (**a**) rhodamine 800, (**b**) rhodamine B. Imaging was done under the in vivo imaging system (IVIS).

3.4. Rhodamine B Uptake and Transmission in Soybean Seedling

In previous studies, RB was found to be well suited for fluorescence imaging in seedlings. The emission and excitation wavelengths had little interference with the fluorescence spectra of chlorophyll and chlorophyll fluorescence [8]. However, RB applied to seeds at 0.1% and 0.5% by weight resulted in phytotoxicity (data not shown); therefore, the 0.05% dosage was used for all seedling-imaging studies.

RB fluorescence was observed in all seedling tissues: root, hypocotyl, cotyledon, epicotyl, and leaf (Figure 4a). The abaxial side of the leaf had greater RB fluorescence than the adaxial side, and the fluorescence was concentrated in the mid-veins (Figure 4b). RB fluorescence was greater in the lower section of the epicotyl (Figure 4c) that is adjacent to the cotyledons, as the cotyledons from the treated seed were the primary source of RB (Figure 3b). RB fluorescence was greater in the lower section of the hypocotyl (Figure 4d) near the roots and was also closer to the soil media surface. RB was observed to diffuse into the silica sand from the treated seed after sowing, and fluorescence was most concentrated in close proximity to the sown seed (not shown).

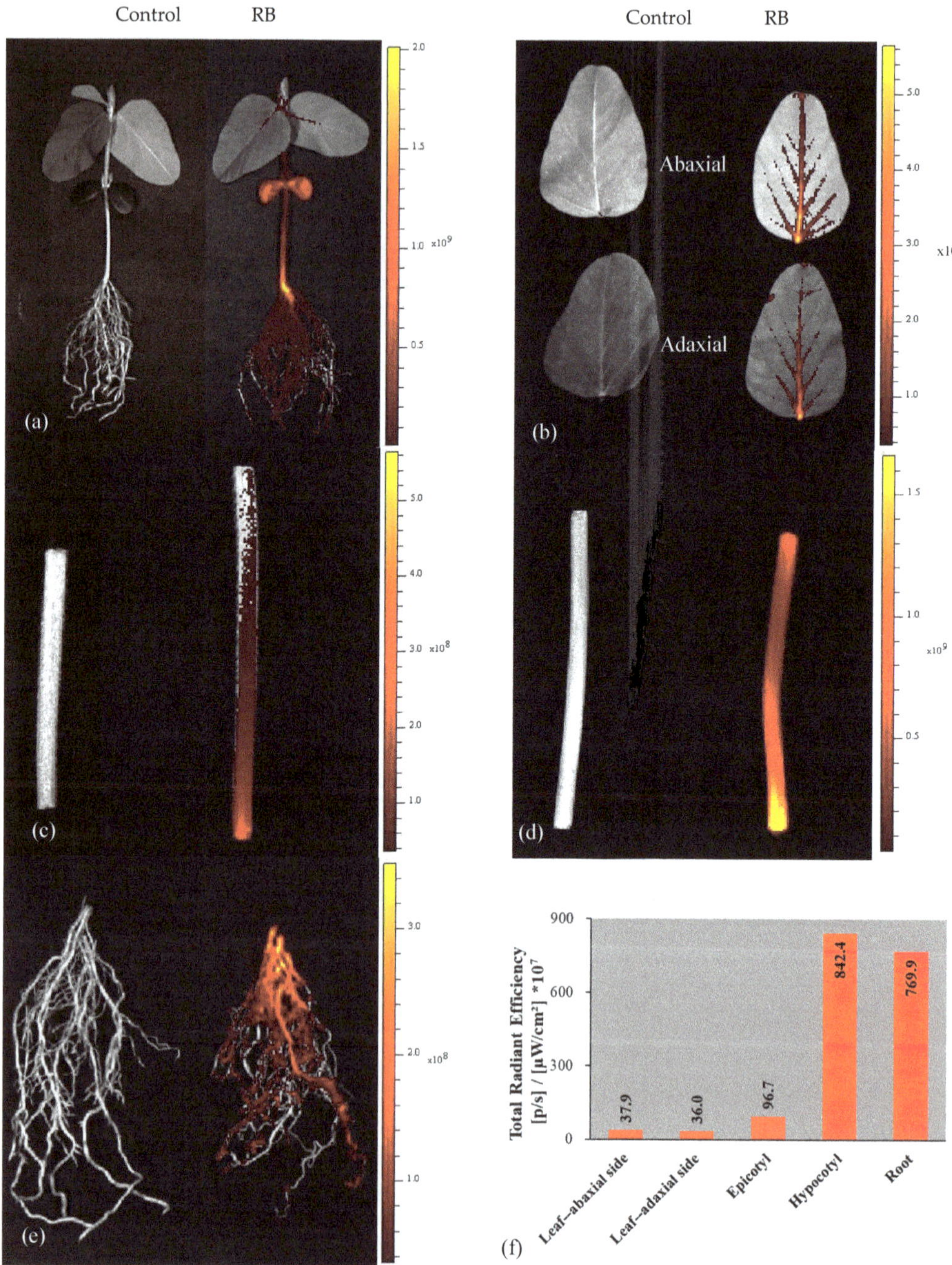

Figure 4. Imaging of rhodamine B (RB) uptake in soybean seedlings under the in vivo imaging system (IVIS): (**a**) whole plant, (**b**) leaf, (**c**) epicotyl, (**d**) hypocotyl, (**e**) root, and (**f**) fluorescence intensity in different parts of soybean seedling.

The intensity of total radiant efficiency in seedling tissues of RB was highest in the hypocotyl and root, 842.4 × 10^7 and 769.9 × 10^7 (p/s)/(μW/cm^2), respectively (Figure 4f). The intensity of RB in the abaxial side of the leaf was slightly higher than the adaxial side of the leaf, but the intensity of RB in the leaves was about 4% compared to the hypocotyl, while the intensity of RB in epicotyl tissue was about 11% compared to the hypocotyl. Similar observations of RB fluorescence distribution were reported on snap bean seedlings [21].

3.5. Relationship of Seed Uptake of Tracers with Systemic Agrochemicals by Roots

An understanding of the uptake of systemic seed treatments is based on the classical work conducted on barley (*Hordeum vulgare* L.) in the 1980s by Briggs and others [22]. The uptake, termed the transpiration stream concentration factor (TSCF), revealed a Gaussian distribution in relation to the log *P* of agrochemicals. This same Gaussian distribution was also documented for other crops, and the Gaussian curve for soybean [23] is shown in Figure 5. Therefore, root uptake by agrochemicals would conservatively be predicted for systemic compounds with log *P* >1 and log *P* <5. However, in the range of log *P* 1–5, 14 tracers revealed seed uptake, while 12 tracers were non-seed uptake, and this differential uptake of tracers was not related to electrical charge (Figure 5). The rule of two that included log *P*, molecular mass, and hydrogen donors only predicted with 55% accuracy those tracers that showed seed uptake (Tables 1 and 3). However, investigating the same fluorochrome, a number of piperonyl amides applied as seed treatments were taken up by soybean seeds in the range of 0 to 4.2 log *P* [24]. Therefore, log *P* is one factor, but not the sole property responsible for seed uptake by chemically diverse tracers used in these experiments. Again, this supported that other physical/chemical properties were responsible for seed uptake and some tracers in this study acted as dyes or stains [25]. Collectively, soybean seed uptake was not related to root uptake with respect to log *P* of the compound.

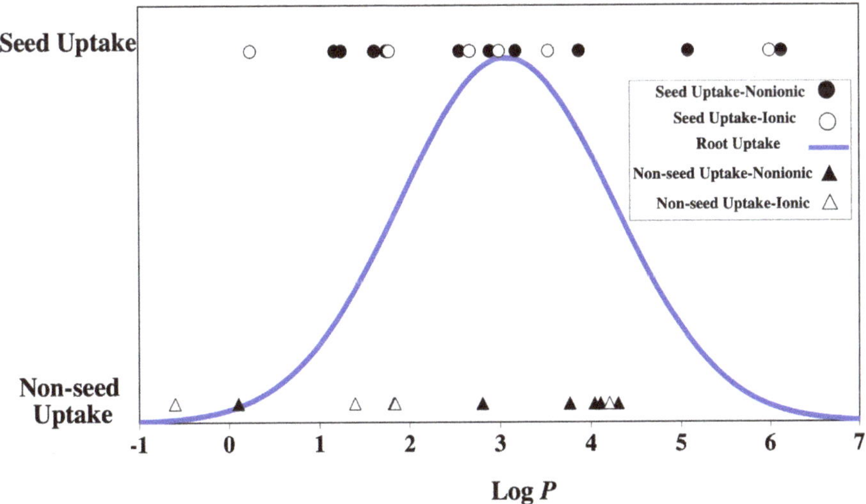

Figure 5. The log *P* and electrical charge of 32 tracers in relation to qualitative seed uptake or nonseed uptake. The blue line represents Gaussian distribution for systemic root uptake adapted from Hsu et al., 1990 [23].

4. Conclusions

Fluorescent tracers are powerful tools to study seed and seedling uptake but can have much broader applications in agriculture and biological science research beyond seed science. The list (Table 1) of 32 fluorescent tracers with identifying CAS number and physical/chemical properties provides a large number of tracers that can be used in future research for both fundamental and applied applications. With respect to uptake of agrochemicals by crop plants, the lipophilicity of a compound, measured as the log *P*, is well documented in the literature to model systemic uptake by roots [22]. Our study reveals that the log *P* is not the only property that can be used to predict tracer seed uptake, thus indicating that other physical/chemical properties are involved in seed uptake. RB has utility as a fluorescent tracer applied as a seed treatment that can be translocated to transpiring leaves, and it has been proposed as a systemic crop signaling system for real-time detection of crop seedlings in the field [21]. Therefore, selected fluorescent tracers have the potential for crop plant detection for enhanced pest management systems for field application. Although RB has many desirable characteristics as a tracer, the problems with phytotoxicity need to be addressed before the technology can be used in agriculture.

Author Contributions: A.G.T. conceived and supervised the project. Z.W., S.A.G.A., D.Y., and M.A. performed the research and analyzed data. Z.W. wrote the article; A.G.T. and M.A. revised the article for all authors. All authors have read and approved the final version of the manuscript.

Funding: This material is based upon work that is supported by the National Institute of Food and Agriculture, US Department of Agriculture, Multi-state Project W-3168, under accession #1007938. Funding for the senior and fourth authors was provided by the China Scholarship Council (grant no. 201808845004 and 201503250009, respectively). The IVIS spectrum optical imager was funded by Cornell University Biotechnology Resource Center (BRC) and BRC Imaging instrument grant NIH S10OD025049.

Acknowledgments: The authors thank Stephen Donovan, Michael Loos, Catharine Catranis, and Yi Qiu for helpful suggestions and for their technical assistance. We further thank Johanna M. Dela Cruz at BRC at Cornell University for her help on the use of the IVIS, and Hilary Mayton for critically reviewing the manuscript. We greatly appreciate recommendations on seed treatment binders from Incotec, and technical information on fluorescent tracers provided by Day-Glo Color Corp., AaKash Chemicals, Milliken Chemical, and Exciton Dye Technologies.

Conflicts of Interest: The authors declare no conflict of interest.

References

1. Taylor, A.G. Seed treatments. In *Encyclopedia of Applied Plant Sciences*, 1st ed.; Thomas, B., Murphy, D.J., Murray, B.G., Eds.; Elsevier Academic Press: Amsterdam, The Netherlands, 2003; pp. 1291–1298. ISBN 9780122270505.
2. Taylor, A.G. Seed Storage, Germination, Quality, and Enhancements. In *Physiology of Vegetable Crops*, 2nd ed.; Wien, H.C., Stuetzel, H., Eds.; CAB International: Wallingford UK, 2020; p. 496. ISBN 978-1786393777.
3. Taylor, A.G.; Eckenrode, C.J.; Straub, R.W. Seed treatments for onions: Challenges and progress. *HortScience* **2001**, *36*, 199–205. [CrossRef]
4. Salanenka, Y.A.; Taylor, A.G. Seed coat permeability: Uptake and post-germination transport of applied model tracer compounds. *HortScience* **2011**, *46*, 622–626. [CrossRef]
5. Taylor, A.; Salanenka, Y. Seed treatments: Phytotoxicity amelioration and tracer uptake. *Seed Sci. Res.* **2012**, *22*, S86–S90. [CrossRef]
6. Salanenka, Y.A.; Taylor, A.G. Uptake of model compounds by soybean, switchgrass and castor seeds applied as seed treatments. In *Symposium Proceeding of Seed Production and Treatment in a Changing Environment, 2 April 2009*; Alton: Hants, UK, 2009; pp. 76–81.
7. Salanenka, Y.; Taylor, A. Seed coat permeability and uptake of applied systemic compounds. *Acta Hortic.* **2008**, *782*, 151–154. [CrossRef]
8. Su, W.H.; Fennimore, S.A.; Slaughter, D.C. Fluorescence imaging for rapid monitoring of translocation behaviour of systemic markers in snap beans for automated crop/weed discrimination. *Biosyst. Eng.* **2019**, *186*, 156–167. [CrossRef]
9. Living Image Software User's Manual (Version 4.0). Available online: https://research.cchmc.org/cic/training/spectrumct_manual.pdf (accessed on 10 May 2020).
10. Collins, J.W.; Meganck, J.A.; Kuo, C.; Francis, K.P.; Frankel, G. 4D Multimodality Imaging of Citrobacter rodentium Infections in Mice. *J. Vis. Exp.* **2013**, *78*, e50450. [CrossRef] [PubMed]

11. Lin, L.C.; Hsu, J.H.; Wang, L.C. Identification of Novel Inhibitors of 1-Aminocyclopropane-1-carboxylic Acid Synthase by Chemical Screening in *Arabidopsis thaliana*. *J. Boil. Chem.* **2010**, *285*, 33445–33456. [CrossRef] [PubMed]
12. Chen, I.-J.; Lo, W.S.; Chuang, J.Y.; Cheuh, C.M.; Fan, Y.S.; Lin, L.C.; Wu, S.J.; Wang, L.C. A chemical genetics approach reveals a role of brassinolide and cellulose synthase in hypocotyl elongation of etiolated *Arabidopsis* seedlings. *Plant. Sci.* **2013**, *209*, 46–57. [CrossRef] [PubMed]
13. Xu, X.; Miller, S.A.; Baysal-Gurel, F.; Gartemann, K.H.; Eichenlaub, R.; Rajashekara, G. Bioluminescence Imaging of Clavibacter michiganensis subsp. michiganensis Infection of Tomato Seeds and Plants. *Appl. Env. Microbiol.* **2010**, *76*, 3978–3988. [CrossRef] [PubMed]
14. Lipinski, C.; Lombardo, F.; Dominy, B.W.; Feeney, P.J. Experimental and computational approaches to estimate solubility and permeability in drug discovery and development settings. *Adv. Drug Deliv. Rev.* **2001**, *46*, 3–26. [CrossRef]
15. Jampílek, J. Potential of agricultural fungicides for antifungal drug discovery. *Expert Opin. Drug Discov.* **2015**, *11*, 1–9. [CrossRef] [PubMed]
16. Clarke, E.D.; Delaney, J. Physical and Molecular Properties of Agrochemicals: An Analysis of Screen Inputs, Hits, Leads, and Products. *Chim. Int. J. Chem.* **2003**, *57*, 731–734. [CrossRef]
17. ChemAxon. Available online: https://chemicalize.com (accessed on 9 May 2020).
18. Kwon, Y. Partition and Distribution Coefficients. In *Handbook of Essential Pharmacokinetics, Pharmacodynamics and Drug Metabolism for Industrial Scientists*; Kluwer Academic/Plenum Publishers: New York, NY, USA, 2001; p. 44. ISBN 978-1-4757-8693-4.
19. Yang, D.; Avelar, S.A.G.; Taylor, A.G. Systemic Seed Treatment Uptake during Imbibition by Corn and Soybean. *Crop. Sci.* **2018**, *58*, 2063–2070. [CrossRef]
20. Diaz, M.; Taylor, A.; Cicero, S.M. Uptake of systemic seed treatments by maize evaluated with fluorescent tracers. *Seed Sci. Technol.* **2014**, *42*, 101–107. [CrossRef]
21. Su, W.H.; Fennimore, S.A.; Slaughter, D.C. Development of a systemic crop signalling system for automated real-time plant care in vegetable crops. *Biosyst. Eng.* **2020**, *193*, 62–74. [CrossRef]
22. Briggs, G.G.; Bromilow, R.H.; Evans, A.A. Relationships between lipophilicity and root uptake and translocation of non-ionized chemicals by barley. *Pestic, Sci.* **1982**, *13*, 495–504. [CrossRef]
23. Hsu, F.C.; Marxmiller, R.L.; Yang, A.Y.S. Study of Root Uptake and Xylem Translocation of Cinmethylin and Related Compounds in Detopped Soybean Roots Using a Pressure Chamber Technique. *Plant. Physiol.* **1990**, *93*, 1573–1578. [CrossRef] [PubMed]
24. Yang, D.; Donovan, S.; Black, B.C.; Cheng, L.; Taylor, A.G. Relationships between compound lipophilicity on seed coat permeability and embryo uptake by soybean and corn. *Seed Sci. Res.* **2018**, *28*, 229–235. [CrossRef]
25. Green, F.J. *The Sigma-Aldrich Handbook of Stains, Dyes and Indicators*; Alderich Chemical company, Inc.: Milwaukee, WI, USA, 1990; ISBN 0-941633-22-5.

Article

Effect of Seed Priming with Potassium Nitrate on the Performance of Tomato

Muhammad Moaaz Ali [1,†], Talha Javed [2,3,†], Rosario Paolo Mauro [4,*], Rubab Shabbir [2,3], Irfan Afzal [3] and Ahmed Fathy Yousef [1,5]

1. College of Horticulture, Fujian Agriculture and Forestry University, Fuzhou, Fujian 350002, China; muhammadmoaazali@yahoo.com (M.M.A.); ahmedfathy201161@yahoo.com (A.F.Y.)
2. College of Agriculture, Fujian Agriculture and Forestry University, Fuzhou, Fujian 350002, China; talhajaved54321@gmail.com (T.J.); rubabshabbir28@gmail.com (R.S.)
3. Seed Physiology Lab, Department of Agronomy, University of Agriculture, Faisalabad 38040, Pakistan; irfanuaf@gmail.com
4. Dipartimento di Agricoltura, Alimentazione e Ambiente (Di3A), Università degli Studi di Catania, Via Valdisavoia, 5-95123 Catania, Italy
5. Department of Horticulture, College of Agriculture, University of Al-Azhar (branch Assiut), Assiut 71524, Egypt

* Correspondence: rosario.mauro@unict.it; Tel.: +39-095-4783314

† Both authors contributed equally to this work.

Received: 12 September 2020; Accepted: 23 October 2020; Published: 25 October 2020

Abstract: The seed industry and farmers have challenges, which include the production of poor quality and non-certified tomato seed, which ultimately results in decreased crop production. The issue carefully demands pre-sowing treatments using exogenous chemical plant growth-promoting substances. Therefore, to mitigate the above-stated problem, a series of experiments were conducted to improve the quality of tomato seeds (two cultivars, i.e., "Sundar" and "Ahmar") and to enhance the stand establishment, vigor, physiological, and biochemical attributes under growth chamber and greenhouse conditions by using potassium nitrate (KNO_3) as a seed priming agent. Seeds were imbibed in 0.25, 0.50, 0.75, 1.0, and 1.25 KNO_3 (weight/volume) for 24 h and then dried before experiments. The results of growth chamber and greenhouse screening show that experimental units receiving tomato seeds primed with 0.75% KNO_3 in both cultivars performed better as compared to other concentrations and nonprimed control. Significant increase in final emergence (%), mean emergence time, and physiological attributes were observed with 0.75% KNO_3. Collectively, the improved performance of tomato due to seed priming with 0.75% KNO_3 was linked with higher activities of total soluble sugars and phenolics under growth chamber and greenhouse screening.

Keywords: *Solanum lycopersicum* L.; crop establishment; potassium nitrate; seed quality

1. Introduction

Tomato (*Solanum lycopersicum* L.) is a major vegetable crop on a global scale and one of the principle sources of phytonutrients [1,2], which makes it one of the preferred targets by researchers for metabolic engineering, as it is easily docile to biotechnological modifications [3]. Globally, being a vegetable of major economic importance, the tomato is a source of minerals and vitamins, as well as an anticancer agent [4]. Ripe tomatoes contain (average values per 100 g of edible portion) water (94.1%), energy (23 calories), calcium (1.0 g), magnesium (7.0 mg), vitamin A (1000 IU), ascorbic acid (22 mg), thiamin (0.09 mg), riboflavin (0.03 mg), and niacin (0.8 mg) [5]. In tomato, germination and crop establishment are the most crucial physiological stages that are affected by seed quality and genetics [6]. Rapid and uniform germination and seedling establishment is essential for increasing tomato yield and quality [7], which is of economic importance in agriculture. Therefore, various seed enhancement approaches, such as coating, pelleting, and priming, can be responsible to a major extent for improved quality of seeds. Among these approaches, seed priming with suitable priming agents and concentrations can induce some physiological and biochemical changes in the seed, which result in improved crop performances in terms of enhanced germination potential, seedling vigor, and final yield [6,8].

Seed priming is a process of regulating the imbibition and active metabolism phases of germination before radical emergence followed by drying and maintenance of near to original moisture content [9]. It increases the ability of radical to protrude rapidly, as the initial stages of germination are already fulfilled even under environmental stresses [10]. Seed priming helps the plants to cope with the adverse effects of unfavorable environmental conditions [11,12]. According to Liu [13], priming improves the activities of anti-oxidative metabolites, such as superoxide dismutase and peroxidase, during seed germination. Priming helps the plants to accelerate cell division, transport stored proteins and hasten the speed of seed germination [14]. Seed priming improved germination and seedling vigor in tomato [9] by activation of antioxidants [15], reduced membrane permeability, and maintenance of tissue water contents [6].

Exogenous application of priming agents to the seeds have remarkable role for pre-sowing accomplishment of germination phases [16]. Sliwinska [17] reported that 42% of primed tomato root tip cells were arrested in the G2-phase of mitosis and did not complete cell division. Previous studies revealed the positive role of potassium nitrate (KNO_3) as a seed priming agent on seedling establishment and vigor [18]. In addition, considerable increase in germination potential and seedling vigor was observed in tomato seed treated with KNO_3 at the concentration of 50 mmol [19]. Similarly, exogenous KNO_3 treatment on rice seeds concurrently improved multiple aspects of germination and physiology. This implies that KNO_3 might play a signaling role in prompting a wide adaptation of rice seedlings [20]. Therefore, the present study was conducted to evaluate the effects of exogenously applied KNO_3 (0.25%, 0.50%, 0.75%, 1.0%, and 1.25%) as seed priming agent in two different tomato cultivars.

2. Materials and Methods

2.1. Seed Source

Six months old seeds of the pure line tomato cultivars "Sundar" and "Ahmar" (oval shaped fruit with regular leaves) were obtained from Ayyub Agricultural Research Institute, Vegetable Research Section, Faisalabad-38000, Punjab, Pakistan. The initial germination and seed moisture content before seed treatment were (86% and 10.5% in "Sundar"; 84% and 11% in "Ahmar"), respectively.

2.2. Seed Priming Treatments

Tomato seeds were primed/imbibed with 0.25%, 0.50%, 0.75%, 1.0%, and 1.25% (weight/volume) KNO_3 for 24 h at 25 °C. Pre-weighed seeds (5 g) were imbibed on two blotter papers in 9-cm diameter Petri dishes with appropriate concentration of KNO_3 solutions, followed by covering of dishes with aluminum foil. For aeration, a hole was provided in the center of each Petri dish. After each treatment, seeds were rinsed thoroughly with distilled water and dried back closer to original moisture level under shaded conditions. Nonprimed tomato seeds were maintained as control for comparison.

2.3. Experimental Site and Conditions

Growth chamber and greenhouse experiments were conducted at the research station of the University of Agriculture, Faisalabad, Punjab, Pakistan (30.37° N, 69.34° E) from 29 October 2019 to 1 December 2019. Well pulverized soil was collected from the field of the research station, and each plastic tray 35 cm × 25 cm × 15 cm in size was filled with 6 kg of soil. The textural class of soil was sandy loam having pH (6.8), electric conductivity (0.396 dS m^{-1}), available phosphorus (17.67 ppm), and potassium (353.96 ppm). After leveling the soil surface in each tray, moisture was applied up to field capacity. In each tray, 30 seeds were sown with equal distance in the soil in both experiments and considered as one replicate. Both experiments were laid out in a completely randomized design with four replications. For growth chamber screening (optimal germination and growth conditions), all the trays were placed in the growth chamber with an optimal temperature of 25 °C and a light period of 12 h. The relative humidity during the complete execution of the growth chamber experiment was maintained at 65%. For the greenhouse experiment (suboptimal conditions), all the trays were placed in the greenhouse under natural environmental conditions. The climate data during the complete execution of greenhouse experiment is given in Figure 1.

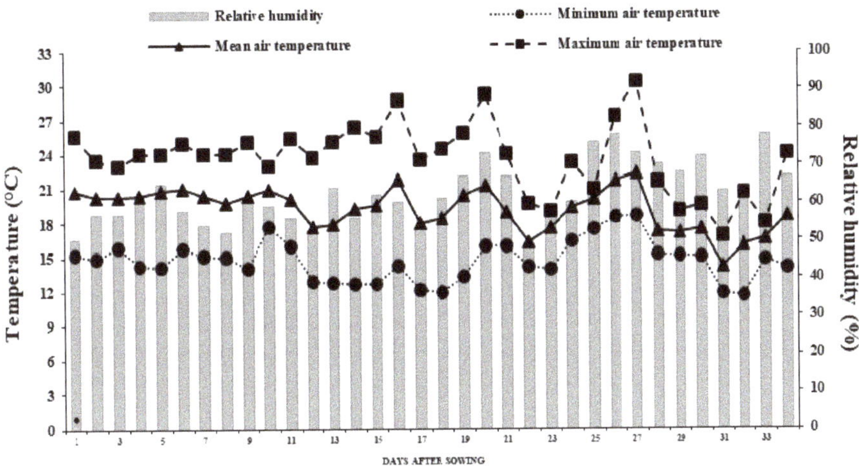

Figure 1. Microclimate conditions inside the greenhouse during the experiment at research station of University of Agriculture, Faisalabad, Punjab, Pakistan.

2.4. Seedling Establishment

Seedling emergence was recorded daily and recorded when the hypocotyl came above the soil surface. Final emergence, expressed on a percentage basis, was calculated as the ratio among number of emerged seedlings and total number of seeds sown at the end of the experiment [21]. Mean emergence time (days) was recorded as per the equation earlier reported by International Seed Testing Association (ISTA) [22]:

$$\text{Mean Emergence Time (MET)} = \frac{\sum \text{Dn}}{\sum \text{n}},$$

where n is the number of seeds which emerged on day D, and D is the number of days counted from the beginning of emergence.

2.5. Seedling Vigor

Thirty days after sowing (DAS) plant height was determined on 5 randomly selected seedlings. On the same date, both fresh and dry weight of tomato plants were recorded. For dry weight, plants were dried at 70 °C till constant weight in an oven.

2.6. Physiological Variables

At 30 DAS, i.e., with the plants at the stage of 6 true leaves, measurements of CO_2 index (μmol mol^{-1}), net photosynthetic rate (μmol CO_2 m^{-2} s^{-1}), and transpiration rate (μmol H_2O m^{-2} s^{-1}) were made on a fully expanded leaf from the top of the plant canopy by using an open system LCA-4 (ADC BioScientific Ltd., Hoddesdon, UK) portable infrared gas analyzer. Measurements were made between 6:00 a.m. and 7:00 a.m., with the following specifications: ambient pressure (P)

99.95 kPa, leaf chamber molar gas flow rate (U) 251 μmol s^{-1}, molar flow of air per unit leaf area (Us) 221.06 mol m^{-2} s^{-1}, temperature of leaf chamber (Tch) varied from 39 to 44 °C, Photosynthetically active radiation (PAR) at leaf surface was maximum up to 918 μmol m^{-2}, and leaf chamber molar gas flow rate (U) 251 μmol s^{-1}.

2.7. Biochemical Variables

To determine total phenolics, leaves were ground in liquid nitrogen by using pestle and mortar and a 20 μL sample was mixed with 1.60 mL distilled water, 100 μL Folin-Ciocalteu reagent ($_2$N), and 300 μL sodium carbonate solution in a test tube [23]. After 30 min at 40 °C in water bath, test tubes were immediately moved to an ice box and absorbance recorded at 765 nm with a spectrophotometer (UV 4000). The total soluble sugars (TSS) in leaf samples were determined by the anthrone method [24]. Ground leaf sample (25 mg) was mixed with 5 mL of $_{2.5}$NHCl in a test tube. Tubes were placed in water bath 100 °C for 3 h, followed by cooling of tubes at room temperature. By using distilled water, the volume of tube was made to 100 mL and centrifuged at 4000 rpm for 10 min. After that, 0.5 mL supernatant, 0.5 mL distilled water, and 4 mL anthrone (0.2% v/v anthrone on 95% sulfuric acid) was taken in another tube. The tube was heated again in boiling water bath for 8 min. The tube was cooled rapidly and reading was taken at 630 nm by using spectrophotometer (UV 4000).

2.8. Statistical Analysis

Collected data were subjected to a two-way analysis of variance (ANOVA) (2 genotypes × 6 KNO_3 level) and Tukey's honest significance difference (HSD) test for means comparison at 5% significance level, using the analytical software package 'Statistix 8.1'.

2.9. Greenhouse Microclimate Conditions During the Trial

During the experiment, the average mean temperature was 19.2 °C, with a sharp decrease from 22 to 14 °C (on 27 and 31 DAS, respectively), whereas average minimum and maximum temperatures oscillated between 12–19 °C and 17–30 °C, respectively (Figure 1). The average relative humidity varied between 49% and 78%, with the lowest value recorded at 22 DAS and highest one at 33 DAS (Figure 1).

3. Results

3.1. Growth Chamber Screening

3.1.1. Seedling Establishment

Seedling establishment of tomato includes seedling emergence (%) and the number of days required by seeds to germinate (mean emergence time—MET). The results of the present study indicated that the seed priming with KNO_3 improved the stand establishment of tomato (cv. "Sundar" and "Ahmar") grown in growth chamber. Tomato seeds primed with 0.75% KNO_3 had maximum emergence rate in both cultivars (98% "Sundar"; 99% "Ahmar"), so they were showing better performances when compared to the other treatments. Minimum germination (82% "Sundar"; 84% "Ahmar") was observed in nontreated seeds of tomato, while, in the case of MET, the maximum number of days (4–5) was observed in nontreated seeds of tomato. The seeds treated with 0.75% KNO_3 germinated earlier than all other treatments (Table 1).

Table 1. Final emergence (%) and mean emergence time (days) of tomato seedlings as affected by the cultivar and seed priming with potassium nitrate (KNO_3), under two different growth conditions.

	Treatments	Final Emergence (%)			Mean Emergence Time (Days)		
		"Sundar"	"Ahmar"	KNO_3 Mean	"Sundar"	"Ahmar"	KNO_3 Mean
Growth chamber	Control	82	84	83^{d}	5.3	4.7	5.0^{a}
	0.25% KNO_3	84	90	87^{c}	4.2	3.7	4.0^{b}
	0.50% KNO_3	89	93	$91^{b,c}$	3.3	3.5	$3.4^{b,d}$
	0.75% KNO_3	97	99	98^{a}	3.0	3.0	3.0^{d}
	1% KNO_3	89	93	$91^{b,c}$	3.3	3.1	$3.2^{c,d}$
	1.25% KNO_3	84	90	87^{c}	3.8	3.5	$3.7^{b,c}$
	Cultivar mean	88^{b}	92^{a}		3.8^{a}	3.6^{b}	
	$HSD_{interaction\ (p \leq 0.05)}$	7			0.7		
Greenhouse	Control	79	82	81^{e}	7.2	6.6	6.9^{a}
	0.25% KNO_3	81	85	83^{d}	6.3	5.8	6.0^{b}
	0.50% KNO_3	84	88	86^{c}	5.3	4.9	5.1^{c}
	0.75% KNO_3	93	97	95^{a}	3.9	3.6	3.7^{e}
	1% KNO_3	89	90	89^{b}	4.7	4.1	4.4^{d}
	1.25% KNO_3	82	85	84^{d}	5.2	5.3	5.3^{c}
	Cultivar mean	85^{b}	88^{a}		5.4^{a}	5.0^{b}	
	$HSD_{interaction\ (p \leq 0.05)}$	4			0.9		

Values sharing the same letters are non-significantly different ($p \leq 0.05$).

3.1.2. Seedling Vigor

Statistical analysis of data about seedling vigor revealed that the effect of seed priming treatments was significant in both "Sundar" and "Ahmar". All priming treatments significantly improved the seedling length in both cultivars, whereas the cultivar did not exert any significant effect. Maximum seedling length was achieved in tomato seed primed with 0.75% (8.36 and 8.35 cm in "Sundar" and "Ahmar", respectively), followed by 1% KNO_3 solution (7.5 and 7.8 cm), whereas the lowest values for seedling length in both cultivars was observed in control (5.2 and 5.3 cm). In both cultivars, plants raised from seeds treated with 0.75% KNO_3 showed higher values for seedling fresh weight (35.1 and 37.6 mg in "Sundar" and "Ahmar", respectively) and dry weight (17.9 and 19.1 mg) as compared to other treatments (Table 2).

Table 2. Seedling length, shoot fresh weight, and shoot dry weight of tomato as affected by seed priming with KNO_3 under two different growth conditions.

	Treatments	Seedling Length (cm)			Shoot Fresh Weight (mg)			Shoot Dry Weight (mg)		
		"Sundar"	"Ahmar"	KNO_3 Mean	"Sundar"	"Ahmar"	KNO_3 Mean	"Sundar"	"Ahmar"	KNO_3 Mean
Growth chamber	Control	5.2	5.3	5.2^{d}	25.7	24.4	25.1^{d}	10.9	12.9	11.9^{d}
	0.25% KNO_3	6.2	6.1	6.2^{c}	29.2	28.6	28.9^{c}	13.8	15.7	14.7^{c}
	0.50% KNO_3	7.2	7.4	7.3^{b}	32.5	34.7	33.6^{b}	16.0	16.6	16.3^{b}
	0.75% KNO_3	8.4	8.4	8.4^{a}	35.1	37.6	36.3^{a}	17.9	19.1	18.5^{a}
	1% KNO_3	7.5	7.8	7.7^{b}	31.9	32.4	32.2^{b}	15.4	16.9	16.2^{b}
	1.25% KNO_3	7.3	7.7	7.5^{b}	28.9	28.6	28.8^{c}	13.0	12.4	12.7^{d}
	Cultivar mean	7.0^{b}	7.1^{a}		30.5^{a}	31.1^{a}		14.5^{b}	15.6^{a}	
	$HSD_{interaction\ (p \leq 0.05)}$	0.7			2.5			1.9		
Greenhouse	Control	3.8	3.9	3.8^{d}	21.8	25.1	23.4^{c}	9.3	10.1	9.7^{c}
	0.25% KNO_3	4.5	4.9	4.7^{c}	24.3	24.1	24.2^{c}	10.1	11.2	10.6^{c}
	0.50% KNO_3	6.0	6.1	6.0^{b}	27.6	26.5	27.1^{b}	11.9	13.1	12.5^{b}
	0.75% KNO_3	7.0	7.1	7.0^{a}	30.3	32.1	31.2^{a}	14.6	14.4	14.5^{a}
	1% KNO_3	6.0	5.7	5.9^{b}	26.9	27.7	27.3^{b}	12.9	14.1	$13.5^{a,b}$
	1.25% KNO_3	6.1	6.2	6.1^{b}	23.9	24.9	24.4^{c}	10.4	11.6	11.0^{c}
	Cultivar mean	5.6^{a}	5.7^{a}		25.8^{b}	26.7^{a}		$11.5^{b,c}$	12.4^{a}	
	$HSD_{interaction\ (p \leq 0.05)}$	0.9			3.3			2.2		

Values sharing the same letters are non-significantly different ($p \leq 0.05$).

3.1.3. Physiological and Biochemical Attributes

The results shown in Tables 3 and 4 indicate that seed priming treatments significantly improved the physiological and biochemical attributes of both tomato cultivars, while the genotype effect was also found significant. The highest photosynthesis rate, transpiration rate, and CO_2 index were linked to tomato seeds treated with 0.75% KNO_3, whereas the lowest values were found in the nonprimed seeds (Table 3). A statistical evaluation of data demonstrated that total soluble sugars and phenolic contents were significantly influenced by seed priming treatments. Though all the seed priming treatments proved successful to improving these biochemical attributes, highest values were observed in plants deriving from seeds primed with 0.75% KNO_3 under growth chamber screening (Table 4).

Table 3. Variations in physiological attributes of two tomato cultivars under the influence of seed priming with KNO_3 in two different growth conditions.

	Treatments	Photosynthetic Rate (µmol CO_2 m^{-2} s^{-1})			Transpiration Rate (µmol H_2O m^{-2} s^{-1})			CO_2 Index (µmol mol^{-1})		
		"Sundar"	"Ahmar"	KNO_3 Mean	"Sundar"	"Ahmar"	KNO_3 Mean	"Sundar"	"Ahmar"	KNO_3 Mean
Growth chamber	Control	9.7	10.7	10.2 ^{e}	0.81	0.92	0.86 ^{e}	110	113	111 e
	0.25% KNO_3	11.7	12.3	12.0 ^{d}	0.99	1.09	1.04 ^{d}	121	123	122 ^{d}
	0.50% KNO_3	13.8	14.8	14.3 ^{b,c}	1.21	1.29	1.25 ^{c}	138	141	139 ^{b}
	0.75% KNO_3	16.7	17.3	17.0 ^{a}	1.61	1.70	1.65 ^{a}	162	164	163 ^{a}
	1% KNO_3	14.9	16.3	15.6 ^{a,b}	1.30	1.40	1.35 ^{b}	144	145	145 ^{b}
	1.25% KNO_3	13.0	13.7	13.3 ^{c,d}	1.24	1.30	1.27 ^{c}	133	134	134 ^{c}
	Cultivar mean	13.3 ^{b}	14.2 ^{a}		1.19 ^{b}	1.28 ^{a}		135 ^{a}	137 ^{a}	
	$HSD_{interaction}$ ($p \leq 0.05$)	3.0			0.19			11		
Greenhouse	Control	11.0	12.7	11.8 ^{e}	1.01	1.12	1.06 ^{e}	121	123	122 ^{e}
	0.25% KNO_3	12.7	14.3	13.5 ^{d,e}	1.19	1.28	1.23 ^{d}	131	133	132 ^{d}
	0.50% KNO_3	15.1	17.0	16.1 ^{b,c}	1.41	1.49	1.45 ^{c}	148	152	150 ^{b}
	0.75% KNO_3	18.3	19.3	18.8 ^{a}	1.81	1.88	1.84 ^{a}	172	177	174 ^{a}
	1% KNO_3	16.3	18.3	17.3 ^{a,b}	1.51	1.60	1.55 ^{b}	154	155	155 ^{b}
	1.25% KNO_3	14.7	15.7	15.2 ^{c,d}	1.44	1.50	1.47 ^{c}	144	144	144 ^{c}
	Cultivar mean	14.7 ^{b}	16.2 ^{a}		1.39 ^{b}	1.47 ^{a}		145 ^{a}	147 ^{a}	
	$HSD_{interaction}$ ($p \leq 0.05$)	3.1			0.20			13		

Values sharing the same letters are non-significantly different ($p \leq 0.05$).

Table 4. Variations in biochemical attributes of two tomato cultivars under the influence of seed priming with KNO_3 in two different growth conditions.

	Treatments	Total Soluble Sugars (mg g^{-1})			Phenolics (mg g^{-1})		
		"Sundar"	"Ahmar"	KNO_3 Mean	"Sundar"	"Ahmar"	KNO_3 Mean
Growth chamber	Control	70.2	64.5	67.4 ^{b}	1.29	1.27	1.28 ^{c}
	0.25% KNO_3	70.8	69.9	70.3 ^{b}	1.45	1.34	1.39 ^{c}
	0.50% KNO_3	83.5	82.5	83.0 ^{a}	1.70	1.65	1.67 ^{a,b}
	0.75% KNO_3	85.0	86.3	85.6 ^{a}	1.81	1.74	1.77 ^{a}
	1% KNO_3	80.8	79.0	79.9 ^{a,b}	1.75	1.53	1.64 ^{a,b}
	1.25% KNO_3	79.5	76.0	77.8 ^{a,b}	1.68	1.50	1.59 ^{b}
	Cultivar mean	78.3 ^{a}	76.4 ^{a}		1.61 ^{b}	1.50 ^{a}	
	$HSD_{interaction}$ ($p \leq 0.05$)	9.8			0.29		
Greenhouse	Control	71.3	65.6	68.4 ^{c}	1.33	1.26	1.29 ^{e}
	0.25% KNO_3	71.7	71.2	71.5 ^{b,c}	1.51	1.35	1.43 ^{d,e}
	0.50% KNO_3	84.5	83.0	83.7 ^{a}	1.74	1.63	1.68 ^{b,c}
	0.75% KNO_3	88.8	87.2	88.0 ^{a}	1.86	1.76	1.81 ^{a}
	1% KNO_3	83.8	80.2	82.0 ^{a,b}	1.79	1.52	1.65 ^{a,b}
	1.25% KNO_3	79.7	77.3	78.5 ^{a,c}	1.73	1.49	1.61 ^{c,d}
	Cultivar mean	79.9 ^{a}	77.4 ^{a}		1.66 ^{b}	1.50 ^{a}	
	$HSD_{interaction}$ ($p \leq 0.05$)	11.0			0.24		

Values sharing the same letters are non-significantly different ($p \leq 0.05$).

3.2. Greenhouse Screening

3.2.1. Seedling Establishment

Seed priming treatments improved the final emergence of both tomato cultivars under greenhouse conditions. The highest emergence values were recorded in tomato seed primed with 0.75% KNO_3 (93.29% and 96.68% in "Sundar" and "Ahmar", respectively), whereas the lowest ones were found in the nonprimed seeds. Seed priming with 1% KNO_3 proved to improve the final emergence of both cultivars too (88.7% in "Sundar" and 90.1% in "Ahmar") (Table 1). In both cultivars, no variation in final emergence was observed among experimental units receiving tomato seed primed 0.50% and 1% KNO_3. However, when compared to the other treatments, the lowest MET value was recorded in tomato seeds primed with 0.75% followed by 1% KNO_3. Besides, both nontreated cultivars showed the highest values for MET (Table 1).

3.2.2. Seedling Vigor

Seedling length of both tomato cultivars is presented in Table 2, and data revealed that maximum seedling length in both cultivars was achieved in tomato seed primed with 0.75% (7.0 and 7.1 cm in "Sundar" and "Ahmar", respectively), followed by 1.25% KNO_3 solution (6.1 and 6.2 cm in "Sundar" and "Ahmar", respectively), whereas the lowest ones were recorded in the control (3.8 and 3.8 cm in "Sundar" and "Ahmar", respectively). Seed priming with KNO_3 also proved effective in improving the seedling fresh and dry weight; nonetheless, the effect of different cultivars was not pronounced. Plants in both cultivars raised from seeds treated with 0.75% KNO_3 showed higher values for seedling fresh (30.3 mg in "Sundar" and 32.1 mg in "Ahmar") and dry weight (14.6 and 14.4 mg, respectively) as compared to all other treatments. No significant difference in seedling fresh and dry weight was observed among tomato seed treated with 0.50% and 1.0% KNO_3 in both cultivars (Table 2).

3.2.3. Physiological and Biochemical Variables

Seed priming treatments improved the physiological and biochemical of both tomato cultivars under greenhouse conditions. Higher values for photosynthetic rate, transpiration rate, and CO_2 index were recorded in experimental units receiving tomato seed primed with 0.75% KNO_3 as compared to control (Table 3), while in the case of genotypes, "Ahmar" showed better photosynthetic rate and transpiration rate as compared to "Sundar". Seed priming with 1% KNO_3 improved the physiological attributes of both cultivars, too. No variation in physiological attributes was observed among experimental units receiving tomato seed primed 0.50% and 1% KNO_3 in "Ahmar". In the same way, maximum total soluble sugars were observed in tomato seeds primed with 0.75% followed by 1% KNO_3 (Table 4). The lowest values for phenolic contents were recorded in control in both cultivars (Table 4).

4. Discussion

4.1. Seedling Establishment

Tomato seed priming with KNO_3 affected the emergence of seedling and the speed of seed germination. Major events in other literature on priming includes metabolic changes, such as repair of DNA and increases in the biosynthesis of RNA [25], and enhancement in the respiration process of seed [26]. This indicates that the time of seed imbibition is very important for seed priming. For the study of seed priming of tomato with different levels of KNO_3, it is important to know about the emergence percentage and mean emergence time. The results of the present study indicate that the performance of both tomato cultivars primed with 0.75% KNO_3 was appreciable in growth chamber, as well as in greenhouse screening, meaning that this effect was still appreciable under suboptimal growth conditions. The pattern of seedling emergence and mean emergence time were almost the same in both cultivars, as well as in both growth environments (growth chamber and greenhouse). The time of water intake by the seed during priming can vary within the cultivars, which can affect the performance of the seed priming agent (KNO_3) [27]; similarly, in our study, the difference between the performance of both cultivars were seen.

The data shown in Table 1 indicated that priming of tomato seeds with 0.75% KNO_3 was better than other treatments in terms of final emergence and mean emergence time. Our study is in correspondence with another study that revealed that the emergence percentage of wheat seeds was decreased when they were primed with >1% KNO_3 [28]. This indicates that KNO_3 concentration above a certain threshold may not be appropriate to boost seed germination. Seed priming with 1% KNO_3 was found useful in terms of emergence percentage in sorghum [29] and rice [30]. Besides, soybean seed primed with 1% KNO_3 for 1 day enhanced the emergence percentage as compared to nontreated seeds, both in laboratory and field experiments [18].

4.2. Seedling Vigor

Seedling vigor is the combined result of the emerged seeds under a wide range of biotic and abiotic stresses. Seedling vigor is not a single measurable entity, but it is a sum of many growth parameters, such as seedling length, seedling fresh weight, and seedling dry weight [22]. Maximum vigor was observed when seed priming with 0.75% KNO_3 was done. Our study is in line with another study in which seedling vigor of wheat was improved by priming with KNO_3 [28]. Similar results were found in corn when the priming of seed was done with 1% KNO_3 [31]. Our findings are similar to other studies, in which the shoot length of watermelon and tomato were increased by the seed priming with KNO_3 [32,33]. Seed priming with 0.5% and 1% KNO_3 improved the vegetative growth of watermelon [34] and tomato [35], respectively, under salt stress. Seed priming with KNO_3 can cause a significant increase in seedling vigor of the wheat crop as compared to hydro-priming or dry broadcasting [36].

4.3. Physiological and Biochemical Attributes

Plant growth is based mainly upon photosynthesis, while its performance is mostly dependent on the opening/closing of stomata, which modulates photosynthetic rate, respiration rate, and CO_2 index [37–39]. The results of the present study revealed that the maximum photosynthesis rate, transpiration rate, and CO_2 index was observed in tomato plants grown by seeds primed with 0.75% KNO_3, compared to other priming treatments. Our study is in corroboration with another study in which the increased photosynthetic rate, respiration rate, and CO_2 index of cucumber seedlings as the result of seed priming with KNO_3 were reported. The photosynthesis rate of the seedlings has a positive correlation with the growth of seedling [40]. The results of the present study revealed that the biochemical attributes, e.g., total soluble sugars and phenolic content, of tomato plants were enhanced by seed priming with KNO_3. The maximum increase was observed when seeds were treated with 0.75% KNO_3, while minimum values were seen in nonprimed seedlings. Previous studies expressed that seed priming with KNO_3 significantly improved the biochemical indices of chicory [41] and rice [20].

5. Conclusions

The performance of tomato is diminished by the poor quality of seed. Therefore, the present study was conducted to improve the quality of tomato seed by priming with KNO_3. The results presented in this paper revealed that tomato seeds of both cultivars primed with 0.75% KNO_3 proved to be successful for improving seedling establishment and vigor, as well as physiological and biochemical attributes, under growth chamber and greenhouse conditions. The present study provides the direction towards further molecular investigation related to the seed priming of tomato.

Author Contributions: M.M.A. and T.J. conceptualization, conceived and data analysis and original draft preparation. R.S., I.A. and A.F.Y. helped in data analysis. R.P.M. conceptualization, data curation and editing the manuscript. All authors have read and approved the final manuscript.

Funding: This research received no external funding.

Acknowledgments: The authors thank the valuable contributions for data collection provided by Ahmed Mukhtar. Helpful suggestions were provided by Hafiz Sohaib Ahmed Saqib for data analysis.

Conflicts of Interest: The authors declare no conflict of interest.

References

1. Mauro, R.P.; Rizzo, V.; Leonardi, C.; Mazzaglia, A.; Muratore, G.; Distefano, M.; Sabatino, L.; Giuffrida, F. Influence of harvest stage and rootstock genotype on compositional and sensory profile of the elongated Tomato cv. "Sir Elyan". *Agriculture* **2020**, *10*, 82. [CrossRef]
2. Mauro, R.P.; Lo Monaco, A.; Lombardo, S.; Restuccia, A.; Mauromicale, G. Eradication of Orobanche/Phelipanche spp. seedbank by soil solarization and organic supplementation. *Sci. Hortic.* **2015**, *193*, 62–68. [CrossRef]
3. Canene-Adams, K.; Campbell, J.K.; Zaripheh, S.; Jeffery, E.H.; Erdman, J.W. The tomato as a functional food. *J. Nut.* **2005**, *135*, 1226–1230. [CrossRef] [PubMed]
4. Verma, A.K.; Tiwari, R.; Karthik, K.; Chakraborty, S.; Deb, R.; Dhama, K. Nutraceuticals from fruits and vegetables at a glance: A review. *J. Biol. Sci.* **2013**, *13*, 38–47.
5. Uddain, J.; Hossain, K.M.A.; Mostafa, M.G.; Rahman, M.J. Effect of different plant growth regulators on growth and yield of tomato. *Int. J. Sustain. Agric.* **2009**, *1*, 58–63.
6. Farooq, M.; Aziz, T.; Basra, S.M.A.; Cheema, M.A.; Rehman, H. Chilling tolerance in hybrid maize induced by seed priming with salicylic acid. *J. Agron. Crop Sci.* **2008**, *194*, 161–168. [CrossRef]
7. Javed, T.; Afzal, I. Impact of seed pelleting on germination potential, seedling growth and storage of tomato seed. *Acta Hortic.* **2020**, 417–424. [CrossRef]
8. Afzal, I.; Hussain, B.; Basra, S.M.A.; Rehman, H. Priming with moringa leaf extract reduces imbibitional chilling injury in spring maize. *Seed Sci. Technol.* **2012**, *40*, 271–276. [CrossRef]
9. Farooq, M.S.M.A.; Basra, S.M.A.; Saleem, B.A.; Nafees, M.; Chishti, S.A. Enhancement of tomato seed germination and seedling vigor by osmopriming. *Pak. J. Agric. Sci.* **2005**, *42*, 3–4.
10. Ibrahim, E.A. Seed priming to alleviate salinity stress in germinating seeds. *J. Plant Physiol.* **2016**, *192*, 38–46. [CrossRef]
11. Ashraf, M.; Foolad, M.R. Pre-sowing seed treatment-A shotgun approach to improve germination, plant growth, and crop yield under saline and non-saline conditions. *Adv. Agron.* **2005**, *88*, 223–271.
12. Chen, K.; Arora, R.; Arora, U. Osmopriming of spinach (*Spinacia oleracea* L. cv. Bloomsdale) seeds and germination performance under temperature and water stress. *Seed Sci. Technol.* **2010**, *38*, 36–48. [CrossRef]
13. Liu, J.; Liu, G.; Qi, D.; Li, F.; Wang, E. Effect of PEG on germination and active oxygen metabolism in wildrye (*Leymus chinensis*) seeds. *Acta Pratacult. Sin.* **2002**, *11*, 59–64.
14. De Castro, R.D.; van Lammeren, A.A.M.; Groot, S.P.C.; Bino, R.J.; Hilhorst, H.W.M. Cell division and subsequent radicle protrusion in tomato seeds are inhibited by osmotic stress but dna synthesis and formation of microtubular cytoskeleton are not. *Plant Physiol.* **2000**, *122*, 327–336. [CrossRef] [PubMed]
15. Farooq, M.; Aziz, T.; Wahid, A.; Lee, D.J.; Siddique, K.H.M. Chilling tolerance in maize: Agronomic and physiological approaches. *Crop Past. Sci.* **2009**, *60*, 501. [CrossRef]
16. Coolbear, P.; McGill, C.R. Effects of a low-temperature pre-sowing treatment on the germination of tomato seed under temperature and osmotic stress. *Sci. Hortic.* **1990**, *44*, 43–54. [CrossRef]
17. Sliwinska, E. Nuclear DNA replication and seed quality. *Seed Sci. Res.* **2009**, *19*, 15–25. [CrossRef]
18. Mohammadi, G.R. The effect of seed priming on plant traits of late-spring seeded soybean (*Glycine max* L.). *Am. Eurasian J. Agric. Environ. Sci.* **2009**, *5*, 322–326.
19. Ebrahimi, R.; Ahmadizadeh, M.; Rahbarian, P. Enhancing stand establishment of tomato cultivars under salt stress condition. *Southwest J. Hortic. Biol. Environ.* **2014**, *5*, 19–42. [CrossRef]
20. Javed, T.; Ali, M.M.; Shabbir, R.; Gull, S.; Ali, A.; Khalid, E.; Abbas, A.N.; Tariq, M. Rice seedling establishment as influenced by cultivars and seed priming with potassium nitrate. *J. Appl. Res. Plant Sci.* **2020**, *1*, 65–75.
21. Association of Official Seed Analysis. Rules for testing seeds. *J. Seed Technol.* **1990**, *12*, 101–112.
22. ISTA. *International Rules for Seed Testing*; The International Seed Testing Association (ISTA): Bassersdorf, Switzerland, 2015; pp. 1–276.
23. Waterhouse, A.L. Determination of total phenolics. *Curr. Prot. Food Anal. Chem.* **2002**, *6*, I1-1.
24. Scott, T.A.; Melvin, E.H. Determination of dextran with anthrone. *Anal. Chem.* **1953**, *25*, 1656–1661. [CrossRef]
25. Bray, C.M. Biochemical processes during the osmopriming of seeds. In *Seed Development and Germination*; Jaime Kigel; CRC Press: Boca Raton, FL, USA, 2017.
26. Singh, G.; Gill, S.S.; Sandhu, K.K. Improved performance of muskmelon (*Cucumis melo*) seeds with osmoconditioning. *Acta Agrobot.* **2013**, *52*, 121–137. [CrossRef]

27. Kiers, E.T.; Leakey, R.R.B.; Izac, A.M.; Heinemann, J.A.; Rosenthal, E.; Nathan, D.; Jiggins, J. Ecology: Agriculture at a crossroads. *Science* **2008**, *320*, 320–321. [CrossRef]
28. Shafiei, A.M.; Ghobadi, M. The effects of source of priming and post-priming storage duration on seed germination and seedling growth characteristics in wheat (*Triticum aestivem* L.). *J. Agric. Sci* **2012**, *4*, 256. [CrossRef]
29. Shehzad, M.; Ayub, M.; Ahmad, A.U.H.; Yaseen, M. Influence of priming techniques on emergence and seedling growth of forage sorghum (*Sorghum bicolor* L.). *J. Anim. Plant Sci.* **2012**, *22*, 154–158.
30. Ruttanaruangboworn, A.; Chanprasert, W.; Tobunluepop, P.; Onwimol, D. Effect of seed priming with different concentrations of potassium nitrate on the pattern of seed imbibition and germination of rice (*Oryza sativa* L.). *J. Integr. Agric.* **2017**, *16*, 605–613. [CrossRef]
31. Hadinezhad, P.; Payamenur, V.; Mohamadi, J.; Ghaderifar, F. The effect of priming on seed germination and seedling growth in *Quercus castaneifolia*. *Seed Sci. Technol.* **2013**, *41*, 121–124. [CrossRef]
32. Demir, I.; Van De Venter, H.A. The effect of priming treatments on the performance of watermelon (*Citrullus lanatus* (Thunb.) Matsum. and Nakai) seeds under temperature and osmotic stress. *Seed Sci. Technol.* **1999**, *27*, 871–876.
33. Mirabi, E.; Hasanabadi, M. Effect of seed priming on some characteristic of seedling and seed vigor of tomato (*Lycopersicon esculentum*). *J. Adv. Lab. Res. Biol.* **2012**, *3*, 237–240.
34. Oliveira, C.E.D.S.; Steiner, F.; Zuffo, A.M.; Zoz, T.; Alves, C.Z.; Aguiar, V.C.B. Seed priming improves the germination and growth rate of melon seedlings under saline stress. *Ciência Rural* **2019**, *49*. [CrossRef]
35. Vaktabhai, C.K.; Kumar, S. Seedling invigouration by halo priming in tomato against salt stress. *J. Pharmacol. Phytochem.* **2017**, *6*, 716–722.
36. Basra, S.; Pannu, I.; Afzal, I. Evaluation of seedling vigor of hydro and matriprimed wheat (*Triticum aestivum* L.) seeds. *Int. J. Agric. Biol.* **2003**, *5*, 121–123.
37. Shu, S.; Tang, Y.; Yuan, Y.; Sun, J.; Zhong, M.; Guo, S. The role of 24-epibrassinolide in the regulation of photosynthetic characteristics and nitrogen metabolism of tomato seedlings under a combined low temperature and weak light stress. *Plant Physiol. Biochem.* **2016**, *107*, 344–353. [CrossRef]
38. Mauro, R.P.; Agnello, M.; Distefano, M.; Sabatino, L.; San Bautista Primo, A.; Leonardi, C.; Giuffrida, F. Chlorophyll fluorescence, photosynthesis and growth of tomato plants as affected by long-term oxygen root zone deprivation and grafting. *Agronomy* **2020**, *10*, 137. [CrossRef]
39. Mauro, R.P.; Occhipinti, A.; Longo, A.M.G.; Mauromicale, G. Effects of shading on chlorophyll content, chlorophyll fluorescence and photosynthesis of subterranean clover. *J. Agron. Crop Sci.* **2011**, *197*, 57–66. [CrossRef]
40. De Castro, F.A.; Campostrini, E.; Netto, A.T.; De Menezes De Assis Gomes, M.; Ferraz, T.M.; Glenn, D.M. Portable chlorophyll meter (PCM-502) values are related to total chlorophyll concentration and photosynthetic capacity in papaya (*Carica papaya* L.). *Exp. Plant Physiol.* **2014**, *26*, 201–210. [CrossRef]
41. Dehkordi, F.S.; Nabipour, M.; Meskarbashee, M. Effect of priming on germination and biochemical indices of chicory (*Cichorium intybus* L.) seed. *Sci. Ser. Data Rep.* **2012**, *4*, 24–33.

Article

Zinc Seed Priming Improves Spinach Germination at Low Temperature

Muhammad Imran [1,2], Asim Mahmood [3], Günter Neumann [3] and Birte Boelt [1,*]

1 Department of Agroecology, Aarhus University, Forsøgsvej 1, 4200 Slagelse, Denmark; miguaf@gmail.com
2 Nouryon, Velperweg 76, 6428BM Arnhem, The Netherlands
3 Institute of Crop Science (340 h), University of Hohenheim, 70599 Stuttgart, Germany; A.Mahmood@uni-hohenheim.de (A.M.); guenter.neumann@uni-hohenheim.de (G.N.)
* Correspondence: bb@agro.au.dk

Abstract: Low temperature during germination hinders germination speed and early seedling development. Zn seed priming is a useful and cost-effective tool to improve germination rate and resistance to low temperature stress during germination and early seedling development. Spinach was tested to improve germination and seedling development with Zn seed priming under low temperature stress conditions. Zn priming increased seed Zn concentration up to 48 times. The multispectral imaging technique with VideometerLab was used as a non-destructive method to differentiate unprimed, water- and Zn-primed spinach seeds successfully. Localization of Zn in the seeds was studied using the 1,5-diphenyl thiocarbazone (DTZ) dying technique. Active translocation of primed Zn in the roots of young seedlings was detected with laser confocal microscopy. Zn priming of spinach seeds at 6 mM Zn showed a significant increase in germination rate and total germination under low temperature at 8 °C.

Keywords: spinach; Zn priming; multispectral imaging; Zn localization; abiotic stress

Citation: Imran, M.; Mahmood, A.; Neumann, G.; Boelt, B. Zinc Seed Priming Improves Spinach Germination at Low Temperature. *Agriculture* **2021**, *11*, 271. https://doi.org/10.3390/agriculture11030271

Academic Editor: Alan G. Taylor

Received: 20 February 2021
Accepted: 17 March 2021
Published: 22 March 2021

1. Introduction

Spinach (*Spinacia oleracea* L.) is an annual crop, usually sown in early spring. Low soil temperature during early spring is one of the major factors affecting seed germination of various crops. In spinach, seed germination and early seedling establishment are inhibited at low temperature [1]. Imbibition and rehydration of dry seeds is a critical process during germination, and rapid absorption of water can cause severe membrane damage, leading to leakage of electrolytes, sugar, and amino acids [2]. Wuebker et al. [3] and Bochicchio et al. [4] reported embryo membrane damage, directly linked to the speed of imbibition at low seed moisture. Under low temperature conditions, these problems can be even more severe due to limited ability of cell membranes to maintain the integrity that is required during imbibition [5] and may compromise germination performance, leading to low germination rate, low uniformity, and final stand establishment. Spinach seeds germinate best between a range of temperatures between 15 and 24 °C. The germination speed and/or rate varies with the change in temperature. Germination speed is very slow at a temperature just above freezing. It may take up to three weeks for germination at 5 °C, compared to one week at 20 °C.

Seed priming is a pre-sowing seed treatment, in which seeds are soaked in water and dried back to storage moisture contents for later use. According to Harris et al. [6], the 'on-farm seed priming' method has become very popular in developing countries. With the 'on-farm seed priming' method, seeds are soaked in a water or nutrient solution and air dried (not to storage moisture contents) prior to sowing. This can speed up germination, improve the tolerance to various stress conditions, and increase crop yield [7].

Chen and Arora [8] have proposed a hypothetical model demonstrating the cellular physiology of priming-induced stress-tolerance, likely achieved via two strategies. First, seed priming activates germination-related processes (e.g., respiration, endosperm weakening, and gene transcription and translation, etc.) that facilitate the transition of quiescent dry seeds into the germinating state, which improves germination potential. Secondly, priming imposes abiotic stress on seeds that repress radicle protrusion but stimulate stress responses (e.g., accumulation of Late Embryogenesis Abundant proteins (LEAs), potentially inducing cross-tolerance. The authors suggest that these two strategies constitute a "priming memory" in seeds, which mediates greater stress-tolerance of germinating primed seeds after the exposure to various stress conditions.

Stored reserves are the primary source of mineral nutrients during seed germination and early growth and should be adequate to sustain the seedling until the root system mediates nutrient uptake from the soil. Stored mineral nutrients are vital, particularly when seedlings are exposed to conditions of nutrient limitation [9]. In barley and wheat, seeds with low Zn contents showed delayed germination and poor seedling vigor, which negatively affected plant growth and final grain yield [10–12]. In wheat, seeds with high Zn concentrations produced better stand establishment, and seedlings were able to take up more Zn under Zn-deficient soil conditions as compared to plants established from seeds low in Zn seed reserves [13]. During germination and early seedling development, particularly under stress conditions, micronutrients are essential. Zinc is a co-factor of various enzymes (superoxide dismutase (SOD)) involved in the detoxification of reactive oxygen species, such as O_2^- (superoxide radical) and H_2O_2 (hydrogen peroxide) [14]. Zn is directly involved in membrane stabilization, biosynthesis of auxins [15] and gibberellins [16] in plant growth regulation, and protein synthesis in general.

In "nutrient seed priming", seeds are soaked in a nutrient solution instead of pure water to improve seed nutrient contents in combination with the priming effect, which improves germination and seedling establishment. Ashraf and Rauf [17] found that priming maize seeds with $CaC1_2$ improved final germination, rate of germination, and fresh and dry biomass of plumules and radicles, compared to untreated control and water-primed seeds under salt stress. Maize seed priming with 1% $ZnSO_4$ enhanced plant growth and increased final grain yield and Zn content of harvested seed from plants grown on soil with low Zn availability [18]. It has also been shown in maize [19] and rice [20] that primed Zn is translocated to growing shoots during germination and early seedling development. Furthermore, Imran et al. [21] also showed increased maize grain yield via Zn seed priming under low Zn-available soils combined with low temperature climatic conditions.

Based on the findings of seed priming memory in invoking seed stress tolerance [8] and the role of Zn seed priming in stress tolerance in crop plants, this study investigated the functions of water- and Zn-priming of spinach seeds under low temperature. The multi-spectral imaging technique was used to monitor the Zn priming of spinach seeds, and confocal-laser microscopic analysis was performed to study Zn translocation in young spinach seedlings.

2. Materials and Methods

2.1. Seed Material and Priming

Commercially available spinach seed (*Spinacia oleracea* L. *cv* Matador) was obtained from the seed company Vikima Seeds A/S, Denmark. Seeds were primed for 24 h with water and $ZnSO_4 \cdot 7H_2O$, according to Imran, Mahmood, Römheld, and Neumann [21], with some modifications, in which seeds after priming were surface dried at room temperature (20 °C) for 24 h before the germination test.

2.2. Optimal Zn Concentration Levels for Seed Priming (Experiment 1)

To determine the optimal Zn concentration for seed priming, 10 g of spinach seeds were soaked in 100 mL of $ZnSO_4 \cdot 7H_2O$ solution. Concentrations of Zn in the priming solutions were e.g., 0 (deionized H_2O), 1, 2, 4, 6, 8, and 10 mM Zn solutions. Unprimed seeds were used as control treatment. A germination test of primed seeds was performed using the top of paper method at 12 h light and 12 h dark periods at 15 °C. Seeds were germinated in petri dishes with four replicates of 25 seeds per treatment. Seed germination data were recorded at Day 7 and Day 14. The seeds with radicle protrusion > 2 mm were considered germinated.

2.3. *Germination Test of Water- and Zn-Primed Seeds at Low Temperature (Experiment 2)*

Based on the results of experiment 1, unprimed control, water-priming, and two levels of Zn concentrations were selected as seed treatments to test seed germination at two different temperatures, 8 °C (low temperature) and 15 °C (optimal temperature). The germination test was performed as mentioned above in experiment 1.

2.4. *Mineral Analysis of Seeds and Young Seedlings*

After the priming treatments, seeds were rinsed with deionized water for 1 min to remove any compounds and nutrients adhering to the seed coat before the analysis of seed mineral nutrients. Furthermore, mineral nutrients were also determined in young spinach seedlings. For this purpose, shoots and roots were separated in all treatments. To measure Zn concentration in seeds and seedlings, after drying at 65 °C, ground samples were ashed in a muffle furnace at 500 °C for 5 h. After cooling, the samples were extracted twice with 2 mL of 3.4 M HNO_3 (v/v) and subsequently evaporated to dryness. The ash was dissolved in 2 mL of 4 M HCl, subsequently diluted 10-fold with hot deionized water, and boiled for 2 min. After adding 0.1 mL of Cs/La buffer to 4.9 mL of ash solution, Zn and Mn concentrations were measured by atomic absorption spectrometry (UNICAM 939, Offenbach/Main, Germany).

2.5. *Confirmation of Zn Accumulation in Spinach Seeds after Zn Priming*

Staining of seed Zn was performed using 1,5-diphenyl thiocarbazone (DTZ), according to Ozturk et al. [22]. For this purpose, water- and Zn- (6 mM Zn) primed seeds were incubated with 500 mg L^{-1} DTZ at room temperature for 30 min. The stained seeds were rinsed with deionized water and images were taken with a high-resolution digital camera. To determine the localization of primed Zn in the different seed tissues, fresh water and Zn-primed (6 mM Zn, rinsed with deionized H_2O for 20 s) seeds were immediately fixed in NEG 50™ gel. The fixed seeds were dissected to 20 µm thick slices with a freezing microtome (MICROM HM 550, Microm International GmbH, Walldorf, Germany) and placed on microscope slides. Afterwards, 2 µL of 500 mg L^{-1} DTZ solution was applied to the specimen to stain with Zn. After 3 min, a few drops of deionized H_2O were applied before placing the coverslip on the thinly sliced sample. Photos were taken with a light microscope (Axiovert 200, Carl Ziess Microscopy GmbH, Göttingen, Germany).

To test the characteristic properties of the red color of DTZ stained spinach seeds (water- and Zn-primed), multi-spectral images of water- and Zn-primed seeds were taken according to Shrestha et al. [23]. Images from each seed sample were acquired using a VideometerLab instrument (Videometer A/S Herlev, Denmark). In this instrument, a top-mounted camera acquires multispectral images with the help of 19 light emitting diodes (LEDs) at 19 wavelengths (375, 405, 435, 450, 470, 505, 525, 570, 590, 630, 645, 660, 700, 780, 850, 870, 890, 940, and 970 nm). Prior to image acquisition, the instrument was calibrated with respect to color, geometry, and self-illumination to ensure directly comparable images. After images were obtained, VideometerLab software (version 2.13.83) was used to extract and transform pixel data.

2.6. *Translocation of Primed Zn in the Roots of Spinach Seedlings*

The localization of Zn in the roots of 10-day old spinach seedlings was examined by using Zinpyr-1 ($C_{46}H_{36}Cl_2N_6O_5$) fluorescence dye, according to Sinclair et al. [24]. For this purpose, unprimed, water- and Zn-primed (6 and 10 mM) spinach seeds were germinated in filter paper towels at 15 °C. Zinpyr-1 was dissolved in dimethyl sulphoxide (DMSO) to make a 1 mM stock solution and stored at −20 °C. For root incubation, a working solution of 20 µM Zinpyr-1 was prepared from the stock solution. From each treatment, 10-day old equally grown spinach seedlings were selected for Zinpyr-1 incubation.

Before immersing into a Zinpyr-1 working solution, seedlings were washed alternatively three times in deionized water and 10 mM ethylene-diamine-tetra-acetic acid (EDTA). Seedlings were incubated in Zinpyr-1 solution for 5 h at room temperature in the dark. Afterwards, roots of the incubated seedlings were rinsed again in deionized water to remove the Zinpyr-1 dye from the root surface, immersed in 75 µM propidium iodide to stain cell walls red, and rinsed again. For negative control, roots of water and Zn-primed seedlings were immersed in a Zn-chelator, N,N,N′,N-tetrakis(2-pyridylmethyl) ethylenediamine (TPEN), for 30 min. Samples were mounted in 0.9% saline, and images were taken on an Olympus (Hamburg, Germany) confocal laser-scanning microscope (CLSM), using excitation at 488 nm with a 100 mW Ar ion.

2.7. Statistical Analysis

Data on final germination were analysed by one-way analysis of variance (ANOVA) using SigmaStat 3.5 Software. Significant differences between the means were calculated at $p < 0.05$ and marked with different letters.

Differences between means of Zn and Mn concentrations, reflectance, and speed of germination were compared using the standard error (SE) of four replicates (25 seeds in each replicate).

3. Results

3.1. Optimal Zn Concentration

At first count after 7 days (Figure 1a), seed germination performance was increased by all priming treatments, as compared to the unprimed control. Water-primed seeds showed 10% higher germination compared to the untreated control. However, at Day 7, none of the Zn-priming and water priming treatments showed statistical difference in germination, but Zn-priming at 1 mM, 6 mM, and 10 mM had almost 22%, 20%, and 18%higher germination, respectively, as compared to the unprimed control.

(a)

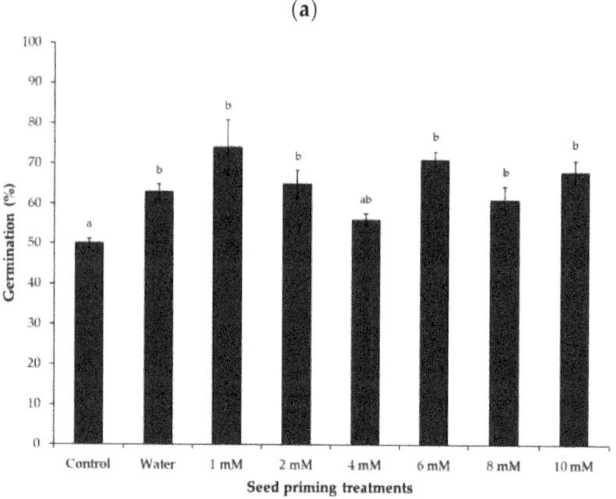

(b)

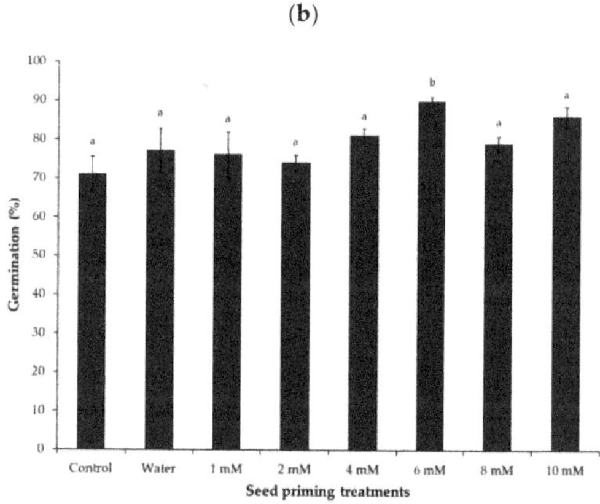

Figure 1. Germination percentage (%) of spinach seeds at Day 7 (**a**) and Day 14 (**b**) after sowing on top of paper in Petri dishes at 15 °C. Seed priming treatments included control (unprimed), water- primed, and Zn-primed at various Zn concentrations (1, 2, 4, 6, 8, and 10 mM Zn) in the priming solution. Bars represent the mean and standard error (SE) of four replicates (25 seeds in each replicate). Significant differences between the means were calculated at $p < 0.05$ and marked with different letters.

At final count on Day 14 (Figure 1b), compared to the first count, differences in % germination between unprimed, water-primed, and most of the Zn-priming treatments were not significant. Compared to other Zn priming treatments, priming at 6 mM Zn concentration showed significantly higher germination as compared with the control and water-primed treatments. Based on the results shown in Figure 1b, seed priming treatments at 6 mM and 10 mM Zn concentrations, together with unprimed and water-priming, were selected for testing germination at a low temperature.

3.2. Zinc Status of Seeds and Young Seedlings

Zinc priming largely increased seed Zn concentration at 6 mM and 10 mM priming treatments compared to unprimed and water-primed treatments. There was approximately an increase of 30 and 48 times in seed Zn concentration, with 6 mM and 10 mM priming solutions, respectively, as shown in Table 1. Concentration of Mn was decreased up to 25% in seeds after Zn priming treatments but concentration of Mn in shoots and roots was not affected by Zn seed priming. Mineral analysis of shoots and roots of young spinach seedlings showed a 4 to 10 time increase in shoot and root Zn concentrations after both Zn-priming treatments compared to unprimed and water primed.

Table 1. Zinc and Mn concentrations in spinach seeds, shoots and roots of young seedlings after water, and Zn seed priming. Values represent the mean and standard error (SE) of four replicates.

Treatments/Plant Parts	Unprimed	Water-Primed	6 mM Zn	10 mM Zn
		Zn µg g^{-1}		
Seeds	56.2 ± 0.1	55.3 ± 0.7	1625.3 ± 87.2	2413.5 ± 50.9
Shoot	163.6 ± 0.8	159.8 ± 0.7	522.2 ± 14.4	653.4 ± 42.9
Root	85.2 ± 2.8	80.6 ± 3.7	603.2 ± 11.5	821.6 ± 15.5
		Mn µg g^{-1}		
Seeds	45.1 ± 1.8	46.5 ± 1.1	36.5 ± 0.9	34.7 ± 0.3
Shoot	45.7 ± 0.3	41.1 ± 0.1	46.6 ± 0.3	44.1 ± 1.2
Root	31.9 ± 2.1	28.6 ± 1.4	26.3 ± 0.5	22.8 ± 0.1

3.3. Detection of Zn Primed Spinach Seeds with VideoMeter Lab and DTZ Staining

To confirm the addition of Zn in spinach seeds via Zn priming, the DTZ staining method was used. Images of water-, 6 mM, and 10 mM Zn-primed seeds (Figure 2a–c, respectively) showed no visible differences in all three treatments. Compared to normal images, multispectral images of the same samples taken with VideometerLab (Figure 2d–f), viewed at a wavelength of 700 nm in hot view mode, revealed more uniform and higher absorption of red color.

Higher development of red color in 6- and 10-mM Zn-primed seeds after DTZ staining indicates a higher level of Zn compared to water primed seeds. Differences in the mean spectrum of DTZ stained water- and Zn-primed seeds are shown in Figure 3.

Other spectrum characteristics data (Table 2), such as CIELab, intensity, hue, and saturation, also revealed the development of a darker red color in Zn-primed seeds after DTZ staining, as compared to water priming.

Table 2. Characteristic color parameters: CIELab L, intensity, hue, and saturation of DTZ-stained spinach seeds after water-, 6 mM, and 10 mM Zn-priming. Values represent the mean and standard error (SE) of four replicates. Significant differences between the means were calculated between seed priming treatments at $p < 0.05$ and marked with different letters.

Priming Treatments	CIELab L	Intensity	Hue	Saturation
Water-primed	62.2 ± 0.9 a	29.3 ± 1.0 a	140.2 ± 0.4 a	18.8 ± 0.7 a
6 mM Zn	55.7 ± 0.7 b	22.8 ± 0.6 b	134.7 ± 0.4 b	16.1 ± 0.6 b
10 mM Zn	53.9 ± 0.9 b	21.4 ± 0.7 b	133.9 ± 0.4 b	15.7 ± 0.7 b

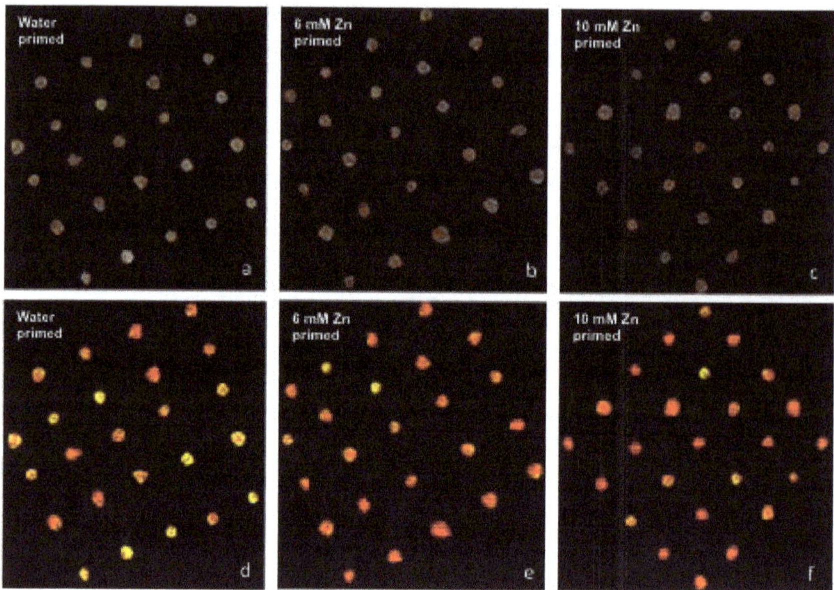

Figure 2. Visible spectrum images of 1,5-diphenyl thiocarbazone (DTZ)-stained water (**a**), 6 mM (**b**), and 10 mM (**c**) Zn-primed seeds. Images (**d–f**) (water-, 6 mM, and 10 mM Zn-primed seeds, respectively) taken with VideometerLab at wavelength 700 nm in hot view mode.

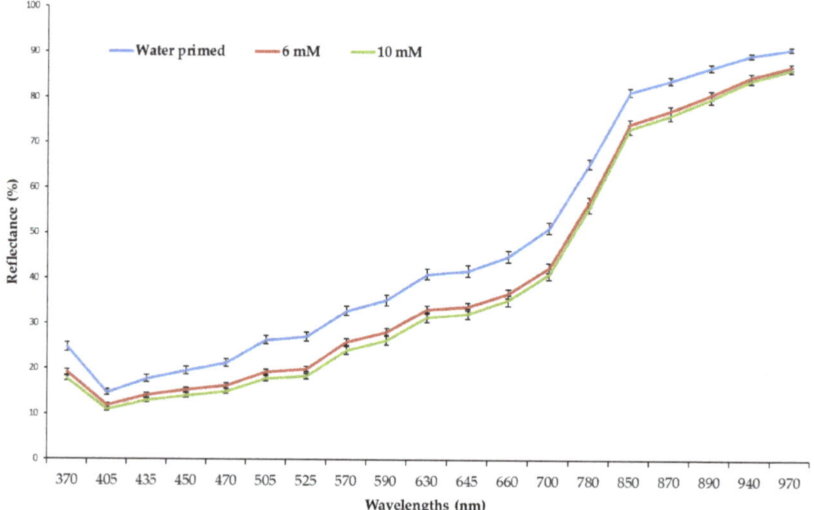

Figure 3. The mean visible spectrum of water-, 6 mM, and 10 mM Zn-primed spinach seeds extracted from the multi-spectral images of seeds at 19 wavelengths (375, 405, 435, 450, 470, 505, 525, 570, 590, 630, 645, 660, 700, 780, 850, 870, 890, 940, and 970 nm). Bars represent the mean and standard error (SE) of four replicates (25 seeds in each replicate).

3.4. Zinc Localization in the Roots of Spinach Seedlings

Laser confocal microscopy was employed to detect Zn (in vascular tissues) in the roots of 10-day-old spinach seedlings, germinated from water-, 6 mM, and 10 mM Zn-primed seeds, (Figure 4a,b). Higher intensities of Zinpyr-1 fluorescence in 6 mM and 10 mM (Figure 5d,e, respectively) compared to unprimed and water-primed seedlings (Figure 5a,b, respectively) indicates higher accumulation of Zn in the roots of Zn-primed seedlings compared to water-primed seedlings, irrespective of Zn concentration in Zn-primed seeds.

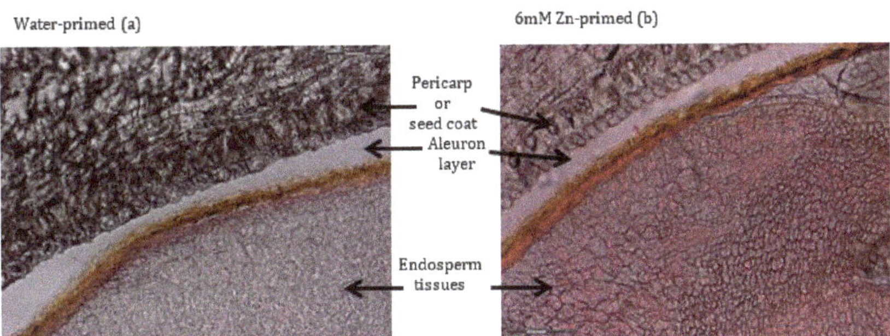

Figure 4. DTZ-staining of Zn in a spinach seed of water-primed (**a**) and Zn-primed (**b**) seedlings with 6 mM $ZnSO_4{\cdot}7H_2O$. Red staining indicates Zn localization, especially in the aleurone layer, endosperm, and pericarp.

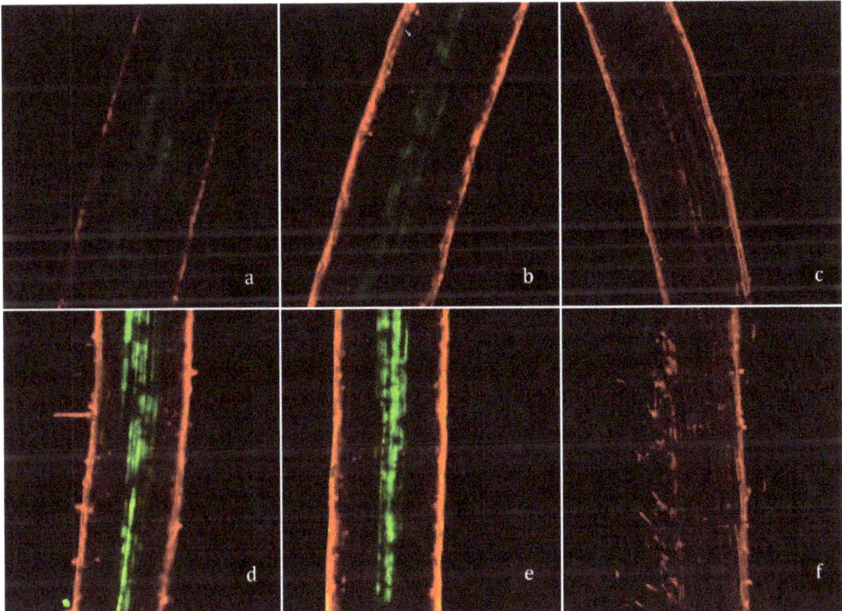

Figure 5. Confocal laser-scanning microscope images of spinach roots of (**a**) control (unprimed) seedlings, (**b**) water-primed seedlings, (**d**,**e**) 6 mM and 10 mM Zn-primed seedlings, respectively, and (**c**,**f**) seedlings treated with 200 μM N,N,N′,N-tetrakis (2-pyridylmethyl) ethylenediamine (TPEN) for 30 min and then exposed to 15 μM Zinpyr-1 for 5 h.

3.5. Seed Germination at 15 °C and 8 °C

Spinach seeds started germination 3 days after sowing. Figure 6a,b represents seed germination at 15 °C and 8 °C, respectively. At 15 °C, there was no statistical difference for germination speed between water- and Zn-primed seeds, but all priming treatments showed a significant increase in germination compared to unprimed seeds. These differences were diminished after Day 6. At the final count (Figure 6a), Day 14, Zn-priming reflected relatively higher total germination but differences were not significant compared to unprimed and water-primed treatments.

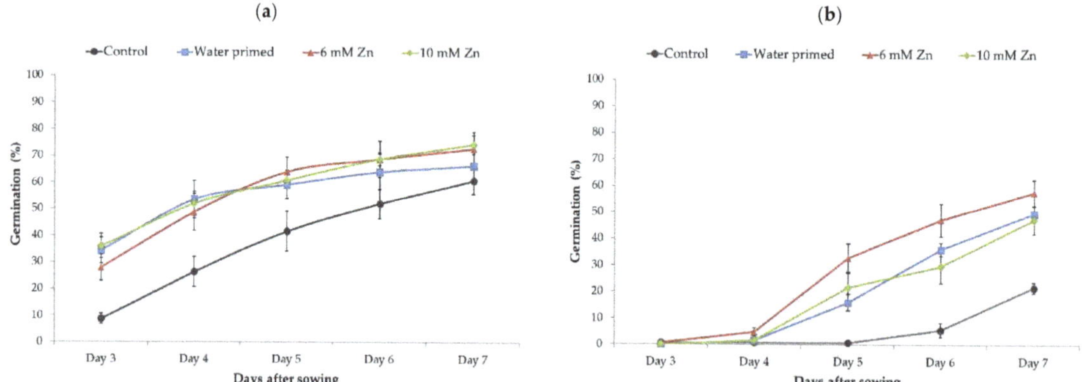

Figure 6. Germination speed of control, water-, 6 mM, and 10 mM Zn-primed spinach seeds germinated at 15 °C (**a**) and 8 °C (**b**) in Petri dishes. Germination was monitored daily up to 7 days after sowing. Data presented are the means of four replicates (25 seeds in each replicate) with standard errors.

At 8 °C, Zn-priming at 6 mM Zn concentration showed a significantly higher germination speed compared to unprimed, water-, and 10 mM Zn-priming treatments (Figure 6b). However, there was no significant difference in germination speed between water- and 10 mM Zn-priming treatments, but both treatments showed significant increases compared to unprimed seeds. At final count (Figure 7b), 6 mM Zn-priming showed a significant increase in total germination (>10%) compared to unprimed and water-primed seeds. There was no significant difference between unprimed, water-, and 10 mM Zn-primed treatments. Final germination recorded on Day 14 showed no significant effect of treatments when seeds were germinated at 15 °C (Figure 7a); however, at 8 °C, treatment with 6 mM Zn improved germination (Figure 7b).

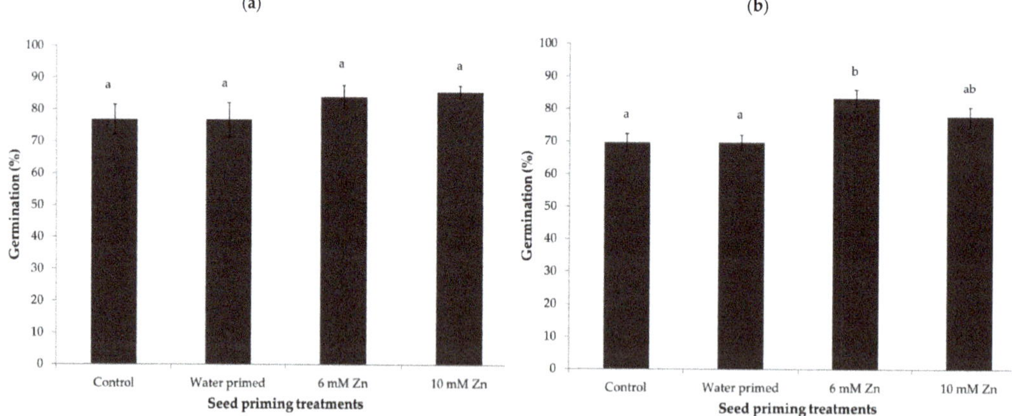

Figure 7. Final germination (%) of spinach seeds after 14 days of sowing, germinated at 15 °C (**a**) and 8 °C (**b**) on top of paper in Petri dishes. Seed priming treatments include control (unprimed), water- primed, 6 mM, and 10 mM Zn in the priming solution. Bars represent the mean and standard error (SE) of four replicates (25 seeds in each replicate). Significant differences between the means were calculated at $p < 0.05$ and marked with different alphabets.

4. Discussion

Adequate Zn contents in the seeds are essential for vigorous seedlings and resistance against different abiotic stress factors during germination and the early seedling development stage [25,26]. Zn seed priming has been used efficiently in various crops, e.g., maize [21,27], barley [28], rice [20], and soybean [29], to improve seed germination and resistance against various abiotic stress factors, such as drought, low root zone temperature, and nutrient deficiencies. However, Zn seed priming improves germination and seedling development, but determining the optimum concentration of Zn in the priming solution to attain the beneficial effects of Zn priming is very important. In the present study, Zn-priming of spinach seeds at 6 mM and 10 mM Zn concentrations showed the best effects on germination (Figure 1b). However, in the beginning (Figure 1a), Zn-priming at 1 mM Zn concentration showed significant increase in seed germination compared to unprimed seeds but that effect had disappeared at the final germination count. Previously, Ajouri et al. [28] and Prom-u-Thai et al. [20] also reported similar findings in barley and rice, respectively, where Zn seed priming showed beneficial effects on seed germination and early seedling development at certain Zn concentrations. Higher levels of Zn concentration in the priming solution may exert a toxic effect during the germination process and development of young seedlings. The results in the present study indicate the advantageous role of Zn seed priming in improving spinach seed germination, in particular during low temperatures. In the current study, final germination recording was performed on Day 14; however, according to International Seed Testing Associsation (ISTA) guidelines [30], final germination in spinach is scored on Day 21. Standard germination tests are performed using either 15 or 10 °C [30]. Since "low temperature" was chosen as 8 °C in this experiment, an even slower germination would be expected. This implies that seeds in the current study may not have reached full germination capacity, in particular those tested at low temperature.

Development of red color after staining spinach seeds with DTZ revealed an increase in seed Zn levels after Zn-priming treatment. DTZ is a Zn–chelating agent [31,32], which gives red color after binding with Zn. Additionally, it has been used to determine the localization of Zn in different organisms and crop seeds, such as algae [33], wheat seeds [22], and rice seeds [20]. In the present study, in the thick and relatively dark colored seed coat of spinach seeds, it was difficult to differentiate the intensity of the red color developed after DTZ staining in water- and Zn-primed seeds with ordinary camera images (Figure 2a–c). Recently, Shrestha et al. [23,34] successfully employed multi-spectral image analysis on tomato seeds to classify varietal differences based on the variation of spectral characteristics in seeds from different varieties. The mean spectrum of multi-spectral images of DTZ-stained seeds after water- and Zn-priming treatments (Figure 2d–f) showed a clear difference between water- and Zn-primed seeds, but reflectance at each wavelength exhibited a similar trend. The variation in spectra of water- and Zn-primed seeds after DTZ staining can be attributed to increased Zn in the spinach seeds. Development of red color in the inner parts of Zn-primed seeds (Figure 4b) revealed that primed Zn was not only absorbed in the outer pericarp or seed coat tissues, it was also accumulating in the inner tissues of Zn-primed seeds. Interestingly, Freitas et al. [35] found the highest accumulation of Zn in the embryo of the ZnO coated maize seed. This study employed micro-X-ray fluorescence spectrometry and micro-X-ray absorption near-edge spectroscopy for the mapping of Zn distribution and Zn speciation analysis.

Translocation of primed Zn in the young germinating spinach seedlings was determined by using a histochemical technique of Zn visualization based on the formation of the green-fluorescent complex with Zinpyr-1 ($C_{46}H_{36}Cl_2N_6O_5$). This technique has been successfully applied in different plant species to detect Zn in different shoot and root tissues, for example Arabidopsis, [24] *Zea mays* [36], *Noccaea caerulescens*, and *Thlaspi arvense* [37]. As shown in Figure 5a,b, higher intensities of green fluorescence in the roots of Zn-primed seeds indicate active transport of primed Zn to the young growing roots. Previously, various authors [20,21,29] have also reported the translocation of primed Zn to the young growing roots during early seedling development, which also supports the seedling growth under Zn-deficient conditions. Zn is well known for its functions in plants under various biotic and abiotic stress conditions [38]. Our results suggest that active translocation of primed Zn to growing roots can be useful in stress tolerance in spinach grown under various stress conditions.

Zn is important in the physiological functions, during germination and seedling development [25]. It is vital in the processes of protein synthesis and gene expression. For the structural and functional integrity of biological systems, almost 10% of proteins require Zn for their synthesis or functioning [39]. Production of reactive oxygen species (ROS) during seed germination is reported by numerous studies [40–42]. Increased oxidative stress is one of the rapid responses under all kind of stress conditions, including suboptimal or low temperatures. This is associated with increased production of ROS, such as superoxide, hydrogen peroxide, and the hydroxyl radical, involved in membrane damage by lipid peroxidation, protein degradation, enzyme inactivation, and disruption of DNA strands [43]. Micronutrients, such as Zn, are important co-factors of different enzymes involved in the detoxification of ROS, such as superoxide dismutase (SOD) [14,44]. In the present study, compared to unprimed and water-primed seeds, an increase in the germination speed and total germination at 8 °C by Zn priming can be attributed to increased Zn in the seeds. Increased localization of Zn in the roots of Zn-primed seedlings (Figure 5d,e) also suggested an active role of primed Zn against the adverse effects of low temperatures during germination, but the current study did not provide any further detail on this potential effect.

Concluding remarks: Previously, beneficial effects of increased seed Zn levels have shown to improve seed germination and early seedling establishment in stress conditions. This study demonstrates that increased Zn level of spinach seeds via Zn priming can enhance seed germination and seedling establishment under low temperature stress conditions. The physiological role of primed Zn in membrane stability, reduced oxidative stress and performance under field conditions, needs to be further elucidated and studied. Furthermore, the potential of combining Zn seed priming with agrochemicals, e.g., fungicides, should be evaluated as a tool to reduce pesticide use. Finally, the seed priming technique is simple, cost effective, and can be performed on farms before sowing, e.g., by small-scale farmers.

Author Contributions: Conceptualization, M.I. and B.B.; methodology, M.I.; software, M.I.; validation, M.I.; formal analysis, M.I., G.N. and A.M.; investigation, M.I.; resources, B.B.; data curation, M.I.; writing—original draft preparation, M.I.; writing—review and editing, M.I. and B.B.; visualization, M.I.; supervision, B.B.; project administration, B.B.; funding acquisition, B.B. All authors have read and agreed to the published version of the manuscript.

Funding: This material is based on work that was supported by the Innovation Fund Denmark grant number 110-2012-1, SpectraSeed and GUDP (Grønt Udviklings- og Demonstrationsprogram) grant number 34009-12-0528, and the Danish Agricultural Agency under the Ministry of Environment and Food of Denmark.

Institutional Review Board Statement: Not applicable.

Informed Consent Statement: Not applicable.

Conflicts of Interest: The authors declare no conflict of interest.

References

1. Ashraf, M.; Foolad, M. Pre-sowing seed treatment—A shotgun approach to improve germination, plant growth, and crop yield under saline and non-saline conditions. *Adv. Agron.* **2005**, *88*, 223–271.
2. Powell, A.A.; Matthews, S. The Damaging Effect of Water on Dry Pea Embryos during Imbibition. *J. Exp. Bot.* **1978**, *29*, 1215–1229. [CrossRef]
3. Wuebker, E.F.; Mullen, R.E.; Koehler, K. Flooding and Temperature Effects on Soybean Germination. *Crop. Sci.* **2001**, *41*, 1857–1861. [CrossRef]
4. Bochicchio, A.; Coradeschi, M.A.; Zienna, P.; Bertolini, M.; Vazzana, C. Imbibitional injury in maize seed independent of chilling temperature. *Seed Sci. Res.* **1991**, *1*, 85–90. [CrossRef]
5. Bramlage, W.J.; Leopold, A.C.; Parrish, D.J. Chilling stress to soybeans during imbibition. *Plant Physiol.* **1978**, *61*, 525–529. [CrossRef]

6. Harris, D. On-farm seed priming to accelerate germination in rainfed, dryseeded rice. *Int. Rice Res. Notes* **1997**, *22*, 1.
7. Harris, D.; Tripathi, R.; Joshi, A. On-farm seed priming to improve crop establishment and yield in dry direct-seeded rice. In *Direct Seeding: Research Strategies and Opportunities*; International Research Institute: Manila, Philippines, 2002; pp. 231–240.
8. Chen, K.; Arora, R. Priming memory invokes seed stress-tolerance. *Environ. Exp. Bot.* **2013**, *94*, 33–45. [CrossRef]
9. Asher, C. Effects of nutrient concentration in the rhizosphere on plant growth. In Proceedings of the XIIIth Congress International Society of Soil Science's Symposium, Hamburg, Germany, 13 August 1986; Volume 5, pp. 209–216.
10. Genc, Y.; McDonald, G.; Graham, R. The interactive effects of zinc and salt on growth of wheat. In *Plant Nutrition for Food Security, Human Health and Environmental Protection*; Tsinghua University Press: Beijing, China, 2005; pp. 548–549.
11. Longnecker, N.; Marcar, N.; Graham, R. Increased manganese content of barley seeds can increase grain yield in manganese-deficient conditions. *Aust. J. Agric. Res.* **1991**, *42*, 1065–1074. [CrossRef]
12. Rengel, Z.; Graham, R.D. Importance of seed Zn content for wheat growth on Zn-deficient soil. *Plant Soil* **1995**, *173*, 259–266. [CrossRef]
13. Graham, R.D.; Rengel, Z. Genotypic Variation in Zinc Uptake and Utilization by Plants. In *Zinc in Soils and Plants*; Metzler, J.B., Ed.; Springer: Dordrecht, The Netherlands, 1993; pp. 107–118.
14. Cakmak, I. Tansley review no. 111: Possible roles of zinc in protecting plant cells from damage by reactive oxygen species. *New Phytol.* **2000**, *146*, 185–205. [CrossRef]
15. Salami, A.U.; Kenefick, D.G. Stimulation of Growth in Zinc-Deficient Corn Seedlings by the Addition of Tryptophan 1. *Crop. Sci.* **1970**, *10*, 291–294. [CrossRef]
16. Sekimoto, H.; Hoshi, M.; Nomura, T.; Yokota, T. Zinc Deficiency Affects the Levels of Endogenous Gibberellins in *Zea mays* L. *Plant Cell Physiol.* **1997**, *38*, 1087–1090. [CrossRef]
17. Ashraf, M.; Rauf, H. Inducing salt tolerance in maize (*Zea mays* L.) through seed priming with chloride salts: Growth and ion transport at early growth stages. *Acta Physiol. Plant.* **2001**, *23*, 407–414. [CrossRef]
18. Harris, D.; Rashid, A.; Miraj, G.; Arif, M.; Shah, H. 'On-farm' seed priming with zinc sulphate solution—A cost-effective way to increase the maize yields of resource-poor farmers. *Field Crop. Res.* **2007**, *102*, 119–127. [CrossRef]
19. Muhammad, I.; Kolla, M.; Volker, R.; Günter, N. Impact of Nutrient Seed Priming on Germination, Seedling Development, Nutritional Status and Grain Yield of Maize. *J. Plant Nutr.* **2015**, *38*, 1803–1821. [CrossRef]
20. Prom-U-Thai, C.; Rerkasem, B.; Yazici, A.; Cakmak, I. Zinc priming promotes seed germination and seedling vigor of rice. *J. Plant Nutr. Soil Sci.* **2012**, *175*, 482–488. [CrossRef]
21. Imran, M.; Mahmood, A.; Römheld, V.; Neumann, G. Nutrient seed priming improves seedling development of maize exposed to low root zone temperatures during early growth. *Eur. J. Agron.* **2013**, *49*, 141–148. [CrossRef]
22. Ozturk, L.; Yazici, M.A.; Yucel, C.; Torun, A.; Cekic, C.; Bagci, A.; Ozkan, H.; Braun, H.-J.; Sayers, Z.; Cakmak, I. Concentration and localization of zinc during seed development and germination in wheat. *Physiol. Plant.* **2006**, *128*, 144–152. [CrossRef]
23. Shrestha, S.; Deleuran, L.C.; Olesen, M.H.; Gislum, R. Use of Multispectral Imaging in Varietal Identification of Tomato. *Sensors* **2015**, *15*, 4496–4512. [CrossRef]
24. Sinclair, S.A.; Sherson, S.M.; Jarvis, R.; Camakaris, J.; Cobbett, C.S. The use of the zinc-fluorophore, zinpyr-1, in the study of zinc homeostasis in arabidopsis roots. *New Phytol.* **2007**, *174*, 39–45. [CrossRef]
25. Cakmak, I. Enrichment of cereal grains with zinc: Agronomic or genetic biofortification? *Plant Soil* **2008**, *302*, 1–17. [CrossRef]
26. Welch, R.M.; Graham, R.D. A new paradigm for world agriculture: Meeting human needs-productive, sustainable, nutritious. *Field Crop Res.* **1999**, *60*, 1–10. [CrossRef]
27. Harris, D.; Rashid, A.; Miraj, G.; Arif, M.; Yunas, M. 'On-farm' seed priming with zinc in chickpea and wheat in pakistan. *Plant Soil* **2008**, *306*, 3–10. [CrossRef]
28. Ajouri, A.; Asgedom, H.; Becker, M. Seed priming enhances germination and seedling growth of barley under conditions of P and Zn deficiency. *J. Plant Nutr. Soil Sci.* **2004**, *167*, 630–636. [CrossRef]
29. Muhammad, I.; Volker, R.; Günter, N. Accumulation and distribution of Zn and Mn in soybean seeds after nutrient seed priming and its contribution to plant growth under Zn- and Mn-deficient conditions. *J. Plant Nutr.* **2017**, *40*, 695–708. [CrossRef]
30. ISTA. The germination test. In *International Rules for Seed Testing*; International Seed Testing Association: Bassersdorf, Switzerland, 2020; pp. 5–56.
31. McNary, W.F. Zinc-Dithizone Reaction of Pancreatic Islets. *J. Histochem. Cytochem.* **1954**, *2*, 185–195. [CrossRef]
32. López-García, C.; Varea, E.; Palop, J.J.; Nacher, J.; Ramirez, C.; Ponsoda, X.; Molowny, A. Cytochemical techniques for zinc and heavy metals localization in nerve cells. *Microsc. Res. Tech.* **2002**, *56*, 318–331. [CrossRef]
33. Pawlik-Skowrońska, B. Resistance, accumulation and allocation of zinc in two ecotypes of the green alga Stigeoclonium tenue Kütz. coming from habitats of different heavy metal concentrations. *Aquat. Bot.* **2003**, *75*, 189–198. [CrossRef]
34. Shrestha, S.; Deleuran, L.C.; Gislum, R. Classification of different tomato seed cultivars by multispectral visible-near infrared spectroscopy and chemometrics. *J. Spectr. Imaging* **2016**, *5*, 9. [CrossRef]
35. Freitas, M.N.; Guerra, M.B.B.; Adame, A.; Moraes, T.F.; Junior, J.L.; Pérez, C.A.; Abdala, D.B.; Cicero, S.M. A first glance at the micro-ZnO coating of maize (*Zea mays* L.) seeds: A study of the elemental spatial distribution and Zn speciation analysis. *J. Anal. At. Spectrom.* **2020**, *35*, 3021–3031. [CrossRef]
36. Seregin, I.V.; Kozhevnikova, A.D. Histochemical methods for detection of heavy metals and strontium in the tissues of higher plants. *Russ. J. Plant Physiol.* **2011**, *58*, 721–727. [CrossRef]

37. Kozhevnikova, A.D.; Erlikh, N.T.; Zhukovskaya, N.V.; Obroucheva, N.V.; Ivanov, V.B.; Belinskaya, A.A.; Khutoryanskaya, M.Y.; Seregin, I.V. Nickel and zinc effects, accumulation and distribution in ruderal plants Lepidium ruderale and Capsella bursa-pastoris. *Acta Physiol. Plant.* **2014**, *36*, 3291–3305. [CrossRef]
38. Marschner, H.; Marschner, P. *Marschner's Mineral Nutrition of Higher Plants*; Academic Press: Cambridge, MA, USA, 2012.
39. Andreini, C.; Banci, L.; Bertini, I.; Rosato, A. Counting the Zinc-Proteins Encoded in the Human Genome. *J. Proteome Res.* **2006**, *5*, 196–201. [CrossRef] [PubMed]
40. Cakmak, I.; Strbac, D.; Marschner, H. Activities of Hydrogen Peroxide-Scavenging Enzymes in Germinating Wheat Seeds. *J. Exp. Bot.* **1993**, *44*, 127–132. [CrossRef]
41. Bailly, C.; Bogatek-Leszczynska, R.; Côme, D.; Corbineau, F. Changes in activities of antioxidant enzymes and lipoxygenase during growth of sunflower seedlings from seeds of different vigour. *Seed Sci. Res.* **2002**, *12*, 47–55. [CrossRef]
42. Qin, J.; Liu, Q. Oxidative metabolism-related changes during germination of mono maple (Acer mono Maxim.) seeds under seasonal frozen soil. *Ecol. Res.* **2009**, *25*, 337–345. [CrossRef]
43. Allen, D.J.; Ort, D.R. Impacts of chilling temperatures on photosynthesis in warm-climate plants. *Trends Plant Sci.* **2001**, *6*, 36–42. [CrossRef]
44. Cakmak, I.; Marschner, H. Increase in Membrane Permeability and Exudation in Roots of Zinc Deficient Plants. *J. Plant Physiol.* **1988**, *132*, 356–361. [CrossRef]

agriculture

MDPI

Review

The Use of Multispectral Imaging and Single Seed and Bulk Near-Infrared Spectroscopy to Characterize Seed Covering Structures: Methods and Applications in Seed Testing and Research

Anders Krogh Mortensen, René Gislum, Johannes Ravn Jørgensen and Birte Boelt *

Department of Agroecology, Faculty of Technical Sciences, Aarhus University, Forsøgsvej 1, 4200 Slagelse, Denmark; anmo@agro.au.dk (A.K.M.); rg@agro.au.dk (R.G.); jrj@agro.au.dk (J.R.J.)
* Correspondence: bb@agro.au.dk

Abstract: The objective of seed testing is to provide high-quality seeds in terms of high varietal identity and purity, germination capacity, and seed health. Across the seed industry, it is widely acknowledged that quality assessment needs an upgrade and improvement by inclusion of faster and more cost-effective techniques. Consequently, there is a need to develop and apply new techniques alongside the classical testing methods, to increase efficiency, reduce analysis time, and meet the needs of stakeholders in seed testing. Multispectral imaging (MSI) and near-infrared spectroscopy (NIRS) are both quick and non-destructive methods that attract attention in seed research and in the seed industry. This review addresses the potential benefits and challenges of using MSI and NIRS for seed testing with a comprehensive focus on applications in physical and physiological seed quality as well as seed health.

Keywords: fruit morphology; multispectral imaging; near-infrared; pericarp; testa; seed coat; seed testing; image analysis; chemometrics

Citation: Mortensen, A.K.; Gislum, R.; Jørgensen, J.R.; Boelt, B. The Use of Multispectral Imaging and Single Seed and Bulk Near-Infrared Spectroscopy to Characterize Seed Covering Structures: Methods and Applications in Seed Testing and Research. *Agriculture* **2021**, *11*, 301. https://doi.org/10.3390/agriculture11040301

Academic Editor: Alan G. Taylor

Received: 20 February 2021
Accepted: 28 March 2021
Published: 1 April 2021

1. Introduction

Multispectral imaging (MSI) and near-infrared spectroscopy (NIRS) are both quick and non-destructive methods that have received much attention in seed testing and seed research. The fact that it is possible to measure different quality parameters in a non-destructive, quick, and for some methods, automatic way makes it very interesting for seed-testing facilities and the seed industry. Some of the challenges before the methods are fully implemented and integrated are: development and validation of appropriate statistical models to classify future seeds and a better understanding of these models, i.e., why did the seeds belong to the specific group. The latter is probably more interesting from a scientific, research, and development perspective. In some cases, e.g., a commercial setting, a prober model might be sufficient and the deeper understanding of it less important. This review concerns methods and applications in seed testing and research using MSI and single seed and bulk NIRS to characterize the covering structures of seeds used as regeneration material.

1.1. Seed Covering Structure and Chemical Composition

The microstructure and chemical composition of specific seed coat cell layers give rise to species and variety differences in seed coat structure and function. Most morphological features of the seed coat are relatively insensitive to environmental conditions and therefore very useful for taxonomic identification. Seed coat color is influenced by environmental conditions—i.e., climatic conditions during maturation and hence not appropriate for taxonomic purposes [1].

Sugar beet (*Beta vulgaris* subsp. *vulgaris* var. *altissima* Doell.) belongs to the *Amaranthaceae* family, and other important crops in this family are red beet (*Beta vulgaris subsp vulgaris* var. *Conditiva*) and spinach (*Spinacia oleracea* L.). The dry fruit of sugar beet seed is a single achene with the fruit coat (pericarp) composed of lignified cells. The pericarp consists of an outer layer of parenchyma cells and an inner, denser layer of sclerenchyma cells. The fruit coat is a physical and chemical barrier for germination [2]. The seeds of species in this family are characterized by a thick fruit coat consisting of lignified cells.

The typical fruit of the *Poaceae* family (e.g., cereals and grasses) is a caryopsis, comprised by the embryo, the starchy endosperm, and the outer aleurone endosperm, surrounded in turn by the nucellar layer, the testa (seed coat) and the pericarp. In addition, the caryopsis of barley (*Hordeum vulgare* L.) and oats (*Avena sativa* L.) have an adherent outer coat or husk or hull consisting of the glumellae—lemma and palea—or the glumes, which are not removed, enclosing the caryopsis [3–5]. In contrast to species in the *Amaranthaceae* family, the seeds of species in this family are characterized by a thin fruit coat—the husk or hull.

1.2. Seed Coat Function

The seed coat is the seed's primary defense against adverse environmental conditions [6]. The seed coat functions as preserving the integrity of the interior parts of seeds, protects against pests and diseases, regulates gaseous exchanges between the embryo and the external environment and in many families the seed coat plays a role in the control of water absorption during imbibition and germination. Species in the *Fabaceae* family (e.g., beans and forage legumes) have an outer layer consisting of a waxy cuticle [1]. This represents a barrier to imbibition, which may be conferred by waxy or phenolic substances in the epidermis of the seed coat. Many legume species can produce seeds with seed coats temporarily impermeable to water—"hard seeds"—which is a mechanism of physical dormancy.

The intact seed coat protects the embryo from cellular rupture and the leakage of intracellular substances during imbibition. Soybean seeds (*Glycine max* (L.) Merr.) with seed coat epidermal cracking have higher leakage and low viability [7,8] and rapid imbibition of soybean seeds increases the leakage of intracellular substance and decreases seedling survival [7]. Leakage of intracellular substances from imbibing seeds are indicators of low seed vigor and viability.

Damage to seeds by microorganisms occurs by the production of exocellular enzymes which degrade the seed coat, and therefore microorganism infection may also lead to an increase in electrolyte leakage [9].

2. Near-Infrared Spectroscopy

Single seed or bulk seed NIRS is a non-destructive measurement of the seed or seeds in the electromagnetic near-infrared (NIR) spectrum from wavelengths 780 to 2498 nm, equivalent to wavenumbers 12,821 to 4000 cm^{-1}, respectively, with a spectral resolution of 0.5–5 nm (Figure 1) Thus, NIRS radiation is invisible to the human eye in contrast to the shorter wavelengths used in most image analysis systems. The NIR spectrum emerges when monochromatic radiation at a frequency which corresponds to the vibration of a particular chemical bond is absorbed by that bond, while the rest of the radiation is either reflected or transmitted without interacting with other bonds [10]. The C-H, N-H, S-H or O-H bonds absorb the radiation energy and hence it is possible to measure water and organic compounds such as protein, carbohydrates, alcohols and/or lipids [11]. The NIR spectrum consists of overtone bands when radiation energy makes the molecule go from the ground stage (v = 0) to an excited stage (v = 2) defined as the first overtone, or from the ground stage to v = 3 defined as the second overtone. Furthermore, the NIR spectrum consists of combination vibrations, which typically form broad and complex wavebands making it difficult to relate the spectra to individual chemical components [12]. This direct link between spectral information and the chemical compounds makes it obvious to develop a calibration model consisting of single seed NIRS measurements (explainable or X variables) and wet chemical measurement (response or Y variables) of the aforementioned chemical compounds. This model can be used to predict the chemical compounds in other future seeds.

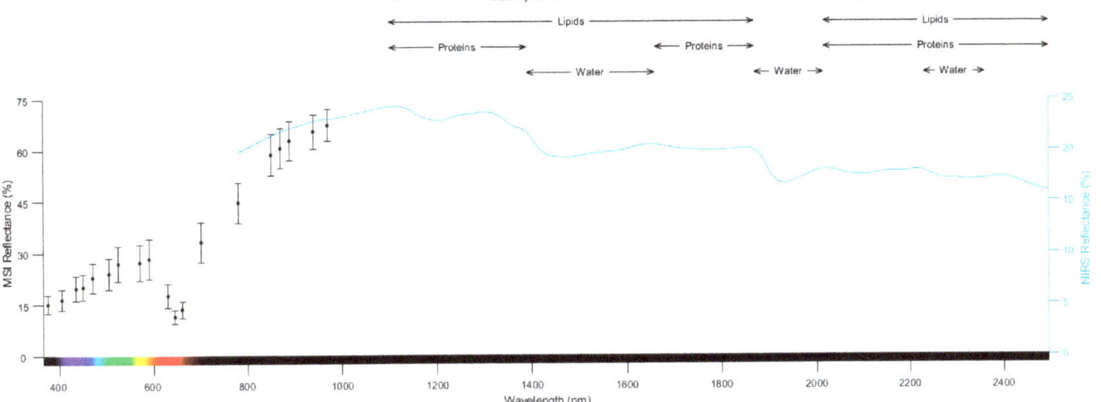

Figure 1. Reflectance of incoming light of a spinach seed lot (N = 70) using MSI (discrete points with error bars) and NIRS (blue continuous line). The MSI reflectance values are the mean and standard deviation of the reflectance of individual seeds at 19 discrete wavebands from a single image. The NIRS reflectance is the mean value of five measurements on the same seed lot. Standard deviation of NIRS reflectance measurements is not shown as it is too small. The color bar below the plot shows the corresponding perceived colors of the human visible spectrum. The ranges above the plot show which chemical compounds contribute to which wavebands [13].

2.1. NIRS Spectra with Good Informative Spectra

The use of NIRS in seed testing and seed research can be through single seed or bulk seed lot measurements. The single seed measurement requires a sample holder with similar form as the seed to reduce the risk of light scatter (light travelling outside the seed to the detector). Near-infrared light can penetrate the seed; however, the depth of the penetration depends on several factors such as the physical proportions of the seed. The NIR light is then reflected, refracted, transmitted, scattered or absorbed in the seed (Figure 2)

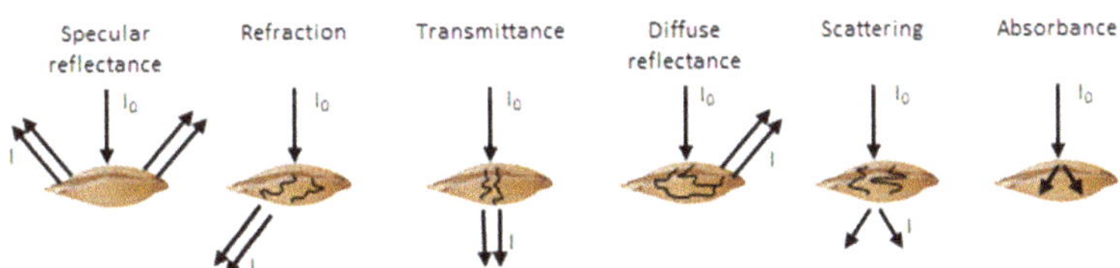

Figure 2. The possible interaction of incident light (I_0) with seed and subsequent reflected, refracted, transmitted, scattered or absorbed light (I).

The method for bulk seed NIRS measurement depends on the available instrumentation, and the output is a mean spectrum of the seeds.

The choice of single seed or bulk seed lot measurement depends on the aim of the project. The main advantage of single seed NIRS is the possibility to obtain a spectral signature, i.e., fingerprint for individual seeds, while bulk analysis is an average spectrum of the measured seeds. The benefit of bulk seed analysis lies in the reduced operation time and the possibility to characterize seed lots with fewer measurements as each spectrum represents the variation within the seed lot.

The raw NIR spectra contain important information in terms of spectral peaks that relate to chemical information. Shrestha et al. [14] showed the NIR spectra of seeds of seven species and even though the trends (spectral peaks) were similar, it was possible to identify spectral differences between the species using principal component analysis (Figure 3).

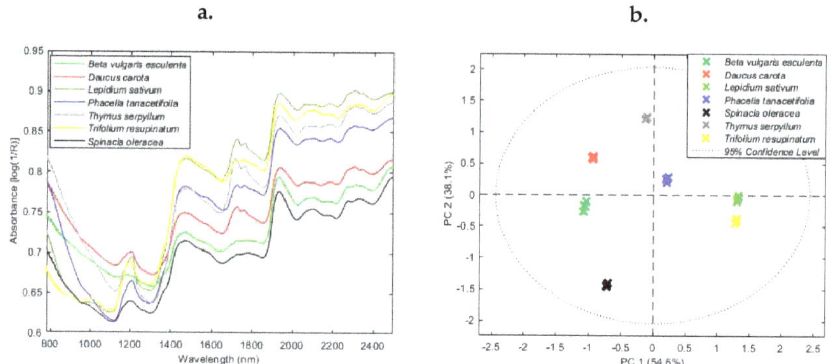

Figure 3. Raw NIR spectra (**a**) and principal component analysis (PCA) analysis (**b**) of seeds of seven species. The NIRS measurements were performed as bulk seed analysis (five repeated measurements on the same samples).

2.2. Spectral Pre-Processing

Pre-processing of the NIR spectra is the first step in developing informative classification models. The purpose of pre-processing is to identify and to remove spectral information that interferes with the desired predictions [15]. If the pre-processing fails, there will be confusion between the information which is sought and the noise which is of no interest [16]. Several pre-processing methods are available and some of them are thoroughly described and shown in Rinnan et al. [17]. In practice, it is important to evaluate the effect of different pre-processing methods on the final models. Another possibility is to use the raw spectra in the subsequent principal component analysis (PCA) as shown in Figure 3. The use of raw spectra will in most cases lead to the usage of more principal components for the final model to reduce noise in the spectra.

2.3. NIRS Model Development and Validation

Models for classification, pattern recognition or clustering developed from NIR spectra for one sample of seeds (either bulk or single seed NIRS) are intended to classify other seeds or seed samples of the same species based on their NIR spectra.

The NIRS data are highly correlated, meaning that data points next to each other are more alike than data points far from each other, and a common method to reduce this dimensionality is through PCA [18]. Subsequently, this reduction in dimensionality is used in different linear and non-linear models as described by [18–20]. The classification models are divided into supervised or non-supervised models where the supervision relates to labelled or non-labelled data. The use of labelled data in supervised classification models will inevitably influence the results and makes proper validation of the models even more important to avoid overfitting. There are a few regression-based classification models, such as partial least squares discriminant analysis [21,22] and extended canonical variates analysis [23].

Validation of models is an essential part of the modelling process to ensure that a model can be used to classify other seeds or seed samples, but also to avoid giving unrealistic (i.e., optimistic) estimates of the ability to classify new samples [24]. Any model should be validated for model performance and prediction ability using either cross-validation or test set-validation. Cross-validation is performed by dividing the full dataset into *G* sample set and using *G-1* sample set as the training set and the remaining segment in the test set. Each segment is successively excluded and used for testing the model based on the remaining samples from the *G-1* segments. Using this method, all samples are used for both calibrating and validating the model. The performance of the model is evaluated by its predictive error in terms of root mean square error of cross-validation. Test set-validation is normally seen as a stronger validation of the obtained models as samples in the test set are not part of the model development. Test set-validation requires the data to be divided into a calibration and a validation set. The calibration set is used to calibrate the model and this model is subsequently tested on the validation set. The model performance using test set-validation is described by root mean square error of prediction.

3. Multispectral Imaging

Multispectral imaging of seeds is a non-destructive technique for simultaneously measuring spectral and spatial information of seeds by imaging their surface reflectance at selected wavelengths from 365 to 970 nm (Figure 1). The combined spectral and spatial measurements provide information about the seed surface chemistry [25] and seed morphology (color, shape, and texture). Multispectral images acquired through MSI is a middle ground between RGB (red green blue) color images and hyperspectral images. RGB images use three wide overlapping wavebands to mimic the human visual perception of colors. In contrast to hyperspectral imaging, which measures the reflectance at hundreds of continuous narrow wavebands across a large spectral range, multispectral imaging measures the reflectance at fewer (<50) and wider discrete wavebands (10–50 nm).

The workflow for MSI of seeds generally includes the following six steps (Figure 4): (1) preparation of seed samples, (2) calibration of multispectral imaging system, (3) acquisition of multispectral images of the seeds, (4) segmentation of regions of interest (ROIs, e.g., the seeds, part(s) of the seeds or foreign matter) in the acquired multispectral images, (5) feature extraction from the segmented ROIs and (6) analysis of the extracted features. If the aim is to study changes in the seeds over time, for example, to follow the imbibition process or radicle emergence, steps 1 to 3 may be repeated multiple times before proceeding with steps 4 to 6.

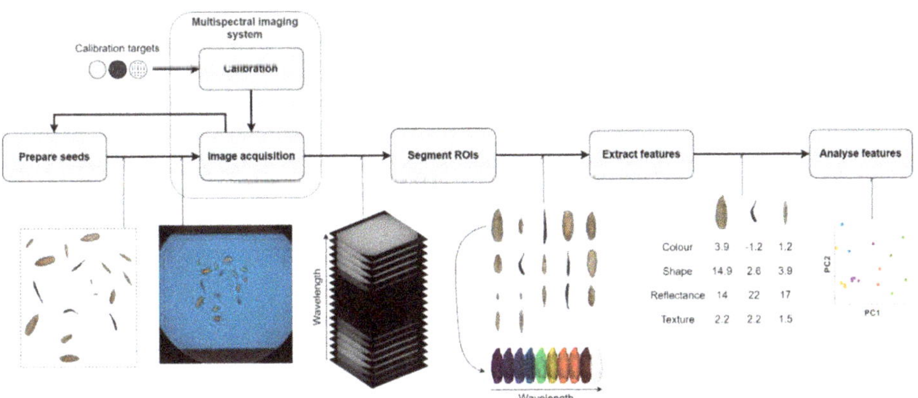

Figure 1. Illustration of the typical workflow in multispectral imaging applications with output examples from each of the six steps.

3.1. Sample Preparation

For MSI, seeds require very little preparation beyond the preparation required for the application or experiment at hand. For example, if the aim was to see if it is possible to identify the presence of particular fungi on seeds, the first step might be to work with sterilized seeds before inoculating them with the fungus/fungi of interest [26–30]. Similarly, it may be necessary to artificially age seeds for different lengths of time, to explore the use of MSI for predicting whether seeds are viable or dead, or parameters related to vigor (e.g. El Masry et al. [31]). On the other hand, in varietal purity applications, the seeds may be imaged without any further preparation.

Due to the spatial nature of the multispectral images, multiple seeds can be imaged simultaneously. Seeds are often placed in a Petri dish and it is important that there is space around each seed. Seeds located too close to each other may touch or even overlap and cause occlusion leading to poorer segmentation and adding noise to the extracted features. To prevent seeds from moving, when placing them in the Petri dish, they may be fixed with double-sided tape [29,32] or placed on an insert with small recesses inside the Petri dish. When placing the seeds, it is important to consider which side is most relevant for the application and thus should be facing the imaging sensor. In applications where multiple sides are equally relevant, such as detection of processing damage, images from multiple sides can be acquired by imaging each seed multiple times [33–35]. For some studies, it may be necessary to keep track of each individual seed through the imaging process to understand the subsequent 'fate' of each seed.

Placing seeds manually in a Petri dish for imaging can be both cumbersome and time-consuming. A conveyer belt can be used to automate the imaging process and increase the number of seeds imaged over time in applications where the seeds do not require any special preparation or manual assessment (e.g., variety or foreign matter identification [36]).

3.2. Calibration of Multispectral Imaging System

The MSI system must be calibrated prior to image acquisition to ensure comparable reflectance measurements across wavebands and images, pixel correspondence between wavebands and to enable spatial measurements in world units [37,38]. This includes both a radiometric calibration and a geometric calibration, which is carried out by imaging calibration targets with known reflectance and geometry [39].

Furthermore, the illumination and exposure times must be set to minimize the number of under- and oversaturated pixels, thereby maximizing the dynamic range and the signal-to-noise ratio of the images [27,37].

3.3. Image Acquisition

After calibration, the MSI system is ready to image the prepared samples. The output of a measurement is a multispectral image or "data cube" consisting of $W \times H$ pixels $\times$ C channels, where W and H are the width and height of the image, respectively, and each pixel contains C channels corresponding to the discrete multispectral bands. When a pixel position overlaps with a seed, the pixel values represent the chemistry on and below the surface of the seed in the small area covered by the pixel [25].

Although multispectral imaging systems can acquire the images through either point scanning, line scanning or area scanning [40], in the vast majority of the applications the images are acquired through area scanning with a charged coupled device (CCD) imaging sensor and sequentially illuminating the seeds using LEDs with the desired wavebands (Table 1). Ideally, these wavebands should be carefully selected to match the application or research question [34]. However, most MSI applications use the same multi-purpose MSI system (all applications with 19 bands in Table 1), where the wavebands and spectral range are selected by the company. However, changing the spectral range will mean changing imaging sensor technology as the spectral range of the current MSI systems is limited by the quantum efficiency of a standard CCD to approximately 400–1000 nm.

Table 1. Summary of selected applications of multispectral imaging for seeds analysis.

Application	Species [1]	Spectral Range (nm)	Number of Bands	Sample Size	Features	Analysis [2]	Accuracy [3] (%)	Reference
Physical Seed Quality								
Varietal identity and purity	Alfalfa (12 cultivars)	365–970	19	2400	Reflectance, color, shape	PCA, CDA, SVM	Color + shape: 42–45 Reflectance: 87–88 All: 92–93	[41]
Varietal identity and purity	Maize (six inbreed lines)	375–970	19	120	Reflectance	CDA	40–100	[42]
Varietal identity and purity	Pepper (three varieties)	365–970	19	1472	Reflectance	k-NN, SVM, CNN	82–98	[43]
Varietal identity and purity	Rice (20 varieties)	365–970	19	598	Reflectance, color	k-NN	93	[25]
Varietal identity and purity	Rice (five varieties)	405–970	19	250	Reflectance, color, shape	PLS, SVM, NN	Reflectance: 62–86 All: 74–94	[44]
Varietal identity and purity	Rice (197 accessions)	365–970	19	3940	Color, shape	PCA, ANOVA	-	[45]
Varietal identity and purity	Soybean (three cultivars)	405–970	19	600	Reflectance, color, shape	PCA, PLS, SVM, NN	Reflectance: 72–93 All: 73–98	[46]
Varietal identity and purity	Tomato (12 varieties hybrid offspring)	375–970	19	2525 / 205	Reflectance, color, shape	PCA, CDA, PLS	79–96	[47]
Varietal identity and purity	Tomato (5 cultivars)	375–970	19	1236	Reflectance	PCA, PLS, SVM	94–100	[14]
Varietal identity and purity	Triticale + Wheat (9+27 varieties)	375–970	19	1728	Reflectance, shape	k-NN	40–97	[48]
Other seeds, inert matter	Alfalfa Sweet clover	365–970	19	2400	Reflectance, color, shape	PCA, CDA, PLS, Adaboost, SVM	Color + shape: 82–94 Reflectance: 54–99 All: 68–100	[39]
Other seeds, inert matter	Maize	375–970	19	910	Reflectance	PCA. PLS	89–100	[34]
Other seeds, inert matter	Mustard (five types of foreign matter [FM])	676–952	25	20 images 395 FM	Reflectance	PCA, SVM, NN	98	[36]
Other seeds, inert matter	Sunflower	405–970	19	118	Reflectance	SDA, CDA, PCA	70–100	[49]
Damage (processing)	Sugar beet (18 varieties)	375–970	19	301 + 200	Color, shape	CDA	67–100	[35]
Damage (insects)	Wheat (one moth species)	405–970	19	600	Reflectance	GLM	-	[28]
Physiological Seed Quality								
Viability	Castor bean	375–970	19	420	Reflectance	CDA	96	[50]
Viability	Cowpea	375–970	20	501	Reflectance	PCA, CDA	Ageing: 87–97 Germination: 81 Germ. time: 68 Viability: 62	[31]

Table 1. *Cont.*

Application	Species [1]	Spectral Range (nm)	Number of Bands	Sample Size	Features	Analysis [2]	Accuracy [3] (%)	Reference
Viability	Jatropha curcas	365–970	19	300	Reflectance	PCA, CDA	96–98	[51]
Viability	Spinach	395–970	19	300	Reflectance, texture	PCA, PLS	51	[52]
Viability	Watermelon	405–970	19	1000	Reflectance, color, shape	PCA, SVM, NN, RF	Reflectance: 72–87 All: 75–92	[53]
Vigor	Legume (six species)	365–970	19	2400	Reflectance, color, shape	PCA, CDA, SVM	78–92	[54]
Vigor	Sugar beet	375–970	19	60	Reflectance, color, shape	CDA	95	[55]
Seed Health								
Fungal infection	Barley (five fungi)	375–970	19	200	Reflectance, color	CDA	-	[55]
Fungal infection	Black oats (one fungus)	365–970	19	800	Reflectance, color, texture	CDA	Reflectance: 73 All: 86	[32]
Fungal infection	Cowpea (three fungi)	365–970	19	240	Reflectance, color, texture	PCA, CDA	Before incubation: 92 After incubation: 100	[29]
Fungal infection	Jatropha curcas (three fungi)	365–970	19	231	Reflectance	PCA, CDA	87	[27]
Fungal infection	Rice (two cultivars, one fungus)	460–940	6	1925	Reflectance	PCA, *k*-NN, CDA, SVM	86–99	[30]
Fungal infection	Spinach (five fungi)	395–970	19	234	Reflectance	CDA	-	[26]
Fungal infection	Triticale + Wheat (9 + 27 varieties, two fungi)	375–970	19	1728	Reflectance	CDA	-	[48]

[1]. Subdivision of species in brackets are made according to the terminology used in the referenced paper. [2]. Adaboost = Adaptive Boosting; ANOVA = Analysis of Variance; CDA = Canonical Discriminant Analysis; CNN = Convolutional Neural Network; GLM = Generalized Linear Model; *k*-NN = *k*-Nearest Neighbors; NN = Neural Network; PCA = Principal Component Analysis; PLS = Partial Least Squares; MLR = Multiple Linear Regression; RF = Random Forest; SDA = Stepwise Discriminant Analysis; SVM = Support Vector Machine. [3]. The accuracy is the number of correctly classified samples with respect to the total number of samples. "-" means the reference did not report the accuracy and most likely used another error metric

Selecting a high contrasting background material on which the seeds are placed can make the segmentation step easier; however, the intensity level of the background should approximately match that of the seeds to fully use the dynamic range of the multispectral imaging system.

3.4. Segmentation of Regions-of-Interest

The multispectral images contain not only ROIs, but also background objects, such as the background material, the Petri dish, a conveyer belt, or other inert matter. In the segmentation step, the ROIs are separated from the background objects and extracted from the image. The ROIs in the multispectral images are often limited to only the seeds, but they may also include other objects such as foreign matter [36]. To ensure that only the correct objects are analyzed, the segmentation method must extract only objects regarded as ROIs. Equally important, the segmentation method must return all pixels related to the ROIs, and only those pixels to reduce noise in the subsequently extracted features.

With a high contrast background material and sufficient space around each seed, the segmentation can often be carried out using a simple threshold in either a single channel [36,40], a sum of the channels [33] or on a score image created through canonical discriminant analysis (CDA) [53] or PCA [36]. Ma et al. [49] used Otsu's algorithm [56] to set the threshold automatically.

Although different methods have been explored, their performance have not been/are seldom quantified (e.g., pixel accuracy or intersection over union) beyond visual inspection as the segmentation step is often seen as an intermediate step towards the final analysis.

3.5. Feature Extraction

In recent applications, several features quantifying the reflectance and morphology of the seed have been explored. These features form four groups related to their characterization of the seed and their relation to the multispectral image: reflectance, color, shape, and texture (Table 1). They are generally extracted from the entire seed; however, they may also focus on only a specific part of the seed such as the endosperm region [42]. The reflectance and color features relate to the spectral dimension (C) of the multispectral image and express the intensity of either reflectance or color of the seed. The reflectance features either treat the wavebands individually by extracting first-order derivatives from the raw wavebands [52] or combine them with a CDA transformation before extracting either a trimmed mean [50] or ratio of pixels above a given threshold [55]. In contrast, the color features combine wavebands overlapping with the human visible spectrum into a well-defined color space, e.g., CIELAB [47], and extracts first-order features from there. The shape features are related to the spatial dimensions ($W \times H$) of the multispectral image and are therefore derived from the binary image created during segmentation. They include simple descriptors, such as area, width and length [57], but also more complex descriptors, such as ellipse fitting parameters and resemblance to known simple shapes (i.e., circle, ellipse, and rectangle). The texture features combine the spatial and spectral dimensions by quantifying the spatial variation in intensity across the seed. This spatial variation in intensity can be caused by both small changes in the surface structure (valleys and hills) as well as changes in color in the seed surface pattern. The color, shape, and texture features describe the morphology of the seed and are therefore jointly referred to as morphological features. Characters of morphologic features of different seed structures play an important role in the delimitation and identification of species [58].

The type of extracted features is somewhat application-dependent (Table 1). Applications related to fungal presence all use reflectance features and to some extent color and texture features. Shape features are, however, not used as the shape of the seed is not affected by the presence of fungus until the fungus has grown significantly. On the other hand, applications related to varietal purity almost all use reflectance, color, and shape features, but do not consider texture features. Applications on seed viability and vigor favor reflectance and to a lesser extent color and shape.

For a given application, it is important to extract features which are expected to correlate well with the desired response variable. Features may be derived from existing knowledge, such as previous work in hyperspectral imaging, NIRS or crop descriptors [59]. However, the selection of features should be well argued.

3.6. Multivariate Data Analysis

The multivariate data analysis of the extracted features often includes a descriptive statistic followed by data modelling. The descriptive statistics compares the mean and variation of the individual features for each class. For the reflectance features, this is often visualized as a mean spectrum for each of the classes [26]. Principle component analysis is also widely used to investigate any trends in the features prior to data modelling.

Several linear and non-linear methods have been used for data modelling in MSI. The most frequently used methods include PCA, CDA, support vector machines (SVM), partial least squares and to a lesser extent neural networks and k-nearest neighbors (Table 1)

Despite a large number of features and correlation between features within feature types (e.g., shape features), dimension reduction [44,57] or feature selection [52] prior to modelling is the exception to the rule. However, several applications evaluate the feature types both individually and combined and show an improvement in accuracy when feature types are combined (Table 1).

In applications related to physiological seed quality and seed health, it may be difficult to ensure an equal number of examples from each class. This leads to an unbalanced dataset, where one or more classes are either over- or underrepresented compared to the remaining classes in the dataset. Unbalanced data in the calibration set can lead to a model with poor generalization on future data, while the model is still reporting misleadingly high values in error metrics such as accuracy. The data imbalance may be handled as a pre-processing step (e.g., resampling) through cost-sensitive learning (assigning different costs to each type of misclassification) or at an algorithm-level [60]. Likewise, error metrics less skewed by an unbalanced dataset should be favored.

4. Applications

The recent applications of multispectral imaging of seeds can be grouped into three categories according to the aspect of seed quality (Table 1): physical seed quality, physiological seed quality, and seed health.

4.1. Physical Seed Quality

The physical seed quality applications include (a) varietal identity and purity, (b) presence of other seeds and inert matter and (c) seed coat integrity.

4.1.1. Varietal Identity and Purity

The microstructure and chemical composition of specific seed coat cell layers give rise to species and varieties differences. Most morphological features of the seed coat are relatively insensitive to environmental conditions and therefor very useful for taxonomic identification.

Multispectral imaging has been employed for varietal discrimination and identification in several species such as tomato (*Solanum lycopersicum* L.), rice (*Oryza sativa* L.) and soybean (reviewed in Boelt et al. [55]). Color, shape and spectral features have been used in the classification models (Table 1). Since then, studies in alfalfa and pepper (*Capsicum annuum* L.) have been reported [41,43].

In pepper, three commercial varieties were analyzed for varietal identification [43]. Each variety was represented by at least 450 seeds and seed material was harvested at different locations. Samples were divided into training and test set in the ratio 9:1. The study employs different multivariate data analysis and resulting classification accuracies are in the range of 86–98%. The multispectral imaging system used in this study has 19 bands. Interestingly, a successive projection algorithm identified nine bands, which provide a classification accuracy almost identical with the outcome with all 19 bands (97%). Still, the authors suggest that a data analysis with lower classification accuracy (93%) may be used as this is easier to operate and has a sufficiently high accuracy for the purpose. This illustrates how feasibility and ease of operation is of importance in the commercial seed industry.

Twelve alfalfa cultivars (*Medicago sativa* L.) with diverse geographic origin were obtained from a genebank [41]. A total number of 200 seeds were split 70:30 in training and testing set, respectively. Different multivariate data analysis was used to classify cultivars (Table 1). When only morphological features were employed, classification accuracy was low (42–44%) but combined with spectral features, accuracy increased to 92–93%. It is noticed that based on spectral reflectance cultivars were classified into three groups correlating with geographic origin. This may be based on common genetic background or seeds may have been produced in different environments. Seed coat color is influenced by environmental conditions—i.e., climatic conditions during maturation and hence not appropriate for taxonomic purposes [1]. However, variation in texture and chemical composition will also be reflected in the spectral features and they are highly relevant for taxonomic discrimination.

Seed accessions in genebanks may not be as uniform as commercial varieties; however, the description of the seed morphology is very important to manage the large accession numbers (for example during the regeneration procedure). Already Hansen et al. [25] demonstrated high classification accuracy among 20 diverse rice varieties (93%) and suggested MSI as an important tool in management of genebank accessions. A recent study used MSI for the assessment of the genetic diversity in a collection of pigmented rice accessions from the Philippines [45]. Geometric seed traits were quantified (area, length, width, roundness, and seed color parameters). The study identified pigmented rice accessions, which represent a valuable genetic resource for the future improvement of commercial rice varieties.

In conclusion, MSI may both be used to distinguishing among commercial varieties in the test of varietal purity and to describe diversity in seed traits during conservation management of plant genetic resources.

4.1.2. Presence of Other Seeds and Inert Matter

Sendin et al. [34] reported the use of MSI for the determination of other crop seeds and plant debris in white maize (*Zea maize* L.). Seeds of crop species were wheat (*Triticum aestivum* L.), sorghum (*Sorghum bicolor* L.), soybean and sunflower (*Helianthus annuus* L.) and all were classified with 100% accuracy. Plant debris was also classified with 100% accuracy and the authors point to the benefit of MSI contra hyperspectral imaging in relation to shorter analysis time and lower cost. Recently Hu et al. [39] published findings on the differentiation of sweet clover (*Melilotus* ssp.) in alfalfa with a classification accuracy of >99% by MSI. Combining morphological features and spectral data in the models increased the accuracy. The survey included six alfalfa varieties and two species of sweet clover: One seed lot of *Melilotus officinalis* and five seed lots of *Melilotus albus*. All seed lots consisted of 200 seeds, divided in training and model testing in the proportion 70:30. Reflection mean intensity showed discrimination both in visible and NIRS wavelength bands.

As indicated in the two above-mentioned studies, very high accuracies may be expected when classifying seeds belonging to different species, and hence this may not attain much consideration in research; however, the determination of other seeds in crop seeds is a very time-consuming task in the seed industry. It appears relevant to develop robust models of crop seeds containing the variability in seed morphology from site to site, year to year for the use in seed testing. There are examples from the food industry in the detection of "foreign matter" which would include some of the same constituents as the inert matter fraction in a seed sample (soil, stones, plant debris) [34,36].

4.1.3. Integrity of Seed Covering Structures

The intact seed coat protects the internal structures of the seed and controls water uptake, but seed coat disrupture may occur due to insect infestation during seed production or storage or mechanical damage during harvest and processing. Seed coat damage negatively affects vigor and viability potential, and the "openings" of the damaged seed coat may be an entrance for pathogenic fungi.

Insect Infestation

When insect infestation occurs during seed production, the damaged seed is often discarded during harvest and processing due to a lower seed mass. Insect infestation occurring in the later developmental phase may not be identified and has the potential to develop during storage. Insect infestation has a direct effect on seed quality by consuming the seed reserves but there is also an indirect effect as it allows the establishment of secondary pests and fungi, for the storage pests lay eggs on the seed surface for the larvae to penetrate the seed coat and the larvae may undergo different larvae stages and finally produce a pupa inside the seed. X-ray and MSI have been tested for the identification of grain moth (*Sitotroga cerealella*) in wheat [28]. The study showed the potential of X-ray for the study of internal structures in the seed, whereas MSI showed the potential for identifying eggs on the seed surface.

Mechanical Damage

Species containing germination inhibitors in the seed coat (for example sugar beet) undergo different treatments during processing to remove these inhibitors. The inner pericarp layer contains crystals of chemical compounds in the sclerenchyma cells [61], and the crystals dissolve in water during washing. This process alters the outer surface structure of sugar beet seed [2]. MSI can detect changes in surface color and reflectance during maturation in sugar beet seed [62], and the study verified a concomitant increase in the content of phenolic compounds. Removal of the pericarp by polishing is another approach for the removal of inhibitory compounds. The polishing process removes most of the large parenchyma cells of the pericarp and hence alters the surface of the sugar beet seed [2]. The ideal treatment will remove the outer pericarp layer, whereas the inner pericarp layer remains intact. Besides removal of germination inhibitors, polishing also makes seed more uniform for pelleting and improves water uptake.

As with any mechanical operation, excessive processing can cause damage to the seed, and this damage can be extended to the interior parts of the seed and affect physiological quality of the seed (Figure 5) Mechanical injuries decrease the seed longevity, expose the seed to the fungal infection and reduce viability.

Due to their sensitivity to water uptake, damaged seeds may result in heterogenous field performance, and there is also evidence from soybean, sweet corn and maize that the damaged seeds are more likely to produce abnormal seedlings [61].

A study by Salimi [35] displayed the potential of MSI in classification of various damage types, without additional analytical evaluation. The study demonstrated MSI as a tool for the identification of mechanical damage from polishing during processing and hence demonstrates MSI as a tool in seed quality assessment. A classification model based on MSI derived information about surface characteristics and multivariate data analysis enabled discrimination into five damage classes with 82% overall accuracy.

Barley (*Hordeum vulgare* L.) grains without hulls will imbibe water and germinate more rapidly than those with firmly adhering husk [63]. During harvest, the hull acts to protect the embryo during the abrasive threshing process in the harvester [64]. However, the husk may be partially or wholly detached at harvest and during post-harvest handling (Brennan, Shepherd et al. 2017). MSI may be a potential tool for the characterization of de-hulled barley grains.

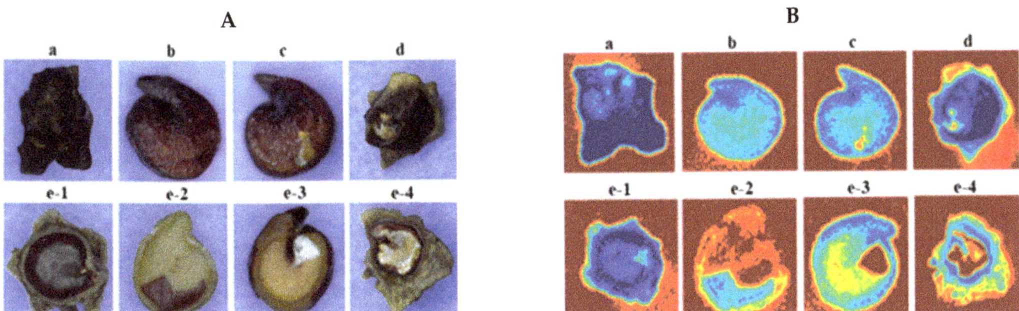

Figure 5. Processing damage in sugar beet seeds. (**A**): RGB images; (**B**): nCDA transformed multispectral images [35]. (**a**). Partially broken pericarp and/or outer testa, (**b**). Completely broken pericarp and outer testa, (**c**). Fractured pericarp and outer testa, partially crushed inner testa with sound embryo, (**d**). partially broken pericarp and/or outer testa, damaged inner testa with intact embryo, (**e1–e4**). Different types of severe damages to the embryo or seeds without any embryo like the pericarp or outer testa. Reproduced with permission from ref. [35]. Copyright 2019, MDPI.

4.2. *Physiological Seed Quality*

Tannins, phenols, waxes, pigments, germination inhibitors and other substances are found in the seed covering structures of different species, and these may influence the function of the seed coat and subsequently the physiological development of the seed.

4.2.1. Viability

Olesen et al. [50] identified viable castor bean (*Ricinus cummunis* L.) seeds with 92% accuracy and showed good correlation between results from tetrazolium tests and MSI. Three seed lots were included in the study. In castor bean, seed coat color was related to the development and the darker seeds were the most developed. In this study, seeds from four ecotypes were studied. The calibration set consisted of 120 seeds from two ecotypes, and they were divided into three groups in depending on seed coat color (visible inspection). The validation set was two other ecotypes, and the seeds of those were also divided into three groups. After acquisition of MSI images, seeds were germinated for the phenotyping of viability, and a tetrazolium test was performed as the viability reference. A high correlation was found (92%). The supervised nCDA model showed 96% precision accuracy in the classification of viable and dead seeds in the validation set. The study showed high differentiation between viable and non-viable seed in mean intensity reflection in the wavelength interval 375–970 nm with the largest difference in the NIR-regions, which is supported by Shetty et al. [52] in a study predicting germination ability in spinach. This latter reference combined the use of single seed NIRS and MSI.

Liu et al. [53] also found a high prediction accuracy (91–92%) for high-quality watermelon (*Citrullus lanatus* (Thunb.)) seed, in two different varieties using both spectral and morphology features in MSI. From each variety 500 seeds were classified into pure, viable; low vigor; other varieties and dead seeds by means of a grow-out trial. Prediction accuracy concerns two classes: pure, viable, and all other seeds for each variety.

4.2.2. Vigor

Several species in the *Fabaceae* family can produce hard seeds (physical dormancy) which are impermeable or semi-permeable and hence do not absorb water. Physical dormancy is often associated with a layer of wax in the outer layers of the seed coat.

Hu et al. [54] examined seeds of six species within the Fabaceae family with MSI for the detection of hard seeds. For each species, 400 seeds were examined 70:30 in training and testing set, respectively, and following image acquisition seeds were imbibed for germination. Hard seeds were identified as un-imbibed, whereas seed which adsorbed water was classified as "soft" non-dormant seed. For three species (sweet clover, alfalfa and galega (*Galega officinalis* L.)) MSI combined with multivariate data analysis has accuracies in the interval of 88–92% in detecting hard seeds, whereas for the other three species they could not be identified. In all three species studied, hard seeds showed a higher reflectance compared to non-hard seeds. Hu et al. [54] used SVM analysis and found that wavelengths in the NIR-region, i.e., 970 nm (water) and 940 nm (lipids, were of highest importance in the separation of the two groups. However, for each species only one seed lot was represented in the analysis and there were proportionally fewer non-hard seeds which made the two groups unbalanced.

Single seed NIRS spectroscopy and MSI have been employed for the assessment of viability after controlled deterioration or artificially seed ageing in spinach [38] and cowpea (*Vigna unguiculata L.*) [31]. In spinach, two seed lots with viability percentages of 90% and 97% were chosen for the examination by single seed NIRS after artificially ageing of both seed lots [38]. In cowpea, variation in germination performance was generated by artificially ageing in four treatments (ageing intervals 24–96 h) [31]. Olesen et al. [38] used Extended Canonical Variates Analysis (ECVA) assigned differences of scatter corrected absorbance spectra from aged and non-aged seeds to CH_2, CH_3 and HC = CH structures which are some of the functional groups in lipids. Lipids play a major role in both ageing and germination. During accelerated ageing lipid peroxidation leads to deterioration of cell membranes and contributes in that way to reducing seed viability of the seed sample. These biochemical changes may be the reason for a clear grouping between aged and non-aged seeds with misclassification in the range of 4–11% when performing the ECVA. In cowpea, the overall correct classification was in the interval 97–98% between aged and non-aged seeds, whereas the classification was lower in the detection of germinated versus non-germinated seed (79–82%). A recent paper reports a strong relationship between X-ray and MSI and seed physiological potential in *Jatropha curcas L.* seed [51]. Both viability and vigor were studied, and the authors find that reflectance data in the NIR wavelength 940 nm showed 96% accuracy.

Ruptured seed coats allow for the diffusion of leachates, which serve as substrates for pathogen growth, and broken seed coats serve as infection sites for seed pathogens. Common measurements of seed leakage in water are the conductivity of electrolytes and ultraviolet (UV) light absorbance (254 and 280 nm) [65,66]. Leaked solutes may be amino acids, proteins, sugars, and phenolics. Brassica seed has a high content of phenolic compounds. One of these is sinapine, the content of which increases under unfavorable storage conditions. Hill et al. [67] found sinapine leakage a more accurate method for the identification of viable cabbage (*Brassica oleracea* var. *capitata* L.) seeds than the conductivity test. Sinapine was measured by the absorbance at 388 nm. The compound fluoresces when irradiated with UV light and has maximum absorbance values of 326 and 388 nm Later work [68] showed that seeds with cracked seed coat leaked faster and that seed coat integrity is a major factor regulating sinapine leakage. Sinapine does not leak from viable seeds [67].

Since leaked solutes have been measured using absorbance of light in the UV region, a future perspective of MSI would be analyzing single dry seeds for diffused solutes often associated with cracks in the seed coat.

4.3. Seed Health

Detection of seeds infected by fungi is traditionally performed by visual inspection of dry seeds, washing tests, incubation methods, embryo count method or seedling symptom tests as well as identification of sporulation [69,70]. These methods require expert knowledge and can be time-consuming. However, the combinations of the features from multispectral images captured by visual light and NIR wavelengths (Figure 1) have proved to be useful in the separation of infected and uninfected seeds (Table 1), but depend on traditional reference methods.

Multispectral imaging for seed health detection has in several studies been based on artificial inoculation of uninfected seeds, with freeze-blotter seed health assay as reference method. First demonstrated in spinach by detection of *Stemphylium botryosum*, *Cladosporium* spp., *Fusarium* spp., *Verticillium* spp. or *Alternaria alternate* [26], and recently by detection of *Drechslera avenae* and *Helminthosporium avenae* in black oat/oats seeds (*Avena strigosa*) [32], *Fusarium pallidoroseum*, *Rhizoctonia solani*, and *Aspergillus* sp. in cowpea [29].

The simplest approach is to use a visual score as reference for fungal infection. However, the method depends on an expert to classify the seeds in healthy and infected seeds as well as determine the species of the fungi. Weng et al. [30] used artificial inoculation of uninfected seeds by *Ustilaginoidea virens* in rice with a visual scoring as reference method. The seeds used in this study were divided into healthy, slightly infected, and infected seeds. However, the healthy and the slightly infected seeds were difficult to separate by a PCA. It was suggested that this was due to only minor changes in seed surface features or chemical components of the slightly infected seeds.

DNA-based data may be used as reference in combination with MSI. Boelt et al. [55] used next generation sequencing (NGS) of the ITS (Internal Transcribed Spacer) from total DNA as reference method on naturally infected barley seeds collected from a wide range of environments. NGS is highly sensitive and gives information to species level as well as the fungal composition and quantities. This is particularly useful as several fungi may infect seeds simultaneously. NGS made it possible to separate seeds infected by *Alternaria infectoria*, *Dothidomycetes* sp., *Fusarium graminearum*, *F. avenaeum* and *Mycosphaerella tassiana* by multispectral imaging.

Magnetic resonance imaging (MRI) to identify anatomical changes in artificial inoculation of *Jatropha curcas* L. was used in combination with MSI by Barboza da Silva et al. [27]. The proposed MRI and MSI methodology allowed the identification of different damage patterns in the endosperm tissues due to infections by *Lasiodiplodia theobromae*, *Colletotrichum siamense*, and *Colletotrichum truncatum*.

5. Summary and Perspectives

Multispectral imaging and single seed or bulk seed NIRS are non-destructive techniques for quality assessment both in research and in seed testing. In contrast, hyperspectral imaging requires more resources for operation and is therefore most relevant in seed testing and seed research. Since recent reviews of multispectral imaging [40,55] there has been a growing evidence of the application of MSI in particular in physical seed quality evaluation and in seed health.

For physical seed quality, focus has been to distinguish genetic purity among varieties (alfalfa and pepper) or between crop species and inert matter (alfalfa versus sweet clover; mustard versus foreign and inert matter). In general, high classification accuracies have been obtained but often the number of samples or sample sizes have been limited or even unbalanced. Future studies ought to include more robust training and validation datasets by including higher and more diverse samples. Exploring seed produced at different sites (years and environmental conditions) would strengthen validation of the models by including variation in seed size and seed coat color and eventually lead to robust global models.

A relevant application for MSI is the characterization of the stored seed samples for the preservation of plant genetic resources. For this application features such as shape, texture, reflectance, and color are highly relevant, but they may be combined with a focus on specific parts of the seed, for example, the morphology of the hilum region, which is a relevant feature in the crop descriptors of legume seeds.

In conclusion, MSI may be used both to distinguish among commercial varieties in the test of varietal purity and to describe diversity in seed traits during conservation of plant genetic resources. For the assessment of physical seed quality, very little sample preparation is required, but a large diversity from each species or variety ought to be included by representing different production sites and climatic environments.

In seed research and seed testing, electrolyte leakage is an established method for vigor evaluation, where seeds are imbibed for a certain period of time, and the imbibition water is analyzed by spectrophotometer. Solutes are measured using the absorbance of light in the UV region. A future perspective of MSI would be analyzing single seeds? for diffused solutes in the UV band. The information acquired on the single seed level may even be combined with other features such as color, physical damage, or cracks in the seed coat. The determination of physiological seed quality will often require more sample preparation depending on the physiological process in question, and sample sizes may be unbalanced, for example, there are far fewer non-viable seeds in a commercial seed lot.

Physiological seed quality is often reflected in the chemistry of the seed and therefore information from the NIR-wavelength regions is often very informative. The region of interest for the chemical information defines which method to apply, where MSI will inspect the seed covering surface and single seed or bulk NIRS will inspect the seed beyond the surface cover. However, none of these methods provides information on internal morphological seed structures.

The use of MSI and single seed and bulk NIRS to characterize seed covering structures is only at the beginning, and there is a future potential for the development of specific applications in seed testing. Cross disciplinary studies between seed research and data science may combine the required insight in seed biology and data analysis to provide relevant seed samples for inspection and optimize feature extraction, data analysis, and model validation.

Author Contributions: Conceptualization, A.K.M., R.G., J.R.J. and B.B.; writing—original draft preparation, A.K.M., R.G., J.R.J. and B.B.; writing—review and editing, A.K.M., R.G., J.R.J. and B.B. All authors have read and agreed to the published version of the manuscript.

Funding: This material is based upon work that is funded by The Ministry of Higher Education and Science, Denmark.

Conflicts of Interest: The authors declare no conflict of interest.

References

1. Souza, F.H.D.D.; Marcos-Filho, J. The seed coat as a modulator of seed-environment relationships in Fabaceae. *Rev. Bras. De Botânica* **2001**, *24*, 365–375. [CrossRef]
2. Ignatz, M.; Hourston, J.E.; Tureckova, V.; Strnad, M.; Meinhard, J.; Fischer, U.; Steinbrecher, T.; Leubner-Metzger, G. The biochemistry underpinning industrial seed technology and mechanical processing of sugar beet. *Planta* **2019**, *250*, 1717–1729. [CrossRef]
3. Rodríguez, M.V.; Barrero, J.M.; Corbineau, F.; Gubler, F.; Benech-Arnold, R.L. Dormancy in cereals (not too much, not so little): About the mechanisms behind this trait. *Seed Sci. Res.* **2015**, *25*, 99–119. [CrossRef]
4. Brennan, M.; Shepherd, T.; Mitchell, S.; Topp, C.F.E.; Hoad, S.P. Husk to caryopsis adhesion in barley is influenced by pre- and post-anthesis temperatures through changes in a cuticular cementing layer on the caryopsis. *BMC Plant Biol.* **2017**, *17*, 169. [CrossRef]
5. Cerri, M.; Reale, L. Anatomical traits of the principal fruits: An overview. *Sci. Hortic.* **2020**, *270*, 109390. [CrossRef]
6. Mohamed-Yasseen, Y.; Barringer, S.A.; Splittstoesser, W.E.; Costanza, S. The role of seed coats in seed viability. *Bot. Rev.* **1994**, *60*, 426–439. [CrossRef]
7. Duke, S.H.; Kakefuda, G.; Harvey, T.M. Differential Leakage of Intracellular Substances from Imbibing Soybean Seeds. *Plant Physiol.* **1983**, *72*, 919–924. [CrossRef] [PubMed]
8. Duke, S.H.; Kakefuda, G.; Henson, C.A.; Loeffler, N.L.; Van Hulle, N.M. Role of the testa epidermis in the leakage of intracellular substances from imbibing soybean seeds and its implications for seedling survival. *Physiol. Plant.* **1986**, *68*, 625–631. [CrossRef]
9. Halloin, J.M. Deterioration resistance mechanisms in seeds. *Phytopathology* **1983**, *73*, 335–339. [CrossRef]
10. Workman Jr., J.; Shenk, J. Understanding and Using the Near-Infrared Spectrum as an Analytical Method. *Near-Infrared Spectrosc. Agric.* **2004**, *44*, 1–10. [CrossRef]
11. Agelet, L.E.; Hurburgh, C.R. Limitations and current applications of Near Infrared Spectroscopy for single seed analysis. *Talanta* **2014**, *121*, 288–299. [CrossRef]
12. Osborne, B.G.; Fearn, T.; Hindle, P.H. *Practical NIR Spectroscopy with Applications in Food and Beverage Analysis*; Longman Scientific and Technical: Harlow, UK, 1993; p. 227.
13. Lequeue, G.; Draye, X.; Baeten, V. Determination by near infrared microscopy of the nitrogen and carbon content of tomato (*Solanum lycopersicum* L.) leaf powder. *Sci. Rep.* **2016**, *6*, 33183. [CrossRef]
14. Shrestha, S.; Deleuran, L.; Gislum, R. Classification of different tomato seed cultivars by multispectral visible-near infrared spectroscopy and chemometrics. *J. Spectr. Imaging* **2016**, *5*, a1. [CrossRef]

15. Boulet, J.-C.; Roger, J.-M. Pretreatments by means of orthogonal projections. *Chemom. Intell. Lab. Syst.* **2012**, *117*, 61–69. [CrossRef]
16. Rinnan, Å. Pre-processing in vibrational spectroscopy—When, why and how. *Anal. Methods* **2014**, *6*, 7124–7129. [CrossRef]
17. Rinnan, Å.; Berg, F.V.D.; Engelsen, S.B. Review of the most common pre-processing techniques for near-infrared spectra. *TrAC Trends Anal. Chem.* **2009**, *28*, 1201–1222. [CrossRef]
18. Wold, S.; Esbensen, K.; Geladi, P. Principal component analysis. *Chemom. Intell. Lab. Syst.* **1987**, *2*, 37–52. [CrossRef]
19. McLachlan, G.J. *Discriminant Analysis and Statistical Pattern Recognition*; John Wiley & Sons: Hoboken, NJ, USA, 2004; Volume 544.
20. Coomans, D.; Massart, D.L. Alternative k-nearest neighbour rules in supervised pattern recognition: Part 2. Probabilistic classification on the basis of the kNN method modified for direct density estimation. *Anal. Chim. Acta* **1982**, *138*, 153–165. [CrossRef]
21. Barker, M.; Rayens, W. Partial least squares for discrimination. *J. Chemom.* **2003**, *17*, 166–173. [CrossRef]
22. Brereton, R.G.; Lloyd, G.R. Partial least squares discriminant analysis: Taking the magic away. *J. Chemom.* **2014**, *28*, 213–225. [CrossRef]
23. Nørgaard, L.; Bro, R.; Westad, F.; Engelsen, S.B. A modification of canonical variates analysis to handle highly collinear multivariate data. *J. Chemom.* **2006**, *20*, 425–435. [CrossRef]
24. Westad, F.; Marini, F. Validation of chemometric models—A tutorial. *Anal. Chim. Acta* **2015**, *893*, 14–24. [CrossRef] [PubMed]
25. Hansen, M.A.E.; Hay, F.R.; Carstensen, J.M. A virtual seed file: The use of multispectral image analysis in the management of genebank seed accessions. *Plant Genet. Resour. Charact. Util.* **2016**, *14*, 238–241. [CrossRef]
26. Olesen, M.H.; Carstensen, J.M.; Boelt, B. Multispectral imaging as a potential tool for seed health testing of spinach (*Spinacia oleracea* L.). *Seed Sci. Technol.* **2011**, *39*, 140–150. [CrossRef]
27. Barboza Da Silva, C.; Bianchini, V.D.J.M.; Medeiros, A.D.D.; Moraes, M.H.D.D.; Marassi, A.G.; Tannús, A. A novel approach for Jatropha curcas seed health analysis based on multispectral and resonance imaging techniques. *Ind. Crop. Prod.* **2021**, *161*, 113186. [CrossRef]
28. França-Silva, F.; Rego, C.H.Q.; Gomes-Junior, F.G.; Brancaglioni, V.A.; Hirai, W.Y.; Rodrigues, D.B.; Almeida, A.d.S.; Martins, A.B.N.; Tunes, L.V.M.D. Determination of Sitotroga cerealella infestation in wheat seeds by radiographic and multispectral images. *Agron. J.* **2020**, *112*, 3695–3703. [CrossRef]
29. Rego, C.H.Q.; Franca-Silva, F.; Gomes, F.G.; de Moraes, M.H.D.; de Medeiros, A.D.; da Silva, C.B. Using Multispectral Imaging for Detecting Seed-Borne Fungi in Cowpea. *Agriculture* **2020**, *10*, 361. [CrossRef]
30. Weng, H.; Tian, Y.; Wu, N.; Li, X.; Yang, B.; Huang, Y.; Ye, D.; Wu, R. Development of a Low-Cost Narrow Band Multispectral Imaging System Coupled with Chemometric Analysis for Rapid Detection of Rice False Smut in Rice Seed. *Sensors* **2020**, *20*, 1209. [CrossRef] [PubMed]
31. ElMasry, G.; Mandour, N.; Wagner, M.H.; Demilly, D.; Verdier, J.; Belin, E.; Rousseau, D. Utilization of computer vision and multispectral imaging techniques for classification of cowpea (*Vigna unguiculata*) seeds. *Plant Methods* **2019**, *15*, 24. [CrossRef]
32. França-Silva, F.; Rego, C.H.Q.; Gomes-Junior, F.G.; Moraes, M.H.D.D.; Medeiros, A.D.D.; Silva, C.B.D. Detection of Drechslera avenae (Eidam) Sharif [Helminthosporium avenae (Eidam)] in Black Oat Seeds (Avena strigosa Schreb) Using Multispectral Imaging. *Sensors* **2020**, *20*, 3343. [CrossRef]
33. Jaillais, B.; Roumet, P.; Pinson-Gadais, L.; Bertrand, D. Detection of Fusarium head blight contamination in wheat kernels by multivariate imaging. *Food Control* **2015**, *54*, 250–258. [CrossRef]
34. Sendin, K.; Manley, M.; Williams, P.J. Classification of white maize defects with multispectral imaging. *Food Chem.* **2018**, *243*, 311–318. [CrossRef] [PubMed]
35. Salimi, Z.; Boelt, B. Classification of Processing Damage in Sugar Beet (*Beta vulgaris*) Seeds by Multispectral Image Analysis. *Sensors* **2019**, *19*, 2360. [CrossRef] [PubMed]
36. Li, M.; Huang, M.; Zhu, Q.; Zhang, M.; Guo, Y.; Qin, J. Pickled and dried mustard foreign matter detection using multispectral imaging system based on single shot method. *J. Food Eng.* **2020**, *285*, 110106. [CrossRef]
37. Dissing, B.S.; Nielsen, M.E.; Ersbøll, B.K.; Frosch, S. Multispectral Imaging for Determination of Astaxanthin Concentration in Salmonids. *PLoS ONE* **2011**, *6*, e19032. [CrossRef] [PubMed]
38. Olesen, M.H.; Shetty, N.; Gislum, R.; Boelt, B. Classification of Viable and Non-Viable Spinach (*Spinacia Oleracea* L.) Seeds by Single Seed near Infrared Spectroscopy and Extended Canonical Variates Analysis. *J. Near Infrared Spectrosc.* **2011**, *19*, 171–180. [CrossRef]
39. Hu, X.W.; Yang, L.J.; Zhang, Z.X.; Wang, Y.R. Differentiation of alfalfa and sweet clover seeds via multispectral imaging. *Seed Sci. Technol.* **2020**, *48*, 83–99. [CrossRef]
40. ElMasry, G.; Mandour, N.; Al-Rejaie, S.; Belin, E.; Rousseau, D. Recent Applications of Multispectral Imaging in Seed Phenotyping and Quality Monitoring-An Overview. *Sensors* **2019**, *19*, 1090. [CrossRef]
41. Yang, L.; Zhang, Z.; Hu, X. Cultivar Discrimination of Single Alfalfa (Medicago sativa L.) Seed via Multispectral Imaging Combined with Multivariate Analysis. *Sensors* **2020**, *20*, 6575. [CrossRef]
42. De la Fuente, G.N.; Carstensen, J.M.; Edberg, M.A.; Lubberstedt, T. Discrimination of haploid and diploid maize kernels via multispectral imaging. *Plant Breed.* **2017**, *136*, 50–60. [CrossRef]
43. Li, X.; Fan, X.; Zhao, L.; Huang, S.; He, Y.; Suo, X. Discrimination of Pepper Seed Varieties by Multispectral Imaging Combined with Machine Learning. *Appl. Eng. Agric.* **2020**, *36*, 743–749. [CrossRef]

44. Liu, W.; Liu, C.; Ma, F.; Lu, X.; Yang, J.; Zheng, L. Online Variety Discrimination of Rice Seeds Using Multispectral Imaging and Chemometric Methods. *J. Appl. Spectrosc.* **2016**, *82*, 993–999. [CrossRef]
45. Mbanjo, E.G.N.; Jones, H.; Caguiat, X.G.I.; Carandang, S.; Ignacio, J.C.; Ferrer, M.C.; Boyd, L.A.; Kretzschmar, T. Exploring the genetic diversity within traditional Philippine pigmented Rice. *Rice* **2019**, *12*, 27. [CrossRef] [PubMed]
46. Liu, C.; Liu, W.; Lu, X.; Chen, W.; Chen, F.; Yang, J.; Zheng, L. Non-destructive discrimination of conventional and glyphosate resistant soybean seeds and their hybrid descendants using multispectral imaging and chemometric methods. *J. Agric. Sci.* **2016** *154*, 1–12. [CrossRef]
47. Shrestha, S.; Deleuran, L.C.; Olesen, M.H.; Gislum, R. Use of multispectral imaging in varietal identification of tomato. *Sensors* **2015**, *15*, 4496–4512. [CrossRef] [PubMed]
48. Vresak, M.; Olesen, M.H.; Gislum, R.; Bavec, F.; Ravn Jorgensen, J. The Use of Image-Spectroscopy Technology as a Diagnostic Method for Seed Health Testing and Variety Identification. *PLoS ONE* **2016**, *11*, e0152011. [CrossRef]
49. Ma, F.; Wang, J.; Liu, C.; Lu, X.; Chen, W.; Chen, C.; Yang, J.; Zheng, L. Discrimination of Kernel Quality Characteristics for Sunflower Seeds Based on Multispectral Imaging Approach. *Food Anal. Methods* **2015**, *8*, 1629–1636. [CrossRef]
50. Olesen, M.H.; Nikneshan, P.; Shrestha, S.; Tadayyon, A.; Deleuran, L.C.; Boelt, B.; Gislum, R. Viability prediction of Ricinus cummunis L. seeds using multispectral imaging. *Sensors* **2015**, *15*, 4592–4604. [CrossRef]
51. Bianchini, V.d.J.M.; Mascarin, G.M.; Silva, L.C.A.S.; Arthur, V.; Carstensen, J.M.; Boelt, B.; Barboza da Silva, C. Multispectral and X-ray images for characterization of Jatropha curcas L. seed quality. *Plant Methods* **2021**, *17*, 9. [CrossRef]
52. Shetty, N.; Olesen, M.H.; Gislum, R.; Deleuran, L.C.; Boelt, B. Use of partial least squares discriminant analysis on visible-near infrared multispectral image data to examine germination ability and germ length in spinach seeds. *J. Chemom.* **2012**, *26*, 462–466 [CrossRef]
53. Liu, W.; Xu, X.; Liu, C.; Zheng, L. Rapid Discrimination of High-Quality Watermelon Seeds by Multispectral Imaging Combined with Chemometric Methods. *J. Appl. Spectrosc.* **2019**, *85*, 1044–1049. [CrossRef]
54. Hu, X.; Yang, L.; Zhang, Z. Non-destructive identification of single hard seed via multispectral imaging analysis in six legume species. *Plant Methods* **2020**, *16*, 116. [CrossRef]
55. Boelt, B.; Shrestha, S.; Salimi, Z.; Jorgensen, J.R.; Nicolaisen, M.; Carstensen, J.M. Multispectral imaging—A new tool in seed quality assessment? *Seed Sci. Res.* **2018**, *28*, 222–228. [CrossRef]
56. Otsu, N. A threshold selection method from gray-level histograms. *IEEE Trans. Syst. ManCybern.* **1979**, *9*, 62–66. [CrossRef]
57. Liu, C.; Liu, W.; Lu, X.; Chen, W.; Yang, J.; Zheng, L. Nondestructive determination of transgenic Bacillus thuringiensis rice seeds (*Oryza sativa* L.) using multispectral imaging and chemometric methods. *Food Chem.* **2014**, *153*, 87–93. [CrossRef] [PubMed]
58. Gabr, D.G. Seed morphology and seed coat anatomy of some species of Apocynaceae and Asclepiadaceae. *Ann. Agric. Sci.* **2014**, *59*, 229–238. [CrossRef]
59. IBPGR. *Descriptors for Cowpea*; IBPGR (International Board for Plant Genetic Resources): Rome, Italy, 1983; p. 30.
60. Haixiang, G.; Yijing, L.; Shang, J.; Mingyun, G.; Yuanyue, H.; Bing, G. Learning from class-imbalanced data: Review of methods and applications. *Expert Syst. Appl.* **2017**, *73*, 220–239. [CrossRef]
61. Chomontowski, C.; Podlaski, S. Impact of sugar beet seed priming using the SMP method on the properties of the pericarp. *BMC Plant Biol.* **2020**, *20*, 32. [CrossRef] [PubMed]
62. Salimi, Z.; Boelt, B. Optimization of Germination Inhibitors Elimination from Sugar Beet (*Beta vulgaris* L.) Seeds of Different Maturity Classes. *Agronomy* **2019**, *9*, 763. [CrossRef]
63. Hoad, S.P.; Brennan, M.; Wilson, G.W.; Cochrane, P.M. Hull to caryopsis adhesion and grain skinning in malting barley: Identification of key growth stages in the adhesion process. *J. Cereal Sci.* **2016**, *68*, 8–15. [CrossRef]
64. Olkku, J.; Kotaviita, E.; Salmenkallio-Marttila, M.; Sweins, H.; Home, S. Connection between Structure and Quality of Barley Husk. *J. Am. Soc. Brew. Chem.* **2005**, *63*, 17–22. [CrossRef]
65. Murphy, J.B.; Noland, T.L. Temperature Effects on Seed Imbibition and Leakage Mediated by Viscosity and Membranes. *Plant Physiol.* **1982**, *69*, 428–431. [CrossRef] [PubMed]
66. Parrish, D.J.; Leopold, A.C. Transient Changes During Soybean Imbibition. *Plant Physiol.* **1977**, *59*, 1111–1115. [CrossRef] [PubMed]
67. Hill, H.J.; Taylor, A.G.; Huang, X.L. Seed Viability Determinations in Cabbage Utilizing Sinapine Leakage and Electrical Conductivity Measurements. *J. Exp. Bot.* **1988**, *39*, 1439–1447. [CrossRef]
68. Taylor, A.G.; Paine, D.H.; Paine, C.A. Sinapine Leakage from Brassica Seeds. *J. Am. Soc. Hortic. Sci.* **1993**, *118*, 546–550. [CrossRef]
69. Mathur, S.; Kongsdal, O. *Common Laboratory Seed Health Testing Methods for Detecting Fungi*; International Seed Testing Association: Bassersdof, Switzerland, 2003; 425p.
70. Lievens, B.; Thomma, B.P.H.J. Recent Developments in Pathogen Detection Arrays: Implications for Fungal Plant Pathogens and Use in Practice. *Phytopathology* **2005**, *95*, 1374–1380. [CrossRef]

Article

Using Multispectral Imaging for Detecting Seed-Borne Fungi in Cowpea

Carlos Henrique Queiroz Rego [1,*], Fabiano França-Silva [1], Francisco Guilhien Gomes-Junior [1], Maria Heloisa Duarte de Moraes [2], André Dantas de Medeiros [3] and Clíssia Barboza da Silva [4]

1 Department of Crop Science, College of Agriculture "Luiz de Queiroz", University of São Paulo, Piracicaba 13418-900, SP, Brazil; fabiano.francads@usp.br (F.F.-S.); francisco1@usp.br (F.G.G.-J.)

2 Department of Plant Pathology and Nematology, College of Agriculture "Luiz de Queiroz", University of São Paulo, Piracicaba 13418-900, SP, Brazil; helo.d.moraes@gmail.com

3 Department of Agronomy, Federal University of Viçosa, Viçosa 36570-900, MG, Brazil; andre.d.medeiros@ufv.br

4 Laboratory of Radiobiology and Environment, Center for Nuclear Energy in Agriculture, University of São Paulo, Piracicaba 13416-060, SP, Brazil; clissia_usp@hotmail.com

* Correspondence: carlosqueirozagro@gmail.com

Received: 13 July 2020; Accepted: 8 August 2020; Published: 17 August 2020

Abstract: Recent advances in multispectral imaging-based technology have provided useful information on seed health in order to optimize the quality control process. In this study, we verified the efficiency of multispectral imaging (MSI) combined with statistical models to assess the cowpea seed health and differentiate seeds carrying different fungal species. Seeds were artificially inoculated with *Fusarium pallidoroseum*, *Rhizoctonia solani* and *Aspergillus* sp. Multispectral images were acquired at 19 wavelengths (365 to 970 nm) from inoculated seeds and freeze-killed 'incubated' seeds. Statistical models based on linear discriminant analysis (LDA) were developed using reflectance, color and texture features of the seed images. Results demonstrated that the LDA-based models were efficient in detecting and identifying different species of fungi in cowpea seeds. The model showed above 92% accuracy before incubation and 99% after incubation, indicating that the MSI technique in combination with statistical models can be a useful tool for evaluating the health status of cowpea seeds. Our findings can be a guide for the development of in-depth studies with more cultivars and fungal species, isolated and in association, for the successful application of MSI in the routine health inspection of cowpea seeds and other important legumes.

Keywords: *Vigna unguiculata* (L.) Walp; seed health; spectroscopy

1. Introduction

Cowpea (*Vigna unguiculata* L. Walp) is a leguminous species which is of nutritional and social importance in underdeveloped regions due to the high protein content in its grains [1,2]. For example, most of the production in Brazil comes from family farming, especially in the North and Northeast regions, but it has currently aroused the interest of farmers in the Midwest region who practice commercial agriculture [3].

Despite its adaptability and rusticity, cowpea seeds are susceptible to several fungal diseases. According to Biemond et al. [4], the contamination of cowpea seeds by *Aspergillus flavus*, *Macrophomina phaseolina*, *Fusarium oxysporum* and *Penicillium* sp. contribute to a marked reduction in germination and seed weight, in addition to acting on accelerating deterioration by producing aflatoxins, thus limiting commercialization of its seeds and consumption of grains.

In agricultural industry, the seed health is mainly monitored by detecting fungi species and their percentages present in the sample, which contributes to make decision regarding the suitability of a lot destined for sowing or marketing [5]. The blotter test is the most well-known and used method for detecting seed-borne fungi. However, it is a time-consuming and subjective test since it depends on visual inspections and requires highly trained specialists [6].

Innovative, accurate and rapid light-based methods have been developed to meet the growing demands of the food and agricultural industries, which can produce a consistent assessment of seed health, overcoming the intrinsic subjectivity of conventional techniques [6]. Unaltered samples can be analyzed with non-destructive and real time visualization of the pathological attributes of seeds, with optimization in the quality control process [7,8]. In addition, these tools produce complementary information related to the energy-matter interaction in the context of seed quality.

The multispectral imaging (MSI) technology is based on the use of spectral bandwidths in the ultraviolet, visible and infrared regions in order to obtain spatial and spectral information from the objects under evaluation [9,10], therefore, it can be a useful tool to distinguish healthy seeds from seeds that are carrying important pathogens [11–13]. For instance, there are reports that the use of MSI showed over 80% separation of healthy spinach seeds from those with different fungi species [14].

Considering that each material has intrinsic spectral characteristics that vary according to chemical or physical attributes, this study had two main objectives. Firstly, to evaluate the efficiency of the MSI technique in the evaluation of cowpea seed health. The second objective was to evaluate whether different fungal species can be discriminated using MSI associated with statistical models, before and after seed incubation.

2. Materials and Methods

2.1. Seed Samples and Fungi Inoculation

Cowpea seeds from BRS Tucumanque cultivar were used in this study. Seeds were inoculated with three fungi species isolates (*Fusarium pallidoroseum*, *Rhizoctonia solani* and *Aspergillus* sp.). Each species of fungus was grown in three 9-cm Petri dishes containing potato dextrose agar (PDA) medium and kept in a growth chamber with a temperature adjusted to 20 ± 2 °C with a 12-h photoperiod of white fluorescent light for a period of 15 days.

The seeds were disinfected for inoculation in sodium hypochlorite solution (1% concentration for 3 min), washed in distilled water and then dried on paper towels at room temperature for 24 h. After drying, 100 seeds per plate were added in order to be in contact with the fungus colony, kept in a growth chamber under the conditions described above for 24 h. After the contact period, the seeds were removed from the plates and placed in a single layer on paper towels at room temperature for 24 h to dry. Afterwards, seeds were divided into two groups for the image acquisition; the first group was called 'Inoculated seeds' (dry seeds), for which 30 seeds were distributed into three Petri dishes (10 seeds per plate), fixed with double tape facing the bottom, positioned one by one in a single layer and equidistant from each other. The second group was called 'Incubated seeds', and a deep-freezing blotter method was used to kill the seeds. Three subsamples of ten seeds were placed in three Petri dishes containing three filter paper sheets moistened with 3.5 mL of distilled water, kept at 20 ± 2 °C for 24 h. After this period, the plates were transferred to a freezer at −20 °C for 24 h and, subsequently, incubated at 20 ± 2 °C with a photoperiod of 12 h with fluorescent lamps, for 4 days; seeds were positioned equidistantly from each other in a single layer.

2.2. Multispectral Imaging Application

The Petri dishes were positioned under the sphere of integration of the VideometerLab4® instrument (Videometer A/S, Herlev, Denmark) and, after successive illumination of the samples at 19 contiguous light emitting diodes (LEDs), a monochrome charge-coupled chip (CCD) recorded the reflectance of the seeds and generated 19 images (2192 × 2192 pixels) corresponding to the 19 wavelengths (365, 405, 430, 450, 470, 490, 515, 540, 570, 590, 630, 645, 660, 690, 780, 850, 880, 940, 970 nm) of the electromagnetic spectrum.

Data analysis were performed with VideometerLab4 software version 3.14.9 (Videometer A/S, Herlev, Denmark). The multispectral images were transformed using normalized canonical discriminant analysis (nCDA) to minimize the distance within classes and to maximize the distance among classes. Each seed was identified as a region of interest (ROI), and it was built a mask to segment the seeds from the background, which was based on an nCDA transformation of seeds and Petri dish and a simple threshold. The seeds were collected in a blob database, and 36 variables were extracted from the individual seeds, including tristimulus components of color as hue (angular specification for color perceived as red, yellow, blue or green) and saturation (degree of difference between the color and neutral gray).

MultiColorMean feature extracts the reflectance mean of each seed for the 19 spectral bands (from 365 to 970 nm). To eliminate the influence of outliers at both the high and low ends, a trimmed mean excludes 10% of the lowest and highest values before calculating the mean. RegionMSI_Mean calculates a trimmed mean of transformed pixel values within the blob (each single seed), and RegionMSIthresh measures the percentage of blob region with transformation value higher than threshold, based on the nCDA model (derived from all the classes).

A gray level run length matrix (GLRLM), was generated to identify and distinguish texture patterns. GraylevelRunStatistics feature captures the coarseness of a texture in specified directions according to algorithm described by Galloway [15] and Albregtsen and Nielsen [16]: (0) = Short Run Emphasis (SRE) measures the distribution of short runs, and higher values indicate fine textures; (1) = Long Run Emphasis (LRE) measures the distribution of long runs, and higher values indicate coarse textures; (2) = Gray Level Non-Uniformity (GLN) measures the similarity of gray level values in the image, and GLN values are lower if gray level values are similar throughout the image; (3)= Run Length Non-Uniformity (RLN) expresses the similarity of run lengths throughout the image, with lower values if the run lengths are the same throughout the image; (4) = Run Percentage (RP) determines the distribution and homogeneity of runs in an image in a particular direction. The texture features described by Chu, et al. [17] were also measured: (5) = Low Grey Level Run Emphasis (LGRE) and (6) = High Grey Level Run Emphasis (HGRE). Short run emphasis measures the short run distribution and it is large for fine textures. Long run emphasis calculates the long run distribution and it is large for coarse structural textures.

The CIE color spaces were measured for the axes of lightness (L^*) and chromaticities (a^* and b^*), where CIELab L^* represents lightness from black to white, CIELab a* the color appearance from green to red and CIELab b^* the color appearance from blue to yellow. The CIELab system is a simplified mathematical approximation to a uniform color space composed of perceived color differences [18]. It was defined by the International Commission on Illumination (CIE), and comprises all perceivable colors of the spectrum, even outside the human vision gamut [19]. An intensity-hue-saturation transformation was applied to map the standardized RGB (sRGB) image into intensity, which is independent of color hue that is the dominant wavelength, and saturation which is the colorfulness or the prominence of the dominant color.

2.3. Unsupervised Analysis

The data obtained from multispectral images were exported to Excel and subsequently subjected to unsupervised multivariate analysis. Multivariate principal component analysis (PCA) was used in this study as an exploratory technique to identify hidden patterns in the data obtained from the MSI analysis. The data obtained for each seed were normalized, and the eigenvalues and eigenvectors were calculated from the covariance matrices. The results were plotted on two-dimensional graphs using the R 4.0.0 software program [20].

2.4. Supervised Discriminant Analysis

Two models were developed based on the Linear Discriminant Analysis (LDA) algorithm to classify different fungal species associated with cowpea seeds. The first model was developed based on the MSI information obtained from the inoculated seeds, while the second model used data from the incubated seeds. The classes used in both models were: Class (1) Control—seeds without fungal infestation; Class (2) *Aspergillus*—Seeds infested with *Aspergillus* sp. fungus; Class (3) *F. pallidoroseum*—Seeds infested with *Fusarium pallidoroseum* fungus; Class (4) *R. solani*—Seeds infested with *Rhizoctonia solani* fungus. The data was partitioned so that 70% was used to train the models and 30% was used for independent validation. In addition, a 10-fold cross-validation was applied. The metrics of general accuracy, Cohen's Kappa coefficient, sensitivity and specificity were used to evaluate the performance of the models. The R 4.0.0 software program (R core Team, 2020) was used to develop the models with the LDA algorithm.

3. Results

The reflectance patterns in classes of healthy seeds, before incubation, and after incubation with *Fusarium pallidoroseum*, *Rhizoctonia solani* and *Aspergillus* sp. were different in images captured at 780 nm (Figure 1). The nCDA method revealed a slight distinction among classes before incubation (Figure 1a) compared to seeds after incubation (Figure 1b): the intense colonization of the fungi after incubation showed greater separation between healthy and unhealthy seeds and also among the different fungal species.

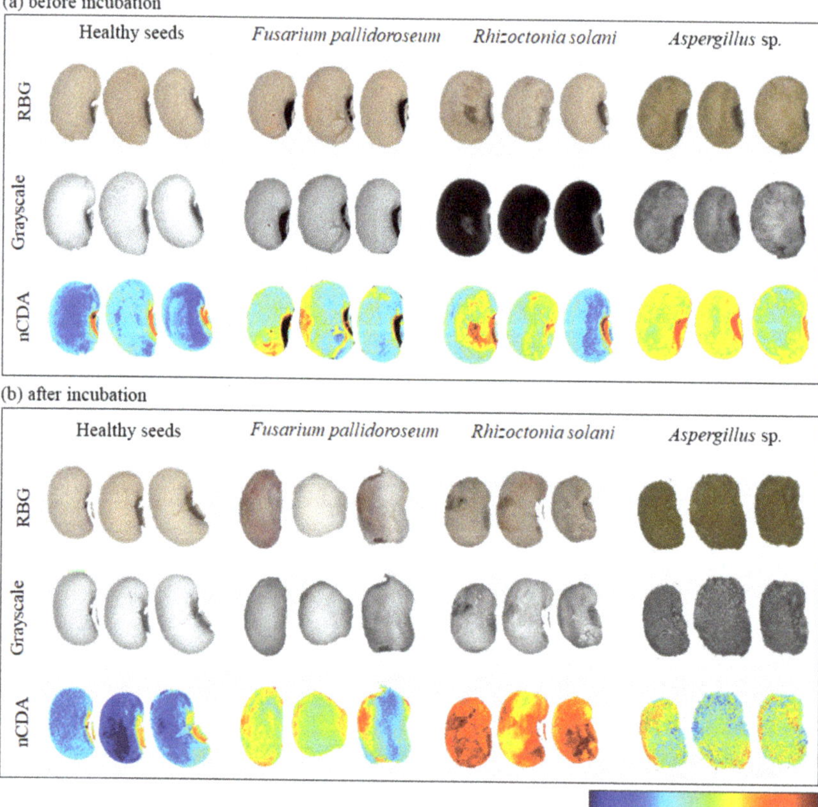

Figure 1. Raw RGB images of cowpea seeds and corresponding transformed images into grayscale and by canonical discriminant analysis (nCDA) captured at 780 nm, with reflectance patters in classes of healthy seeds, *Fusarium pallidoroseum*, *Rhizoctonia solani* and *Aspergillus* sp. before incubation (**a**), and after incubation (**b**).

Figure 2 shows the mean reflectance spectra at 19 wavelengths in a range from 365 to 970 nm. Before incubation (Figure 2a) all classes showed similar spectral signature with the exception of the '*Aspergillus*' class. However, there was an expressive discrimination among classes after incubation (Figure 2b), especially at wavelengths from 365 to 645 nm, and the '*Aspergillus*' class showed a higher distinction from the other classes across the spectrum. At longer wavelengths, there was a difficulty in distinguishing '*F. pallidoroseum*' from healthy seeds, particular in the NIR region.

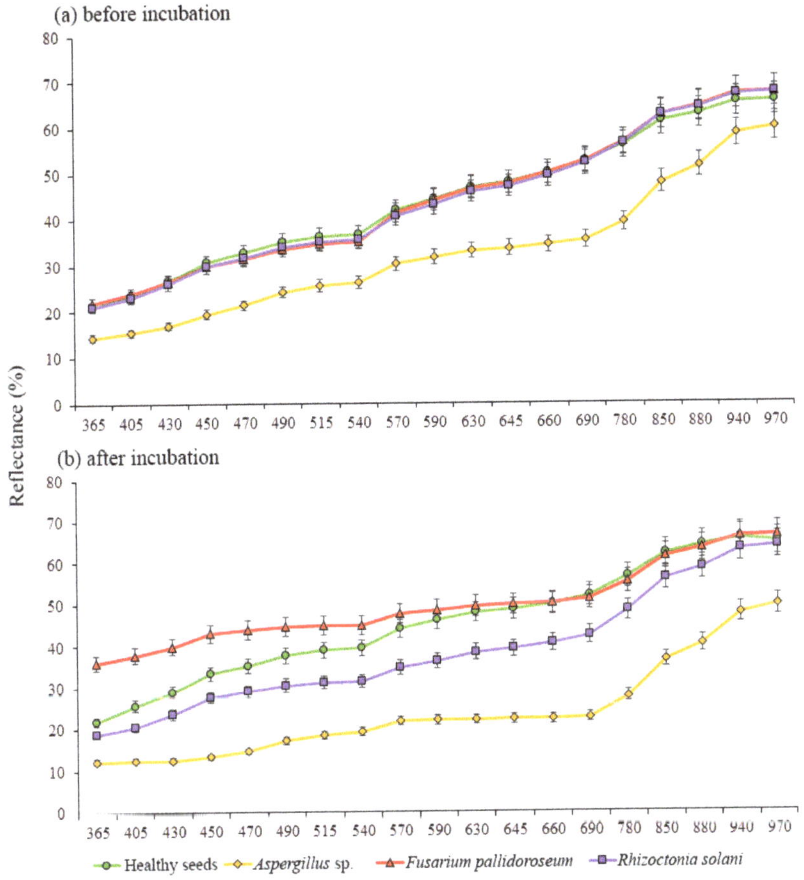

Figure 2. Spectral signature for classes of healthy seeds, *Fusarium pallidoroseum*, *Rhizoctonia solani* and *Aspergillus* sp. at 19 wavelengths in a range from 365 to 970 nm before incubation (**a**) and after incubation (**b**).

The reflectance data of the 19 spectrum bands and the color and texture resources were submitted to PCA analysis (Figure 3). Before incubation, components 1 (PC1) and 2 (PC2) were responsible for 73.4% and 11.3% of the total variation, respectively (Figure 3a). The contribution of components after incubation was 77.8% in PC1 and 6.8% in PC2 (Figure 3b). In this context, there was similar behavior of the vectors originating from the spectra reflectance (represented in green), indicating that the 'healthy seeds', '*F. pallidoroseum* 'and '*R. solani*' classes had higher reflectance values compared to '*Aspergillus*'. Meanwhile, the '*Aspergillus*' class showed higher values mainly for CIELab, before and after incubation. The classes of 'healthy seeds', '*F. pallidoroseum* 'and '*R. solani*' showed strong interaction before incubation (Figure 3a), but there was less interaction among them after seed incubation (Figure 3b).

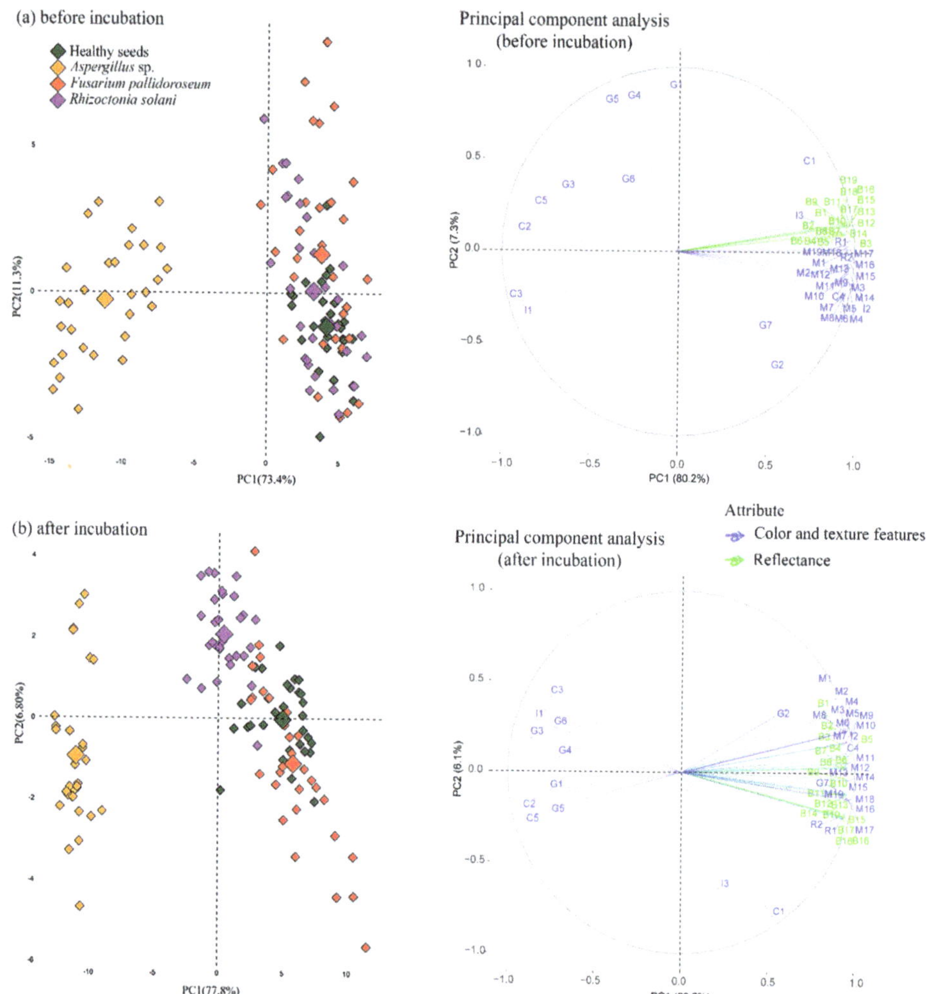

Figure 3. Biplots of principal component analysis for multispectral reflectance, color and texture features for classes of healthy seeds, *Fusarium pallidoroseum*, *Rhizoctonia solani* and *Aspergillus* sp. at 19 wavelengths (365 to 970 nm) before incubation (**a**) and after incubation (**b**). Attributes: Color and texture features – (R2): RegionMSIThresh; (R1): RegionMSI_Mean; (M19): MultiColorMean1_(18); (M18): MultiColorMean1_(17); (M17): MultiColorMean1_(16); (M16): MultiColorMean1_(15); (M15): MultiColorMean1_(14); (M14): MultiColorMean1_(13); (M13): MultiColorMean1_(12); (M12): MultiColorMean1_(11); (M11): MultiColorMean1_(10); (M10): MultiColorMean1_(9); (M9): MultiColorMean1_(8); (M8): MultiColorMean1_(7); (M7): MultiColorMean1_(6); (M6): MultiColorMean1_(5); (M5): MultiColorMean1_(4); (M4): MultiColorMean1_(3); (M3): MultiColorMean1_(2); (M2): MultiColorMean1_(1); (M1): MultiColorMean1_(0); (I3): IHSSaturationMean; (I2): IHSIntensityMean;(I1): IHSHueMean; (G7): GraylevelRunStatistics_(6); (G6): GraylevelRunStatistics_(5); (G5): GraylevelRunStatistics_(4); (G4): GraylevelRunStatistics_(3); (G3): GraylevelRunStatistics_(2); (G2): GraylevelRunStatistics_(1); (G1): GraylevelRunStatistics_(0); (C5): CIELab_Saturation; (C4): CIELab_L; (C3): CIELab_Hue; (C2): CIELab_B; (C1): CIELab_A. Reflectance – (B19): Band_19; (B18): Band_18; (B17): Band_17; (B16): Band_16; (B15): Band_15; (B14): Band_14; (B13): Band_13; (B12): Band_12; (B11): Band_11; (B10): Band_10; (B9): Band_9; (B8): Band_8; (B7): Band_7; (B6): Band_6; (B5): Band_5; (B4): Band_4; (B3): Band_3; (B2): Band_2; (B1): Band_1.

Next, models were developed based on the LDA algorithm using reflectance, color and texture features of the seeds. An overall accuracy of 100% and 92% was observed for the training and test set, respectively, in the first model developed before incubation (Table 1). In testing set for class membership of 'healthy seeds' and '*Aspergillus*', the hit rate was achieved with 100% sensitivity, while other classes showed less individual precision. There was confusion between '*F. pallidoroseum*' and '*R. solani*' (Table 1), since the spectral patters of these classes were very similar (Figure 2).

Table 1. Confusion matrices of the LDA model in training and testing set using reflectance, color and texture features of cowpea seeds at 19 wavelengths (365 to 970 nm) for class membership of healthy seed and inoculated seed with *Fusarium pallidoroseum*, *Rhizoctonia solani* and *Aspergillus* sp.

Before Incubation				
Treatment	Training set (n = 84)			
	Healthy seeds	*Aspergillus* sp.	*F. pallidoroseum*	*R. solani*
Healthy seed	21	0	0	0
Aspergillus sp.	0	21	0	0
F. pallidoroseum	0	0	21	0
R. solani	0	0	0	21
Overall acurracy	1.00			
Cohen's Kappa	1.00			
Sensitivity	1.00	1.00	1.00	1.00
Specificity	1.00	1.00	1.00	1.00
	Cross validation (fold = 10)			
Overall acurracy	0.89 ± 0.01			
Cohen's Kappa	0.85 ± 0.14			
Sensitivity	0.97 ± 0.08			
Specificity	1.00 ± 0.00			
Treatment	Testing set (n = 36)			
	Healthy seeds	*Aspergillus* sp.	*F. pallidoroseum*	*R. solani*
Healthy seed	9	0	1	0
Aspergillus sp.	0	9	0	0
F. pallidoroseum	0	0	7	1
R. solani	0	0	1	8
Overall acurracy	0.92			
Cohen's Kappa	0.89			
Sensitivity	1.00	1.00	0.78	0.89
Specificity	0.96	1.00	0.96	0.96

The second model was created using multispectral data after seed incubation, with an overall accuracy of 100% for both training and testing set (Table 2). The metrics also showed high accuracy of the classification model, with values equal to or greater than 97% in cross-validation, pointing out that the multispectral data can be used to distinguish healthy seeds from seeds carrying different fungal species.

Table 2. Confusion matrices of the LDA model in training and testing set using reflectance, color and texture features of cowpea seeds at 19 wavelengths (365 to 970 nm) for class membership of healthy seed and incubated seed with *Fusarium pallidoroseum*, *Rhizoctonia solani* and *Aspergillus* sp.

After Incubation				
Treatment	Training set (n = 84)			
	Healthy seeds	*Aspergillus* sp.	*F. pallidoroseum*	*R. solani*
Healthy seed	21	0	0	0
Aspergillus sp.	0	21	0	0
F. pallidoroseum	0	0	21	0
R. solani	0	0	0	21
Overall acurracy	1.00			
Cohen's Kappa	1.00			
Sensitivity	1.00	1.00	1.00	1.00
Specificity	1.00	1.00	1.00	1.00
	Cross validation (fold = 10)			
Acurracy	0.99 ± 0.04			
Cohen's Kappa	0.98 ± 0.07			
Sensitivity	0.97 ± 0.08			
Specificity	1.00 ± 0.00			
Treatment	Testing set (n = 36)			
	Healthy seeds	*Aspergillus* sp.	*F. pallidoroseum*	*R. solani*
Healthy seed	9	0	0	0
Aspergillus sp.	0	9	0	0
F. pallidoroseum	0	0	9	0
R. solani	0	0	0	9
Overall acurracy	1.00			
Cohen's Kappa	1.00			
Sensitivity	1.00	1.00	1.00	1.00
Specificity	1.00	1.00	1.00	1.00

The two models developed based on LDA algorithm are shown in Figure 4. The first two discriminatory factors (LD1 and LD2) explained 99.47% of the total variation in the first model, and 94.87% in the second model (Figure 4a,b). Again, the '*Aspergillus*'class was clearly distinguished from the other classes in both statistical models, and there was high interaction between classes of '*F. pallidoroseum*' and *R. solani*', before incubation (Figure 4a). Figure 4c,d show the importance of each variable obtained from reflectance, color and texture features for discrimination of the different seed classes. Before incubation, 'CIELabHue' (0.99), 'IHSHueMean' (0.98), CIELab A (0.98), Band 19 (0.98) and Band 18 (0.98) were more effective in discriminating seed classes, and after incubation the 'CIELab B', IHSSaturationMean, MultiColorMean1 [14], GraylevelRunStatistics (2)(3)(4), and Bands 8–17 (Figure 4d), with contribution greater than 0.99. These results emphasize the potential of simple features in discriminating different fungi associated with cowpea seeds.

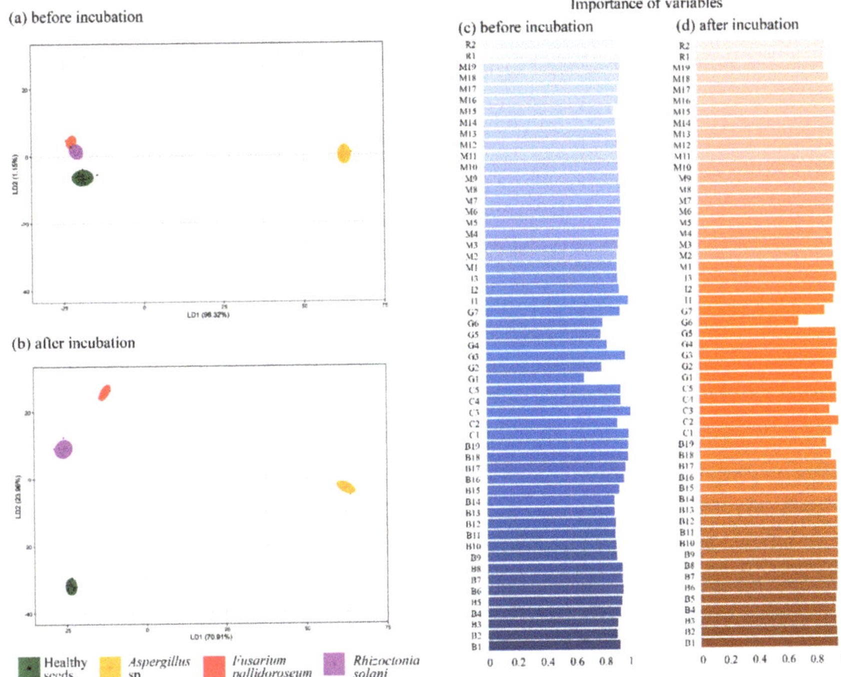

Figure 4. Linear discriminant analysis (LDA) score plot based on reflectance, color and texture features of cowpea seeds at 19 wavelengths (365 to 970 nm) for class membership of inoculated seed (**a**), and incubated seed (**b**) with Healthy seeds, *Fusarium pallidoroseum*, *Rhizoctonia solani* and *Aspergillus* sp. Importance of variables in the models before inoculation (**c**), and after incubation (**d**). Attributes: Color and texture features – (R2): RegionMSIThresh; (R1): RegionMSI_Mean; (M19): MultiColorMean1_(18); (M18): MultiColorMean1_(17); (M17): MultiColorMean1_(16); (M16): MultiColorMean1_(15); (M15): MultiColorMean1_(14); (M14): MultiColorMean1_(13); (M13): MultiColorMean1_(12); (M12): MultiColorMean1_(11); (M11): MultiColorMean1_(10); (M10): MultiColorMean1_(9); (M9): MultiColorMean1_(8); (M8): MultiColorMean1_(7); (M7): MultiColorMean1_(6); (M6): MultiColorMean1_(5); (M5): MultiColorMean1_(4); (M4): MultiColorMean1_(3); (M3): MultiColorMean1_(2); (M2): MultiColorMean1_(1); (M1): MultiColorMean1_(0); (I3): IHSSaturationMean; (I2): IHSIntensityMean; (I1): IHSHueMean; (G7): GraylevelRunStatistics_(6); (G6): GraylevelRunStatistics_(5); (G5): GraylevelRunStatistics_(4); (G4): GraylevelRunStatistics_(3); (G3): GraylevelRunStatistics_(2); (G2): GraylevelRunStatistics_(1); (G1): GraylevelRunStatistics_(0); (C5): CIELab_Saturation; (C4): CIELab_L; (C3): CIELab_Hue; (C2): CIELab_B; (C1): CIELab_A. Reflectance – (B19): Band_19; (B18): Band_18; (B17): Band_17; (B16): Band_16; (B15): Band_15; (B14): Band_14; (B13): Band_13; (B12): Band_12; (B11): Band_11; (B10): Band_10; (B9): Band_9; (B8): Band_8; (B7): Band_7; (B6): Band_6; (B5): Band_5; (B4): Band_4; (B3): Band_3; (B2): Band_2; (B1): Band_1.

4. Discussion

The diagnosis of pathogens transmitted by seeds is an important measure in the quality control program, as it avoids the spread of pathogens to exempt areas, economic losses and the unnecessary use of chemicals, thus reducing costs and environmental contamination. Traditional techniques have the characteristic of requiring considerable time for analysis, in addition to subjectivity for interpreting the test. Thus, the use of techniques which minimize this problem is very desirable; in this sense, technological and computational advances enable new methodologies to be used for this purpose.

This study sought to verify the efficiency of MSI in recognizing different fungal species associated with cowpea seeds; it was possible to observe distinctions in the spectral signature between the different seed classes. Variations in the reflectance spectra can be attributed to changes in color, texture and chemical composition of the surface, thereby enabling separation between the classes of infested and non-infested seeds as evidenced by exploratory data analysis. The differentiation between classes before incubation can be attributed due to the change in color caused by the fungi which were adhered to the seed coat; the '*F. pallidoroseum*' and '*R. solani*' have a simple mycelia formation in their colonies, which could have resulted in less seed covering by the fungi, whereas there is intense spore production by '*Aspergillus*', covering the seeds completely. Therefore, it seems reasonable to assume that the conditions before incubation were not favorable for complete fungus development, only resulting in changes in the seed color.

In addition to the coloration, there are the changes caused by the enzymatic and oxidative activity of the fungi on the seeds from the incubation ('Incubated seeds'), which enabled distinguishing the classes more sharply. According to Williams et al. [21], the main source of variation in chemical alteration is an alteration to the starch and protein content which constitute the seed reserves; after incubating the seeds, the fungus starts to consume these compounds, directing them for their growth. The ability of the MSI technique to distinguish fungi-bearing seeds based on physical-chemical changes has already been proven in studies with other species [12–14,21].

Spectral data made it possible to separate the seed classes in the 'Incubated seeds', however this distinction did not always occur in the same region, making the selection of a spectral band common to the class complex. This can be attributed to the overlap and complexity of continuous data, making it difficult to clearly identify the positions of the characteristic bands that represent the different components to be evaluated [22]. In this context, the color and texture parameters were quite expressive in distinguishing the classes, as seen in the determination coefficients (Figure 4). In this context, Boelt et al. [6] point out that the CIELab resource is efficient in distinguishing between fungi species present in barley seeds; this feature is an interesting alternative, as it enables distinguishing color variations which are not perceptible to the human eye [23], eliminating subjectivity from visual inspections. Another color and texture resource, the 'RegionMSImean', has also proved to be efficient in classifying seeds and has already been applied efficiently in several studies. Olesen et al. [10] used this parameter to distinguish *Ricinus cummunis* L. seeds based on their viability with 92% precision. Likewise, Shrestha et al. [24] concluded that the 'RegionMSImean' parameter was efficient in predicting tomato varieties, with a hit rate above 95% and in some cases reaching 100%.

The contribution of color and texture features in distinguishing the seed classes was evidenced by applying the supervised model. The application of supervised methods such as LDA combined with imaging techniques has already shown promise in several studies [9,23–25]; this is because LDA aims to minimize the distance within classes and maximize the distance between classes, thereby enabling good discrimination between classes.

Despite the satisfactory results, it is important to highlight that this is a preliminary study, which can be a guide for future research, covering a greater number of cultivars and species of fungi, isolated or together with the seeds, bringing results that allow a greater application practical and viable in the evaluation of cowpea seeds and other leguminous species of agricultural importance.

5. Conclusions

The multispectral imaging of cowpea seeds provides the necessary information for quickly distinguishing between different seed classes tested and present accuracy above 92% before incubation and 99% after incubation if associated with a discriminant model; these are promising results, since the amount of data obtained through multispectral imaging is large, and therefore a model capable of selecting the variables which most correlate with a given characteristic, in this case the health status, greatly increases the system's effectiveness, confirming the potential of using technology to assess the seed health.

Author Contributions: Conceptualization, C.H.Q.R. and F.G.G.-J.; Funding acquisition, C.B.d.S.; Investigation, C.H.Q.R., F.G.G.-J. and M.H.D.d.M.; Methodology, C.H.Q.R. and M.H.D.d.M.; Resources, C.H.Q.R.; Software, C.H.Q.R., F.F.-S. and A.D.d.M.; Supervision, F.G.G.-J. and M.H.D.d.M.; Validation, A.D.d.M.; Visualization, C.H.Q.R. and C.B.d.S.; Writing—Original draft, C.H.Q.R.; Writing—Review & editing, C.H.Q.R., F.F.-S., F.G.G.-J., M.H.D.d.M., A.D.d.M. and C.B.d.S. All authors have read and agreed to the published version of the manuscript.

Funding: This research was funded by Fundação de Amparo à Pesquisa do Estado de São Paulo - FAPESP: Grants 2017/15220-7 and 2018/03802-4.

Conflicts of Interest: The authors declare no conflict of interest.

References

1. Akande, S.R. Genotype by environment interaction for cowpea seed yield and disease reactions in the forest and derived savanna agro-ecologies of south-west Nigeria. *Am. Eurasian J. Agric. Environ. Sci.* **2007**, *2*, 163–168. Available online: https://www.idosi.org/aejaes/jaes2(2)/11.pdf (accessed on 3 April 2020).
2. Matos-Filho, C.H.A.; Gomes, R.L.F.; Rocha, M.M.; Freire-Filho, F.R.; Lopes, Â.C.A. Potencial produtivo de progênies de feijão-caupi com arquitetura ereta de planta. *Ciênc. Rural* **2009**, *39*, 348–354. [CrossRef]
3. Freire-Filho, F.R. *Feijão-Caupi no Brasil: Produção, Melhoramento Genético, Avanços e Desafios*; Embrapa Meio-Norte: Teresina, PI, Brazil, 2011. Available online: https://ainfo.cnptia.embrapa.br/digital/bitstream/item/84470/1/feijao-caupi.pdf (accessed on 6 April 2020).
4. Biemond, P.C.; Oguntade, O.; Lava Kumar, P.; Stomph, T.J.; Termorshuizen, A.J.; Struik, P.C. Does the informal seed system threaten cowpea seed health? *Crop Prot.* **2013**, *43*, 166–174. [CrossRef]
5. Brasil, Ministério da Agricultura, Pecuária e Abastecimento. *Regras Para Análise de Sementes*; Brasil, Ministério da Agricultura, Pecuária e Abastecimento: Brasilia, Brazil, 2009.
6. Boelt, B.; Shrestha, S.; Salimi, Z.; Ravn Jørgensen, J.; Nicolaisen, M.; Jens Michael Carstensen, E. Multispectral imaging–a new tool in seed quality assessment? *Seed Sci. Res.* **2018**, *28*, 222–228. [CrossRef]
7. Marcos-Filho, J. *Fisiologia de Sementes de Plantas Cultivadas*; Abrates: Londrina, PR, Brazil, 2015.
8. Mondo, V.H.V.; Cicero, S.M. Análise de imagens na avaliação da qualidade de sementes de milho localizadas em diferentes posições na espiga. *Rev. Bras. Sementes* **2005**, *27*, 9–18. [CrossRef]
9. Huang, M.; Wang, Q.G.; Zhu, Q.B.; Qin, J.W.; Huang, G. Review of seed quality and safety tests using optical sensing technologies. *Seed Sci. Technol.* **2015**, *43*, 337–366. [CrossRef]
10. Olesen, M.H.; Nikneshan, P.; Shrestha, S.; Tadayyon, A.; Deleuran, L.C.; Boelt, B.; Gislum, R. Viability prediction of *Ricinus cummunis* L. seeds using multispectral imaging. *Sensors* **2015**, *15*, 4592–4604. [CrossRef]
11. Bodevin, S.; Larsen, T.G.; Lok, F.; Carstensen, J.M.; Jørgensen, K.; Skadhauge, B. A rapid non-destructive method for quantification of fungal infection on barley and malt. In Proceedings of the 32nd EBC Congress, Hamburg, Germany, 10–14 May 2009.
12. Jaillais, B.; Roumet, P.; Pinson-Gadais, L.; Bertrand, D. Detection of Fusarium head blight contamination in wheat kernels by multivariate imaging. *Food Control* **2015**, *54*, 250–258. [CrossRef]
13. Vrešak, M.; Olesen, M.H.; Gislum, R.; Bavec, F.; Jørgensen, J.R. The use of image-spectroscopy technology as a diagnostic method for seed health testing and variety identification. *PLoS ONE* **2016**, *11*, e0152011. [CrossRef]
14. Olesen, M.H.; Carstensen, J.M.; Boelt, B. Multispectral imaging as a potential tool for seed health testing of spinach (*Spinacia oleracea* L.). *Seed Sci. Technol.* **2011**, *39*, 140–150. [CrossRef]
15. Galloway, M.M. Texture analysis using gray level run lengths. *Comput. Graph. Image Process.* **1975**, *4*, 172–179. [CrossRef]
16. Albregtsen, F.; Nielsen, B. Texture classification based on cooccurrence of gray level run length matrices. *Aust. J. Intell. Inf. Process. Syst.* **2000**, *6*, 38–45. Available online: http://ajiips.com.au/papers/V6.1/V6N1.6%20%20Texture%20Classification%20based%20on%20Cooccurrence%20of%20Gray%20Level%20Run%20Length%20Matrices.pdf (accessed on 12 June 2020).
17. Chu, A.; Sehgal, C.M.; Greenleaf, J.F. Use of gray value distribution of run lengths for texture analysis. *Pattern Recognit. Lett.* **1990**, *11*, 415–419. [CrossRef]
18. Hill, B.; Roger, T.H.; Vorhhagen, F.W. Comparative analysis of the quantization of color spaces on the basis of the CIELAB color-difference formula. *ACM Trans. Graph.* **1997**, *16*, 109–154. [CrossRef]

19. Martín, F.; Miró, J.V.; Moreno, L. Towards exploiting the advantages of colour in scan matching. In Proceedings of the ROBOT2013: First Iberian Robotics Conference, Madrid, Spain, 28–29 November 2013; pp. 217–231. [CrossRef]
20. R Core Team. R: A Language and Environment for Statistical Computing. R Foundation for Statistical Computing, Vienna, Austria. 2019. Available online: http://www.r-project.org/index.html (accessed on 25 May 2020).
21. Williams, P.J.; Geladi, P.; Britz, T.J.; Manley, M. Investigation of fungal development in maize kernels using NIR hyperspectral imaging and multivariate data analysis. *J. Cereal Sci.* **2012**, *55*, 272–278. [CrossRef]
22. Su, W.H.; Sun, D.W. Multispectral imaging for plant food quality analysis and visualization. *Compr. Rev. Food Sci. Food Saf.* **2018**, *17*, 220–239. [CrossRef]
23. Medeiros, A.D.; Pinheiro, D.T.; Xavier, W.A.; Silva, L.J.; Dias, D.C.F.S. Quality classification of Jatropha curcas seeds using radiographic images and machine learning. *Ind. Crops Prod.* **2020**, *146*, 112162. [CrossRef]
24. Shrestha, S.; Deleuran, L.C.; Olesen, M.H.; Gislum, R. Use of multispectral imaging in varietal identification of tomato. *Sensors* **2015**, *15*, 4496–4512. [CrossRef] [PubMed]
25. França-Silva, F.; Rego, C.H.Q.; Gomes-Junior, F.G.; Moraes, M.H.D.; Medeiros, A.D.; Silva, C.B. Detection of Drechslera avenae (Eidam) Sharif [*Helminthosporium avenae* (Eidam)] in Black Oat Seeds (*Avena strigosa* Schreb) Using Multispectral Imaging. *Sensors* **2020**, *20*, 3343. [CrossRef]

 agriculture

MDPI

Article

Relationships of *Brassica* Seed Physical Characteristics with Germination Performance and Plant Blindness

Pedro Bello and **Kent J. Bradford** *

Seed Biotechnology Center, Department of Plant Sciences, University of California, Davis, CA 95616, USA; pbello@ucdavis.edu
* Correspondence: kjbradford@ucdavis.edu

Abstract: *Brassica oleracea* is an important crop species that at early growth stages may exhibit failure of the apical growing point, an abnormality called "blindness". The occurrence of blindness is promoted by exposure to low temperatures during imbibition and germination, but the causes of sensitivity to such conditions are unknown. We combined three analytical seed technology instruments to explore seed physical properties that are highly correlated with quality parameters and might be used directly for grading or sorting seed lots into subpopulations varying in potential susceptibility to blindness. For image analysis, we used the VideometerLab instrument, which can scan 19 wavelengths from ultraviolet to infrared and utilize that information in any combination to potentially identify unique criteria related to seed quality. The iXeed CF Analyzer was utilized to obtain chlorophyll fluorescence values for individual seeds. Chlorophyll contents of many seeds can be used as an indicator of seed maturity, a major contributor to seed quality. Finally, oxygen consumption measurements of individual seeds as obtained with the Q2 instrument are highly correlated with their performance under a wide variety of conditions. Six Brassica seed lots differed in their susceptibility to induction of blindness or loss of viability due to 48 h hydrated incubation at 1.5 °C. Analysis of physical and respiratory parameters identified some measurements that were highly correlated with the occurrence of blindness. Higher chlorophyll content, as detected by the CF-Mobile and certain wavelengths in the Videometer, was associated with greater occurrence of blindness or death following the induction treatment, suggesting that more immature seeds may be susceptible to blindness. Further research is required, but methods to detect and sort such seeds based on physical characteristics appear to be feasible.

Keywords: *Brassica oleracea*; blindness; multispectral; chlorophyll content; seed respiration; seed vigor

Citation: Bello, P.; Bradford, K.J. Relationships of *Brassica* Seed Physical Characteristics with Germination Performance and Plant Blindness. *Agriculture* **2021**, *11*, 220. http://doi.org/10.3390/agriculture11030220

Academic Editor: Alan G. Taylor

Received: 26 January 2021
Accepted: 2 March 2021
Published: 8 March 2021

1. Introduction

Brassica oleracea is a morphologically diverse species that has been selected and bred for its leaves (cabbage, collards and kale), stems (kohlrabi), flower shoots (broccoli and cauliflower) and buds (Brussels sprouts). During early seedling growth, plants of all of these crops may lose the apical growing point, an abnormality called "blindness", which usually occurs at low incidence but can cause major losses in the field for growers under some conditions. The occurrence of blind *B. oleracea* plants was described already in the 1940s. It is characterized by termination of leaf primordia initiation and disorganization in the shoot apical meristem (SAM) [1]. The occurrence of blindness is promoted by low temperature combined with low light conditions, and seed production conditions can play a role in the seed lot sensitivity as well [2]. Recent studies have confirmed that both genetics and seed production environment contribute to the occurrence of blindness [1]. Seed treatments that reduce susceptibility to blindness also have been developed [3]. Early identification of affected plants before transplanting them into the field has not been possible, resulting in high economic losses that can be up to 95% in broccoli under some conditions [2].

Seed production for these crops can be complex due to their indeterminate flowering habit [4]. At a given time during production, these indeterminate species will have immature, mature, over-mature and shattering seeds present simultaneously. Early seed harvest can result in poor seed quality and low germination due to immaturity [5], while delayed harvest may sacrifice up to 50% of seed yield under adverse conditions [6]. In addition to losses due to shattering, a significant fraction of the seed lot is discarded during seed processing due to the removal of smaller, immature seeds. This variability in maturity levels can impact seed lot quality, as immature seeds will lose vigor and viability at a faster rate than mature seeds [5]. These problems can prevent sale of seed lots that do not reach minimum germination levels, with economic losses from discarded lots.

Here we combine three analytical seed technology instruments to explore seed physical and physiological properties that could be highly correlated with quality parameters and potentially used for grading or sorting seed lots to remove lower quality subpopulations. For image analysis, we used the VideometerLab instrument, which can scan 19 wavelengths from ultraviolet to infrared and utilize that information in any combination to measure seed size, detect microbes, classify damage, and potentially identify unique criteria for assessing seed quality [7–11]. The iXeed CF Analyzer was utilized to obtain chlorophyll fluorescence values for individual seeds. Seed chlorophyll contents of many species can be used as an indicator of seed maturity, a major contributor to seed quality [5,12–16]. Finally, oxygen consumption measurements of individual seeds as obtained with the Q2 instrument are highly correlated with their performance under a wide variety of temperature, water potential, hormonal, priming, aging, and other conditions [17–19]. We used these methods to explore the possibility of identifying early indicators of susceptibility to induction of blindness in kohlrabi seeds.

2. Materials and Methods

2.1. Seed and Plant Materials

Six kohlrabi (*B. oleraceae* L. var. *gongylodes*) seed lots comprised of three F1 varieties (A, B, C) with two lots of each (1 and 2, 3 and 4, 5 and 6, respectively) exhibiting different susceptibilities for blindness were provided by Bejo Zaden (Warmenhuizen, The Netherlands).

2.2. Blindness Induction

After initial measurements of physical characteristics of dry seeds, a blindness induction treatment was performed on the seeds preceding respiration measurements. We adapted a published protocol that demonstrated the ability of low temperature treatments to cause shoot apical meristem arrest in *Brassica oleracea* seedlings [1]. Seeds were imbibed in microtiter plate wells in 60 uL of water and incubated at 1.5 °C in a foil-covered incubator (Benchmark IS-1010R placed in a 4 °C room) for 48 h in darkness and then transferred to respiration tests, maintaining individual seed positioning and identities from previous seed imaging throughout.

2.3. Physical Characteristics Measurements

Chlorophyll content. The iXeed CF Analyzer (Figure 1; CF-Mobile; SeQso B.V., The Netherlands) was utilized for chlorophyll fluorescence (CF) measurements [5]. A set of 46–48 seeds per lot were placed on blue metal trays with proper-sized pockets, organized in six rows by eight columns, corresponding to the plate layout and capacity for seed respiration measurements. Three measurements were captured on each tray and seed parameters registered by the CF software were recorded and exported to Microsoft Excel (version 16). The average and standard deviation of CF level and CF size per seed were calculated and combined with parameters gathered subsequently. CF information was captured for a total of 174 seeds for each lot. The parameters derived from the CF-Analyzer data are defined in Table 1.

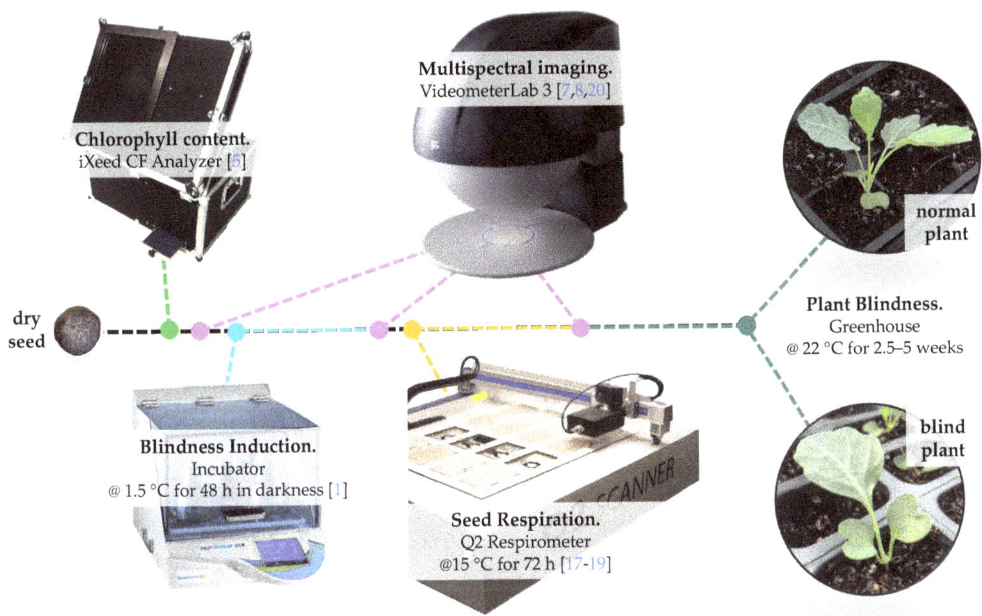

Figure 1. Visual workflow for every seed measured in the study. Starting with chlorophyll content and multispectral imaging taken of the dry seeds (left), followed by the blindness induction treatment, additional multispectral imaging, then seed respiration measurements, additional multispectral data collection and finally seeds were transferred to the greenhouse for growth and plant blindness evaluation.

Multispectral imaging. The VideometerLab instrument (Videometer A/S, Herlev, Denmark) was used for multispectral analyses [7,8,20]. The instrument is equipped with a camera inside an integrating sphere along with diodes that emit light at the following 19 wavelengths: 375, 405, 435, 450, 470, 505, 525, 570, 590, 630, 645, 660, 700, 780, 850, 870, 890, 940 and 970 nm. The same 174 seeds initially scanned for CF information were imaged and analyzed in the VideometerLab instrument for each lot. Multispectral pictures were taken and grouped for each 46–48 seeds, respecting the CF measurements positioning. These seeds were placed in coded 6 (rows) by 8 (columns) cells with each seed placed slightly lower than the previous one, aiding the Videometer software sequence numbering. Over two experimental repetitions, multispectral images of the same seed were captured in different stages of the experiment (see Figure 1): (1) dry seed (all 174 seeds per lot); (2) after blindness induction treatment (110 seeds per lot); and (3) after 72 h at 15 °C for seed respiration measurements (96 seeds per lot) where these seeds were rapidly transferred to marked microtiter plate lids and Videometer images were acquired to quantify seedling area, respecting the seed positioning from the Q2 equipment. A series of up to eight pictures was taken per Q2 plate to avoid overlapping tissues and data were combined together in the BLOB (Binary Large Objects) collection. Selected parameters derived from the VideometerLab data are defined in Table 1.

Table 1. Parameter Definitions.

Type	Parameter	Description
Multispectral	Area	Average or individual projected area (mm^2) calculated for the seed.
	CIELab	CIELab refers to a color space defined by the International Commission on Illumination (CIE); it expresses color as three numerical values: L for lightness from black (0) to white (100), A from green (−) to red (+) and B from blue (−) to yellow (+).
	Saturation	Average amount of pixels that exceed a maximum value of brightness in the image.
	Hue	Average angular position related to a color space coordinate enclosing all colors.
	395–970 nm	Average reflectance of specific wavelengths (in nanometers) for individual seeds. The specified wavelength followed by "SD" refers to the average standard deviation or pixel reflectance variation for individual seeds.
	Tissue Area	Area of uncoated or visbible tissue present in individual seeds, quantified by the number of pixels.
	White Spots	Area of white coloration in seed coats as a percentage of the individual seed area.
Chlorophyll Fluorescence	CF Value	Average or individual chlorophyll fluorescence measured for the seed (total seed fluorescence divided by fluorescence size).
	CF Size	Average or individual calculated size (mm^2) of the CF area in the seed (sum of the pixels that have a CF level above a threshold, and converted to mm^2).
Seed Respiration	R75.Time, R50.Time and R25.Time	Time in hours for individual seeds to deplete oxygen in vials to 75, 50 and 25%, respectively.
	R75, R50, R25 POD curves	Cumulative population oxygen depletion (POD) time course plotting the percentage of seeds depleting the oxygen level to 75, 50 or 25% of the initial, respectively, at each time.
	R50 (50)	Time in hours to 50% level reached on the R50 POD curve.
	R75.Final, R50.Final and R25.Final	Final percentages of the R75, R50 and R25 POD curves, respectively, of the initial value.
	Final-O_2	Final oxygen concentration in the vials after 72 h of imbibition.
Greenhouse Plant Evaluation	Plant Blindness Score	Plants were ranked as dead, normal or blind. Blind plants included: plants without shoot apical meristem [SAM]; plants with needle shaped (first) leaf; plants with funnel shaped (first) leaf; plants without SAM, but with lateral branches; plants with SAM and lateral shoots that are needle- or funnel-shaped or otherwise malformed; plants with oversized first leaf but lacking SAM; plants with abnormal branch architecture.

2.4. Seed Respiration Measurements

The Q2 instrument (now called Seed Respiration Analyzer; Figure 1; Fytagoras B.V., Leiden, The Netherlands) measures oxygen consumption (respiration) rates of individual seeds repeatedly during imbibition and germination. Individual seeds were placed into 2 mL screw-cap vials containing 1.45 mL of agar (0.4% *w/v*) and 0.2% Plant Preservative Mixture (PPM^{TM}), which were sealed with caps that have a dot of a fluorescent polymer centered on their internal side. The polymer contains a dye that changes its fluorescent properties in response to oxygen concentration [21]. As the seed respires, it depletes the oxygen in the sealed well or vial, which changes the fluorescence intensity of the dye. This change is detected by a light source that shines on the dot and a sensor that measures the fluorescence intensity. A robotic arm sequentially moves the light source/sensor over each well, measuring the oxygen concentration inside the wells. Up to 16 plates of 48 vials can be positioned in the apparatus at a time and automatically measured by the robotic sensor, and the measurements can be repeated frequently to obtain time courses of oxygen consumption activity. Measurements reported here were collected every 30 min. Seeds were transferred individually to 2 mL vials using tweezers after pictures were taken in the CF-Analyzer and the VideometerLab. Sample temperature was controlled to ±0.5 °C using Peltier heating/cooling units and fans. In preliminary tests, 48 non-induced seeds (control) per lot were measured at 20 °C, followed by plant blindness evaluation. Thereafter, all seeds were measured in separate 48-well plates at 15 °C for 50 or 72 h, and placement in the Q2 cells was paired to the Videometer and CF-Mobile codes for each seed when applied. Two experimental repetitions were utilized. The parameters derived from the Q2 data are defined in Table 1.

2.5. Plant Blindness Evaluation

After the blindness induction treatment and the respiration measurements, seeds were transplanted to marked trays with numbered cells and placed in a greenhouse at 22 °C and natural light. The day length during plant growth varied from 13 to 15 h in the initial experiment (carried in April through May 2019) and 13 to 11 h during the experiment repetition (September through October 2020). Individual plants grown for 2.5–5 weeks were evaluated and ranked as dead (no seedling emerged), normal or blind plants (Figure 1). The parameters derived from the plant evaluations are defined in Table 1.

2.6. Data Analyses

Data analyses for Videometer images were performed using VideometerLab and the Classifier Design Tool (CDT) software version 3.18.11 (Videometer A/S). We separated seeds from the image background using normalized canonical discriminant analysis (nCDA) transformation followed by simple threshold segmentation within the Videometer software. Several multispectral images were used to create the nCDA transformation model with selected areas of images representing the 2 classes: areas of seeds or background. Automatic normalization was performed to maximize the Rayleigh quotient and input data received a preprocessing band normalization (at 645 nm) and output data was centered around the overall mean between the classes and scaled with the two classes showing means at +1 or −1. The actual data were visualized in a scaling between −2 and 2.

A similar approach was used to quantify the visible tissue area after seed respiration measurements, but in this case using the input with multispectral images featuring the seed coat or embryo/seedling tissue area as the two classes to be separated. The input normalization in this case was performed using preprocessing band normalization at 470 nm, while output normalization was similar. The seedling or tissue areas were quantified by numbers of pixels.

Data from the CF-Analyzer, Q2 and plant evaluations were exported or compiled using Microsoft Excel (version 16). The compiled data were then analyzed in R version 4.0.3 using RStudio version 1.2.1335. Type I analysis of variance (ANOVA) was run on models with experiment as random effect. Normality and heteroscedasticity of the data was visually inspected with histograms and diagnostic plots for all parameters reported using the linear regression analysis (lm) models in R. Tukey Honest Significant Differences were then calculated using the TukeyHSD() function in R and the HSD.test() function of the R package *agricolae* [22].

Boxplots were made using ggplot from the ggplot2 R package [23]. Correlation matrices were calculated using the rcorr.test function from the psych R package [24] and plotted with the corrplot R package [25]. Family-wise error rate was accounted for with adjusted p-values for multiple comparisons using the Holm method [26]. Multiple factor analysis (MFA) was performed in R with the FactoMineR package [27] and additional tools from the factoextra package [28]. Only quantitative variables without missing values and statistically correlated with blindness ($p < 0.01$) were used for the MFA. All comparisons mentioned were statistically significant at $p < 0.05$ unless otherwise stated.

3. Results

3.1. Brassica Blindness: Initial Assessment

All seed lots were first tested for seed respiration and plant growth to identify blindness present in the seed lots and to characterize their vigor prior to the blindness induction treatment. The initial seed respiration test was conducted at 20 °C and all seed lots displayed largely homogeneous and rapid oxygen depletion rates (Supplemental Figure S1). At least 80% of seeds in all lots depleted oxygen in the vials to the 50% level within two days after imbibition and at 72 h most seeds were in anaerobic conditions in the vials. These seeds were then transferred to the greenhouse for plant growth evaluation, and no blind plants were identified after 2.5–5 weeks (data not shown).

To further investigate the blindness potential and susceptibility in all lots, we introduced moderate temperature stress during germination by lowering the temperature to 15 °C in the Q2 test. As expected, oxygen depletion rates were slower for all lots and larger variation in respiratory patterns within lots was also evident (Supplemental Figure S2). The modest temperature stress did not induce blindness, with only one seed showing some blindness symptoms in Variety B, lot number 3. Additionally, a significant number of seeds in most seed lots (except A-2 and B-4) did not reduce the oxygen within vials to the 50% level following imbibition for 50 h. The results demonstrated that little or no blindness was expressed in the seed lots tested under optimal or moderately low temperatures during imbibition and germination.

3.2. Physical Characteristics Assessment and Brassica Blindness Induction

We tested whether the CF-Analyzer, the VideometerLab or the Q2 were able to detect differences among lots and the potential presence of blind seeds. Chlorophyll fluorescence measurements were initially captured for all dry seeds (Figure 2). The seed lots within B and C varieties displayed a significant difference between each other ($p < 0.001$) while the lots in Variety A did not. Seed lots 5 (Variety C, 90.7) and 3 (Variety B, 78.4) displayed higher median chlorophyll contents when compared to the other lots (27.9–37.6). Additionally, these two lots had higher CF level variation (Figure 2, CF Level—larger bars/standard deviation on lots 3 and 5) in comparison to all other seed lots, which had more homogeneous low CF levels, although some outliers were present in these lots (Figure 2, black dots on lots 1, 2, 4 and 6). These lots with low median CF levels were not statistically different from each other but were significantly different from lots 3 and 5 (Figure 2, $p < 0.001$). Similar differences among lots were detected for CF size (or area) (Figure 2).

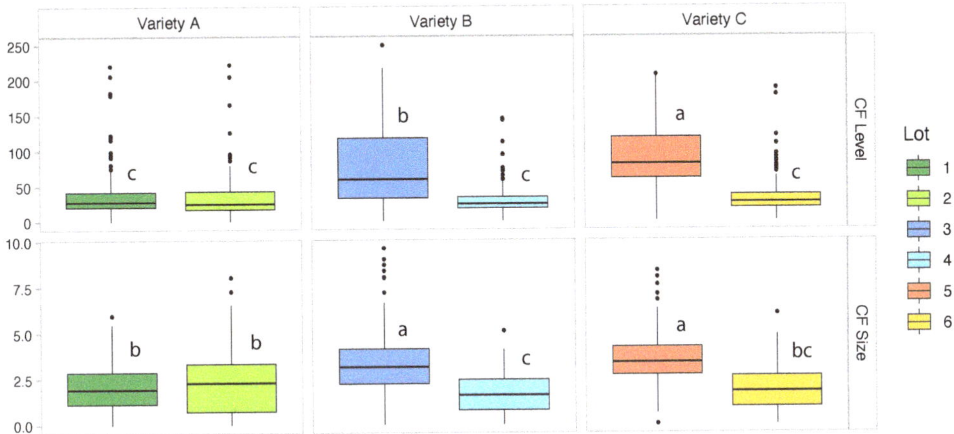

Figure 2. CF-Analyzer parameters (see Table 1) across varieties (columns) and seed lots (color-coded). Letters indicate significant differences among all lots for each parameter (rows) as calculated by ANOVA and Tukey HSD.

Multiple seed features were measured using the Videometer (123 parameters in total over all experimental stages, Supplemental Table S1), including seed size, shape, color and multispectral characteristics. Selected features that displayed some relationship with plant blindness, viability, or parameters gathered by the other analytical equipment used here are described in Table 1. Individual data were obtained for 174 seeds (46 seeds in the first repetition and 128 in the second repetition) for each seed lot (Supplemental Table S1). Several captured seed features differed significantly among lots (Figure 3). The calculated seed area was largest for seed lot 2 (Variety A, 3.84) and smallest for seed lot 4 (Variety B, 2.96). Color space (CIELab L, B) and saturation values together with average reflectance at longer wavelengths, such as red (645 nm) and near-infrared (870 nm) showed consistently and significantly higher ($p < 0.001$) values for seed lots 5 (C) and 3 (B) compared to the other lots (Figure 3), as observed for their CF values. A similar result was obtained for the ultraviolet (375 nm) wavelength, but it also included lot 1 ($p < 0.001$) along with the other two high-valued lots. Color and multispectral values for lot 5 also were significantly higher than for other lots ($p < 0.001$) but comparable with lot 1 for reflectance measured at the indigo color (435 nm) and also similar to the two Variety A seed lots (1 and 2) for Hue values. Wavelengths 435 and 645 nm are indications of chlorophyll A and B levels, respectively.

The spectrum reflectance standard deviation within seeds (or pixel variation) was also calculated on an individual seed basis for all wavelengths. This value quantifies the color or spectrum variation of each seed; seeds with a uniform color will display small values while seeds with a diversity of colors or shades will display larger values. Here we show the reflectance standard deviation for the near-infrared (NIR) 970 nm wavelength, which displayed some relationship with plant performance when measured at the dry seed and after blindness induction phases, although the standard deviation for other wavelengths also displayed similar results (Supplemental Figure S3). The calculated standard deviation for the NIR wavelength (970 nm) showed larger median variation (5.75–6.02) for seed lots 3 and 5 with lowest values for lots 6 (4.86) and 2 (4.50) (Figure 3).

The post-blindness-induction (PBI) seed area had a median increase of about 20% for all lots compared to dry seeds, reflecting expansion due to imbibition (Figure 4). The color space parameter CIELab B (blue (−) to yellow (+)) had a median increase of about 50% in most lots with a smaller increase of 32.2% observed in lot 5 (Variety C), which presented the higher ($p < 0.001$) value for CIELab B before induction (Figure 3). Similar relationships were observed for the saturation, hue and 970 nm-SD values (Figure 4), in which the seed lot with the lower initial value had the largest relative increase after blindness induction.

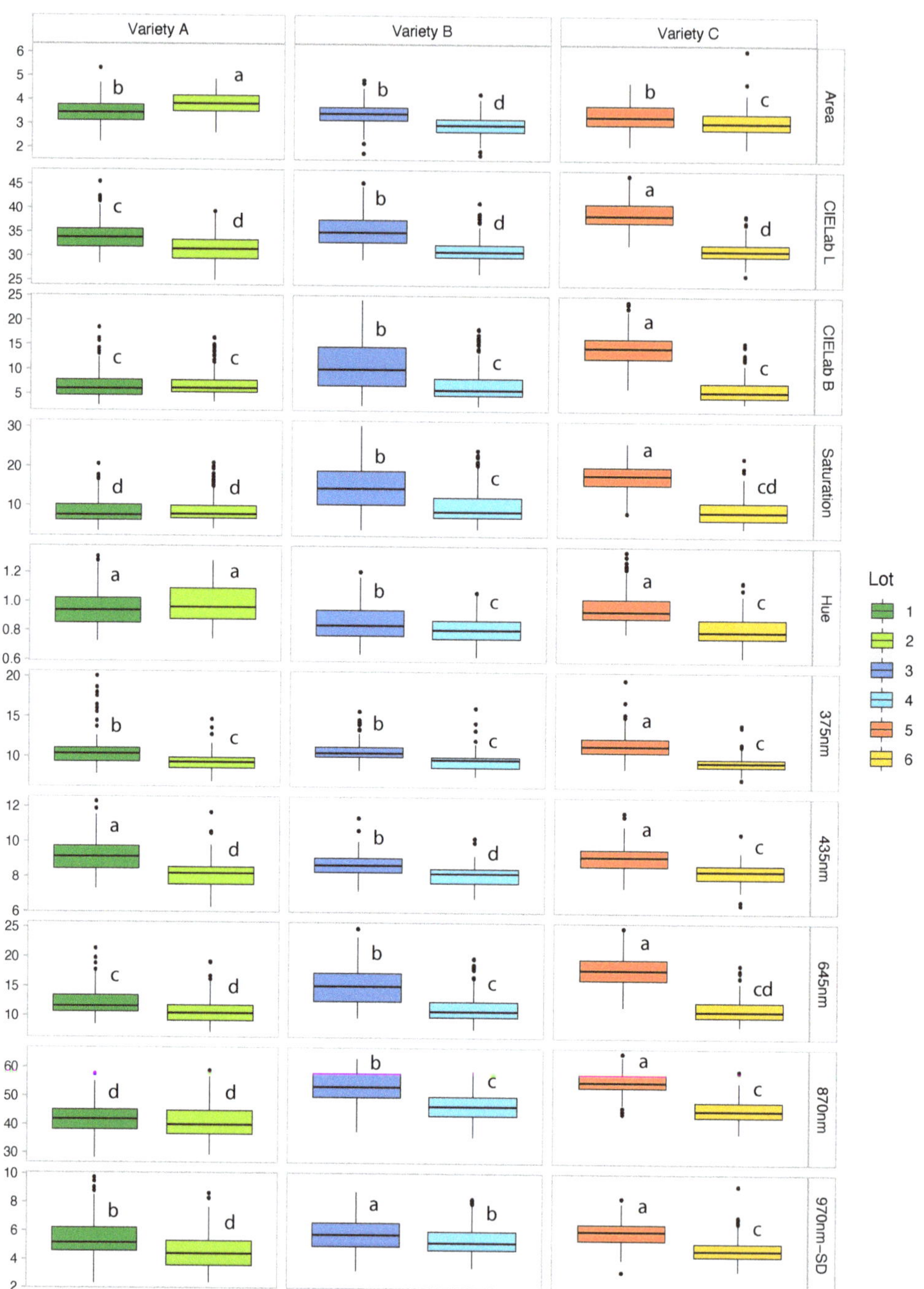

Figure 3. Selected Videometer features (Table 1) across varieties (columns) and seed lots (color-coded). Letters indicate significance between all lots for each parameter (rows) as calculated by ANOVA and Tukey HSD.

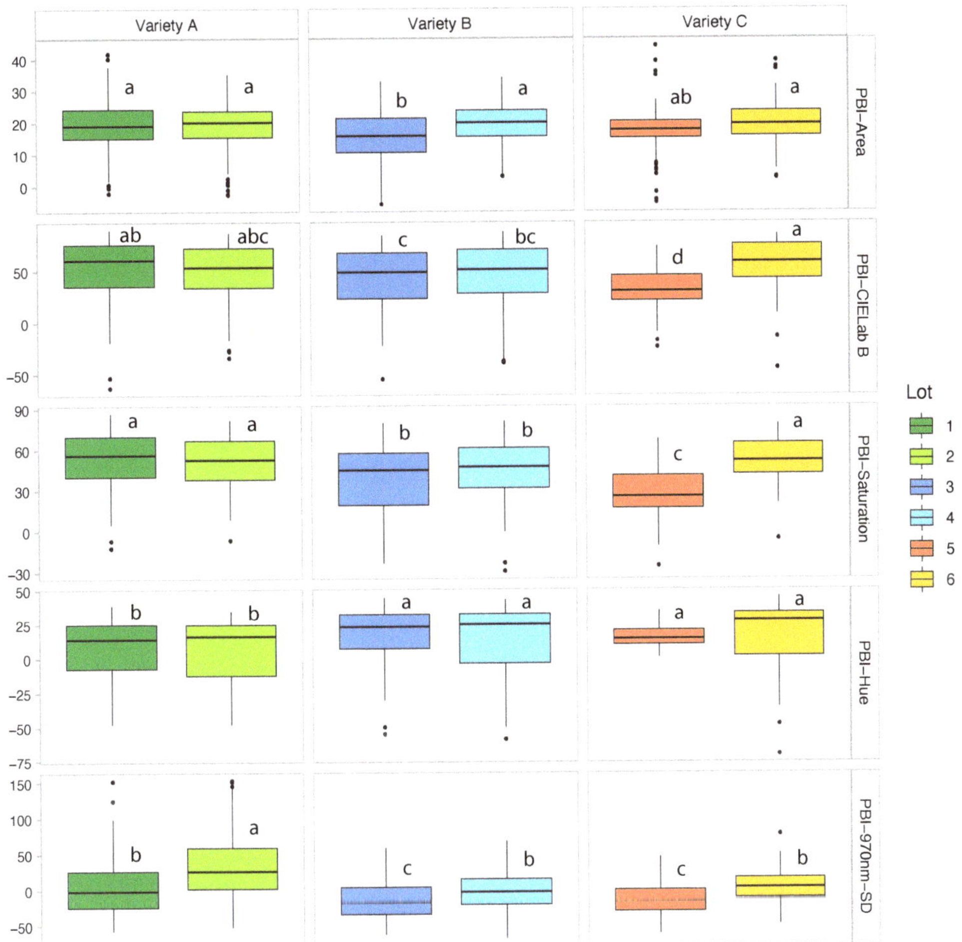

Figure 4. Selected VideometerLab features presented as percentage change in the feature levels post-blindness induction (PBI) relative to initial dry seed analyses for the same seeds. Letters indicate significant differences among all lots for each parameter (rows) as calculated by ANOVA and Tukey HSD.

Q2 measurements at 15 °C on the six seed lots after the cold temperature imbibition treatment indicated an overall delay in respiration rates (Figure 5, Supplemental Table S2—bottom section, mean R75 values ranging from 26.2 to 42 h) compared to measurements at a similar temperature on seeds prior to the induction treatment (Supplemental Figure S2, Supplemental Table S2—middle section, mean R75 values ranging from 22.5 to 30.6 h). In addition, a much larger fraction of induced seeds did not consume oxygen after 72 h (Figure 5, Supplemental Table S2—bottom section, e.g., final R75 POD curves values ranging from41 to 98%) compared to non-induced seeds tested earlier for 50 h (Supplemental Figure S2 and Table S2—middle section, e.g., final R75 POD curves ranging at 80–93%). The mean oxygen level at different time points is also a convenient parameter to quantify the respiration profile and variation (Supplemental Table S2, O_2 at 48 and 72 h, mean and standard deviation, respectively). Lack of oxygen consumption usually indicates lack of seed viability [19], suggesting that the blindness induction treatment had killed some seeds, as was reported previously regarding this treatment [1].

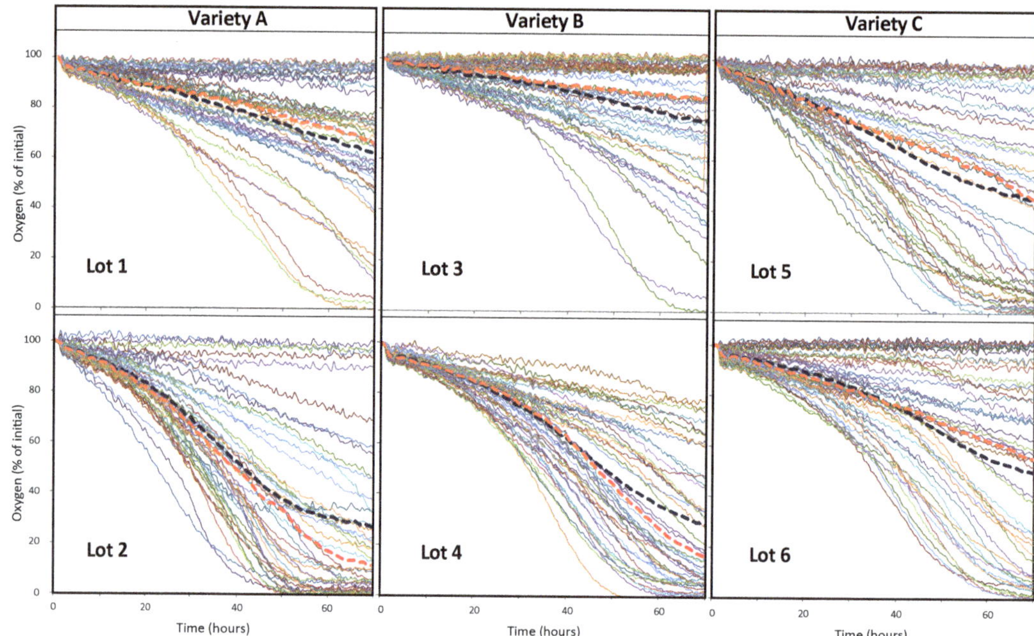

Figure 5. Oxygen depletion curves for individual seeds of kohlrabi seed lots tested at 15 °C for 72 h after blindness induction treatment at 1.5 °C for 48 h. Dashed lines represent the median (red) and average (black) oxygen depletion time courses for the entire seed population.

Based on the median oxygen depletion and R75 POD curves (Figure 6), lots 2 (light green) and 4 (light blue) had overall faster median oxygen depletion rates and higher percentages of seeds depleting the oxygen to at least the 75% level. Lot 5 (orange) contained a fraction of seeds respiring even faster than lot 4, but only around 72% of seeds in that lot consumed more than 75% oxygen in the vials, compared to 89 and 100% in lots 2 and 4, respectively. Lots 1 (dark green) and 3 (dark blue) displayed the lowest oxygen consumption (final medians of 65.5 and 84.5% oxygen remaining, respectively) and slower oxygen consumption (2 lowest R75 POD curves), while lot 6 (yellow) performed somewhat better (final median O_2 depletion curves of 53.7% and slightly faster R75 POD curves).

Some Q2 parameters were selected for comparison among lots (Figure 7). Lots 1, 3 and 6 displayed the slowest median times to 75% (36.2 to 42 h) and 50% (47.6 to 58.2 h) remaining oxygen levels. Area under the curve parameters (R75.Area and R50.Area) are highly correlated with the time to required to lower the oxygen to the same levels (R75.Time and R50.Time) but add more detailed information regarding the oxygen consumption profiles and shapes of depletion curves. As expected, distributions of areas under the curve for the 75% oxygen remaining level also showed lots 1, 3 and 6 as slower ones (larger area values ranging from 30.95 to 36.44), but at the 50% oxygen level lot 6 displayed a somewhat faster but significant ($p < 0.001$) oxygen consumption compared to lot 3 (Figure 7, R50.Area). Seed lots 2, 4 and 5 usually showed lower median times and area values compared to the slower lots and could be considered significantly ($p < 0.001$) faster respirators in most cases. It is important to point out that the lower the remaining oxygen level chosen to compute these values, the smaller the fraction of seeds that are used to calculate them. In this case at least 40% of the seeds were used to calculate parameters generated based on the 75% oxygen level (lot 3 with lowest final percentage in the R75-POD curve, Figure 6) but a little over 20% of the seeds were used to calculate values based in the 50% oxygen level (slow lots 1 and 3—R50 POD curves, Supplemental Table S2—bottom section). To overcome this issue for less vigorous or stressed lots, the remaining oxygen levels at a particular time for all seeds can be used. For example, the final oxygen levels in the vials after 72 h (before seeds were transferred to the greenhouse), clearly showed the wide distributions of respiration rates among seeds in these lots after the blindness induction treatment. Seed lots 2 and 4 had relatively homogeneous low median final oxygen levels (19.4 and 22%, respectively), lot 3 displayed an intermediate variance but at the highest final oxygen level (66.9%), while lots 1, 5 and 6 displayed more intermediate final oxygen levels (41.7, 39.4 and 33.1%, respectively) but with large heterogeneity among seeds (Figure 7, Final-O_2).

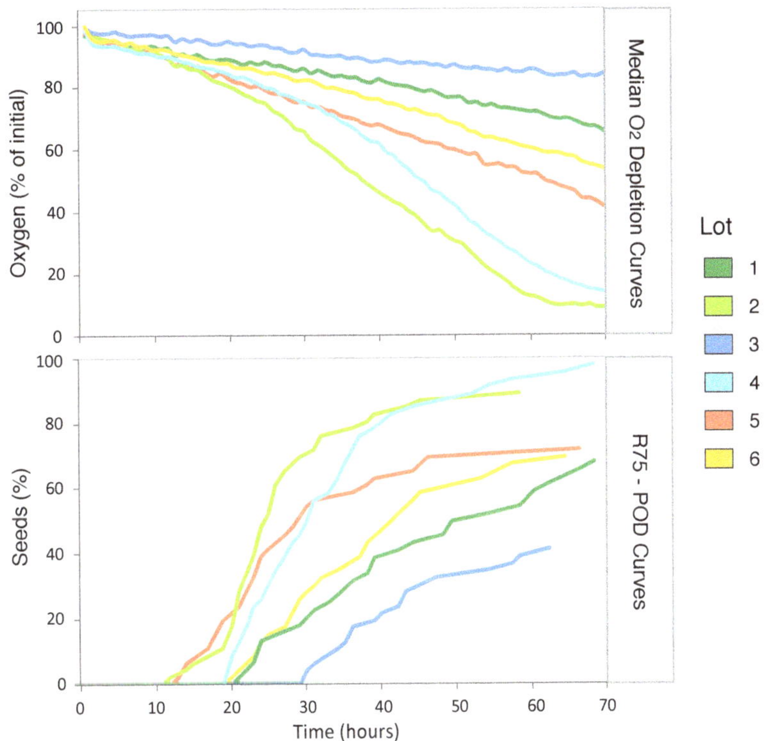

Figure 6. Median oxygen depletion curves (top) and R75 POD curves (bottom) of kohlrabi seed lots tested at 15 °C for 72 h after blindness induction treatment at 1.5 °C for 48 h.

Seedling or tissue area measurements from the Videometer using nCDA models (Figure 8—top sections) after seed respiration measurements exhibited significant differences among seed lots. Seed lot 3 had the smallest exposed tissue area (median at 1602 pixels) with little to no seedling tissue visible in a large fraction of the seeds after the respiration measurements (Figure 8—see bottom panels for BLOB collection with seeds marked with dark blue circles compared to seed lot 4 of the same variety marked with light blue circles). Seed lot 5 displayed an intermediate median tissue area (2809 pixels), while lots 1, 2, 4 and 6 had higher median tissue areas at around 4200 pixels (Figure 9). The largest seedling size variation was present in seed lot 1, where a fraction of seeds showed little to no embryo tissue while another fraction displayed the largest seedlings in the study. This is consistent with the large variation among seeds in this lot for Q2-derived values (Figure 7).

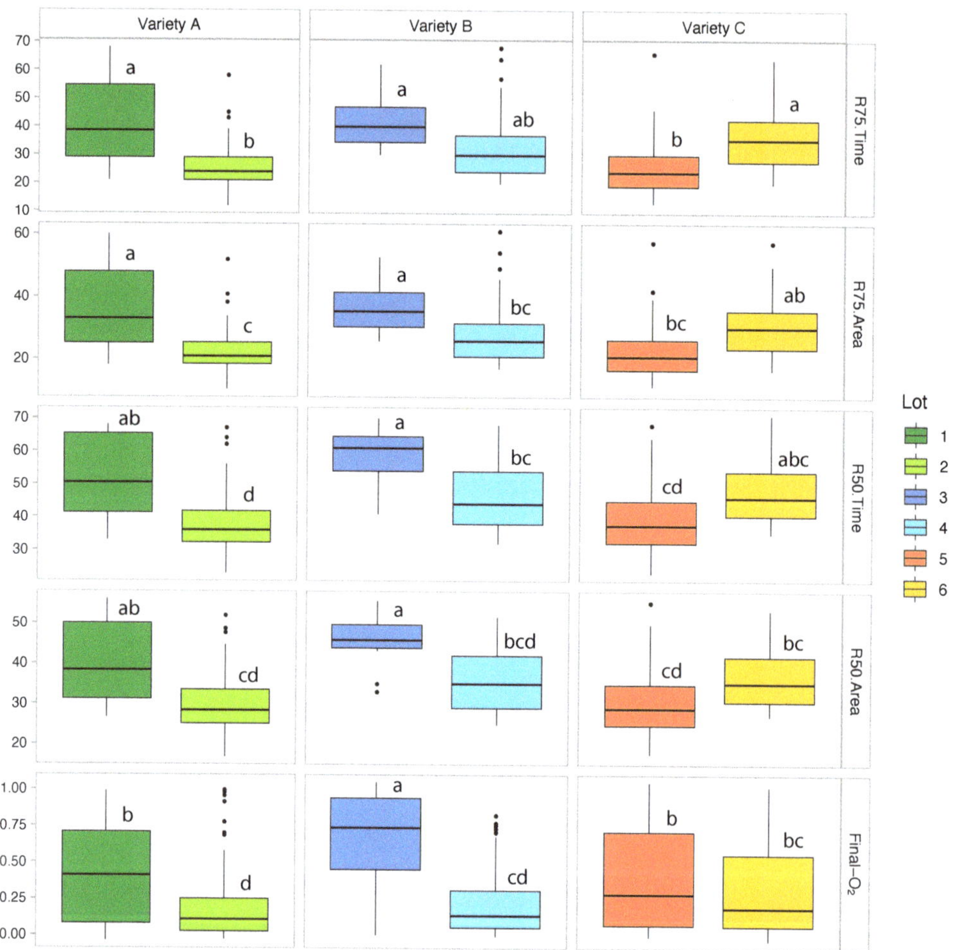

Figure 7. Selected Q2 parameters (see Table 1) of seed lots after blindness induction treatment. Letters indicate significance between all lots for each parameter (rows) as calculated by ANOVA and Tukey HSD.

Finally, seeds were transferred to marked trays and placed in the greenhouse at 22 °C for 2.5–5 weeks, when plants were scored as normal, blind or dead (failed to emerge). Lots 2, 4 and 6 had the largest fractions of normal plants (≥80%) while lots 1, 3 and 5 had smaller fractions of normal plants (>62%) (Figure 10). Additionally, lots 3 (19.1%) and 5 (20.9%) had the largest fractions of blind plants, followed by lot 1 (10.9%), lot 4 (7.3%) and lot 6 (2.7%); lot 2 did not display any blind plants after the induction treatment. The percentages of non-viable plants were higher in lots 3 (43.6%), 1 (27.3%) and 5 (26.4%), while lots 4 (12.7%), 6 (10.9%) and 2 (2.7%) exhibited lower percentages of seed death (Figure 10).

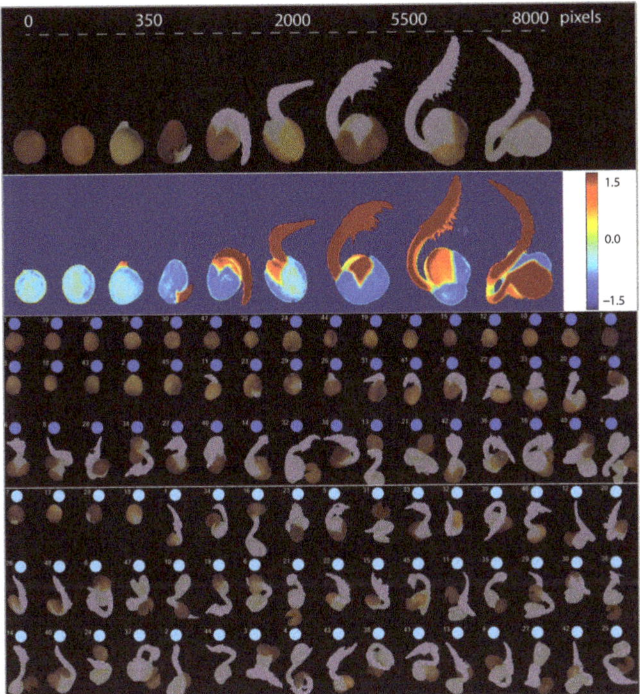

Figure 8. Examples of seedling or plant tissue area in pixels (**top panel**). Area was calculated using nCDA models and proper threshold value to include only relevant tissue and seedlings pixels. Sample of nCDA transformed image where every pixel is scored and only pixels above a certain threshold are counted (**middle panel**). Seedlings from all treatments were isolated and measured, blob (binary large object) collections with samples of these seeds and seedlings sorted by tissue/seedling area for seed lots 3 (dark blue circles) and 4 (light blue circles) are illustrated here (**bottom panels**).

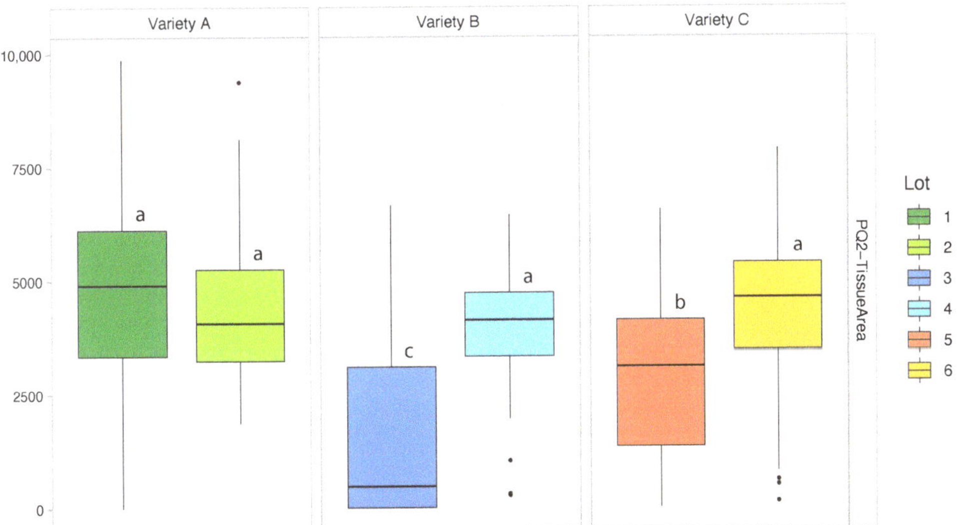

Figure 9. Seedling area (in pixels) of induced seeds after seed respiration measurements and before transfer to the greenhouse for plant growth and evaluation. Letters indicate significance difference among all lots as calculated by ANOVA and Tukey HSD.

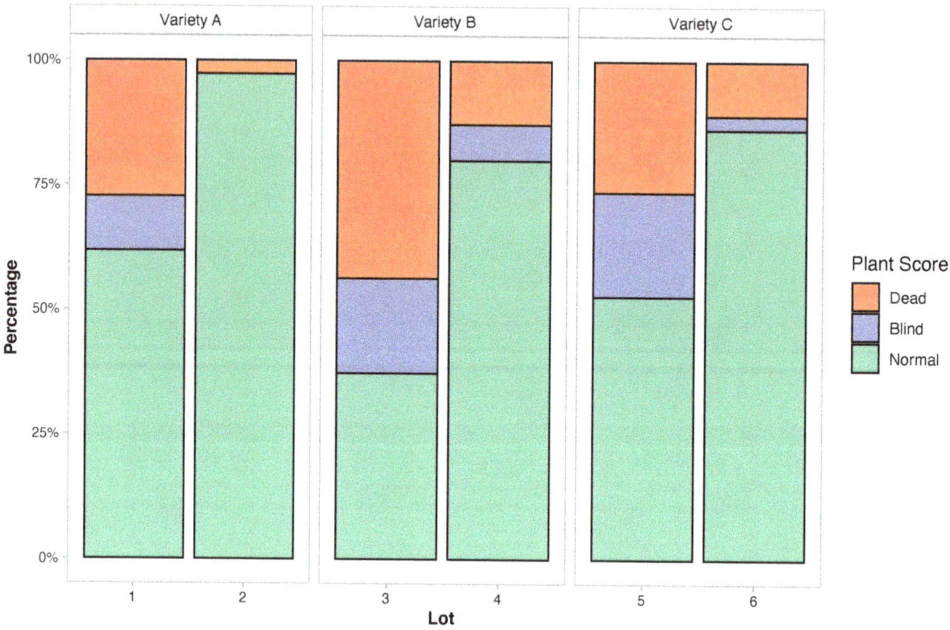

Figure 10. Plant evaluation for induced seeds and scores for dead, blind and normal seedlings after 2.5–5 weeks of growth in the greenhouse following 48 h at 1.5 °C and 72 h at 15 °C for Q2 measurements. Plants were scored per the descriptions in Table 1.

All individual seed data for chlorophyll content, multispectral reflectance, seed respiration and greenhouse plant evaluation scores were consolidated in one data file along with the variety, lot and repetition number (Supplemental Table S1). A full correlation matrix (Supplemental Figures S3 and S4) was constructed using all the data, enabling inspection of relationships among a large number of parameters at once to direct further analyses. To summarize the most critical information and avoid duplicating data, some primary parameters were selected and are presented in a smaller correlation matrix (Figure 11). The correlation numbers presented here provide an indication of their potential for use in individual seed sorting using the intersected parameters.

As expected, some parameters collected in the same instruments throughout all stages of the study were highly correlated with each other (Figure 11; Supplemental Figures S3 and S4, darker blue and red clusters). Correlations between the different analytical instruments used were also expected and observed in some cases. The chlorophyll fluorescence parameters from the CF-Analyzer were correlated with each other and also exhibited a strong correlation with multispectral parameters collected in the VideometerLab at different experimental stages (Figure 11, rows 1 and 2). This relationship was anticipated as both instruments are based on spectral imaging, with the CF-Analyzer targeting only chlorophyll content measurements with specific excitation wavelength and fluorescence wavelength filter while our VideometerLab version measures a broad range of wavelengths but lacks fluorescence filters (although addition of these is possible in the instrument).

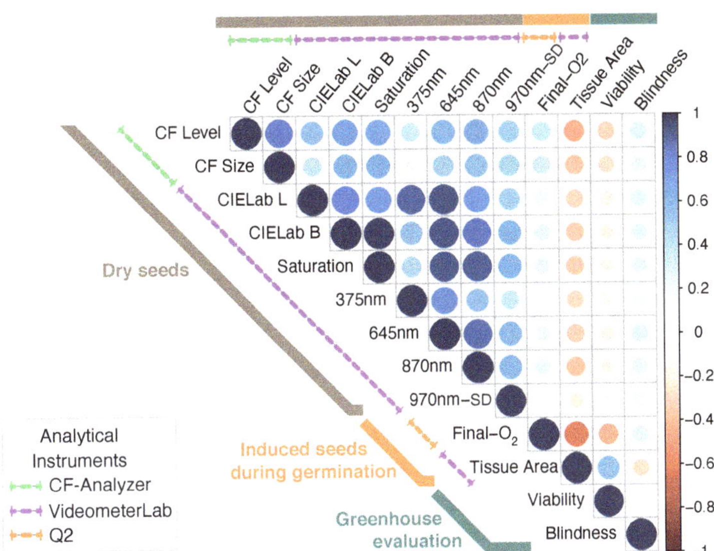

Figure 11. Pearson correlation matrix of selected main parameters (Table 1) from the CF Analyzer (rows and columns 1–2, dashed green), the VideometerLab (rows and columns 3–9 and 11, dashed purple) the Q2 respirometer (row and column 10, dashed orange) and greenhouse tests (rows and columns 12–13) for all induced seeds. Parameters are also organized in the order they were captured in the different stages of the study: dry seeds (rows and columns 1–9, brown bars), induced seeds during germination (rows and columns 10–11, orange bars) and in the greenhouse (rows and columns 12–13, green bars). The Pearson correlation coefficients between the row and column parameters are graphically displayed as circles; their size and color shadings indicate the strength of the correlation coefficients from 0 to 1 for a positive relation (blue color) or 0 to −1 for a negative relation (red color) (only correlation coefficients with significance level for p values below 0.01 are displayed).

The CF level (Figure 11, row 1 and all columns) had a significant relationship with VideometerLab color space parameters collected from dry seeds (CIELab L and B, r = 0.53 and 0.67, respectively, $p < 0.001$), saturation (r = 0.65, $p < 0.001$) and several wavelengths, including the ultraviolet (UV, 375 nm) and the near-infrared (NIR, 875 nm) ranges, with the highest correlation with 780 nm (r = 0.62, $p < 0.001$). This seed maturity indicator was also correlated (positively or negatively) to measurements performed at different stages of the study, such as multispectral imaging after the blindness induction (Supplemental Figure S3), during germination (final oxygen concentration in the Q2 and seedling area after the Q2, r = 0.37/0.45, respectively, $p < 0.001$) and with plant performance in the greenhouse (viability and blindness, r = −0.31/0.29, respectively, $p < 0.001$). Similar relationships were observed for the CF size parameter with weaker but still highly significant ($p < 0.001$) correlations for the dry seed and after induction, germination and greenhouse stages (Figure 11, row 2 and all columns).

The VideometerLab provides numerous parameters to quantify seed characteristics related to size, shape, spectra and others, but it also allows the application of customized features such as the tissue area and white spots/markings that we developed and used in this study (Supplemental Figure S4). In addition to the strong relationship with CF parameters, several of these features collected in dry seeds were correlated with VideometerLab features collected at different stages, Q2 measurements, and plant performance scores in the greenhouse. Some VideometerLab features collected at the dry seed stage were strongly correlated with data collected after blindness induction; these included the color space and saturation parameters from dry seeds and the percentage change in the same parameters after blindness induction (Supplemental Figure S3). The percentage change in the NIR reflectance variation (PBI-970 nm-SD, Figure 4) was the PBI parameter with the highest (negative) correlation coefficients with Q2 parameters (Final-O_2, r = −0.38, $p < 0.001$), seedling area and plant performance (blindness, r = −0.19, $p < 0.001$) (Supplemental Figure S3). These Videometer parameters obtained after the blindness induction were, in most cases, highly correlated with original parameters from dry seeds (e.g., PBI-970 nm-SD with the dry seed 970 nm-SD) and exhibited lower correlations with seed respiration and plant performance. Thus, their relevance for sorting purposes was diminished and they were not included in the correlation matrix.

Videometer color space parameters and saturation values from dry seeds were further associated with the final oxygen measured in the Q2 (r values ranging from 0.23 to 0.27 $p < 0.001$), seedling area after the Q2 (r = −0.31 to −0.34, $p < 0.001$) and viability (r = −0.17 to −0.18, $p < 0.001$) and blindness (r = 0.27 to 0.28, $p < 0.001$) in the greenhouse. Average reflectance of several wavelengths collected in dry seeds also displayed strong correlations with the changes after induction but also with seed respiration, seedling area and plant performance. Some examples of the main wavelengths include UV (375 nm), red (645 nm) and NIR (870 nm) wavelengths that displayed associations with final oxygen level (r = 0.16 to 0.24, $p < 0.001$), seedling area after the Q2 (r = −0.29 to −0.38, $p < 0.001$), viability (r = −0.15 to −0.18, $p < 0.001$) and blindness (r = 0.21 to 0.28, $p < 0.001$). These wavelengths displayed a similar relationships with the quality parameters, but their correlation with CF level was somewhat distinct, with the 375 nm wavelength displaying a lower correlation (r = 0.35, $p < 0.001$) while the red and NIR wavelengths were more closely associated with CF level (r = 0.58 and 0.62, respectively, $p < 0.001$).

Respiration measurements in the Q2 also displayed associations with seedling area and plant performance. The oxygen percentage after 72 h (Final-O_2, Figure 11—row and column 10) was highly negatively correlated with seedling tissue area (r = −0.63, $p < 0.001$) and plant viability (−0.42, $p < 0.001$) and positively with blindness (r = 0.28, $p < 0.001$). These results reinforce that seed respiration is a good indicator for germination timing; seeds with a higher oxygen consumption rate also germinated earlier and had more time for seedling growth and development. Furthermore, the seedling tissue area was highly correlated with plant viability (r = 0.55, $p < 0.001$) and negatively with blindness (r = −0.28, $p < 0.001$).

The correlation among these selected traits across all seed lots (Figure 11) is mostly preserved when this dataset is analyzed separately within each seed lot or within each variety (data not shown), although the strength of correlations varied among lots and varieties. The only exception was found in the seed lots from variety A where the relationship between CF level and blindness was not present, likely due to the absence or limited number of seeds displaying the blindness phenotype (Figure 10).

A multiple factor analysis (MFA) with integrated parameters correlated to blindness from the different stages of the study was utilized to verify whether the distinct plant scores (normal, blind or dead) could be discerned (Figure 12; Supplemental Table S3). The MFA reinforced the complexity of distinguishing among these classes using an unguided approach, but it also revealed a clear trend, with blind and dead classes having considerable overlap but being clearly separated from the normal class (Figure 12, left panel). The first dimension (Dim1 accounting for 57.9% of the total variation, x axis Figure 12) distinguished the majority of blind and dead seeds from the cluster of normal seeds. The main parameters that contributed to this separation were the Videometer parameters 645 nm, saturation, 870 nm, 375 nm, CIELab L and A, 970 nm-SD, followed by CF level and size (Figure 12, right panel). The second dimension (Dim2 accounting for 14.3% of the total variation, y axis Figure 12) was most effective in distinguishing between the normal and dead classes. The top parameters in this dimension were the Final-O_2 and the tissue area (antagonistic), which contributed together more than 80% of the total dimension, followed by smaller contributions from the 970 nm-SD, CIELab A, CF level, 645 nm and the saturation (Figure 12, right panel).

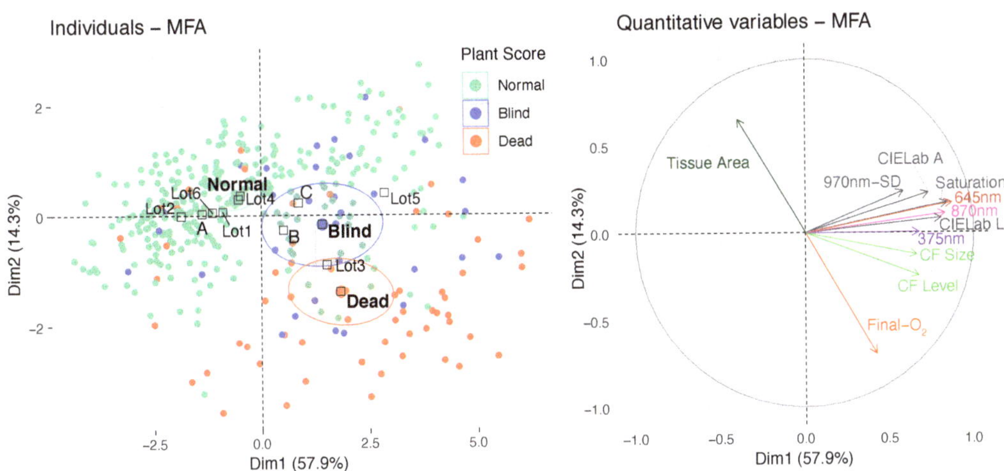

Figure 12. Multiple Factor Analysis (MFA) of features obtained from imaging, respiration and plant growth observations of kohlrabi seeds after blindness induction treatment. (**Left**): Scatter plot of individual seeds identified as normal, blind or dead separated in two dimensions (Dim1 and Dim2) according to their analytical features. Ellipses define confidence areas (95%) for each plant score, while squares represent their corresponding centers of gravity. Additional supplementary qualitative variables for seed variety (A, B or C) and lots (1 to 6) are shown in black. (**Right**): Vector representation of the influence of different measured factors in relation to their contribution to the two principal dimensions. Only complete observations for all parameters shown were used to generate the MFA.

4. Discussion

The potential for kohlrabi (and other *Brassica*) seeds to exhibit blindness, particularly after exposure to low temperatures, creates risks for both seed companies and growers. It has been difficult to identify the specific genetic and/or environmental factors during seed development that result in susceptibility to blindness. While conditions that can promote expression of blindness are known [1] and seed pretreatments can ameliorate susceptibility [3], it would be valuable to identify correlated traits that could be used for prescreening for blindness susceptibility to assign problematic lots for seed treatment or for sorting seed lots to remove the seeds that are susceptible to blindness. Thus, we examined both physical (CF-Analyzer, VideometerLab) and physiological (Q2) approaches to screening individual seeds prior to and after inducing blindness in order to identify whether it would be possible to predict which seeds or lots would be more likely to exhibit blindness.

All of the tested lots exhibited good initial performance at 20 °C, based on seed respiration time courses (Supplemental Figure S1). Lowering the temperature to 15 °C, however, resulted in much larger variances among seeds and discrimination among the seed lots (Supplemental Figure S2), with lots 2 (Variety A) and 4 (Variety B) exhibiting the greatest respiratory capacity at 15 °C (most seeds consuming most of the available oxygen). The subtle temperature stress resulted in few or no blind plants, so a blindness induction treatment was required to reveal the desired phenotype for this study. Following the induction treatment, lots 2 (Variety A) and 4 (Variety B) once more exhibited more active respiratory profiles at 15 °C (Figure 5), and also lower percentages of dead or blind seeds (Figure 10). In contrast, lots 1 (Variety A) and 3 (Variety B) showed greater impairment in respiratory activity at 15 °C and the highest susceptibility to blindness/death due to the induction treatment (Figures 5, 6 and 10). The behaviors of lots 5 and 6 from Variety C were intermediate, as these lots displayed a split respiratory behavior with about half of the seeds consuming most of the oxygen available while the other half consumed little to none (Figures 5 and 6). This result for lot 6 was rather anomalous, as it exhibited relatively poor respiratory capacity at 15 °C (Figure 5), but low susceptibility to blindness/death

(Figure 10). As the effect of the induction treatment increases progressively with longer times of exposure [1], it could be that lot 6 would show greater effects after a longer induction treatment. For the purposes of this experiment, the seed lots exhibited a range of susceptibility to death/blindness from the induction treatment, making it possible to test whether physical measurements would be related to this physiological behavior.

Differences among the seed lots at the dry seed stage were evident from parameters determined by the CF-Analyzer and the VideometerLab. Lots 3 (Variety B) and 5 (Variety C) exhibited higher CF values (Figure 2), as well as higher values for CIELab L, CIELab B, Saturation, 645 nm, 870 nm and the variation of the 970 nm reflectance (970 nm-SD) (Figure 3). The 375 nm and 435 nm values also identified lot 1 within the same group (exhibiting high values) as lots 3 and 5 (Figure 3), in agreement with these lots having more dead and blind seeds (Figure 10). Thus, the shorter wavelength parameters, specially 375 nm, suggest a new possibility for sorting, as that wavelength was not as highly correlated with CF Level as the 645 nm and 870 nm measurements (Figure 11). The 375 nm measures may detect another factor that could also be related to seed maturity. As higher values for all these measurements were associated with greater blindness and fewer normal seedlings (Figure 11), more immature seeds, indicated by higher chlorophyll levels, therefore appear to be associated with greater susceptibility to damage by low temperature imbibition.

Measurements performed after the low temperature induction treatment overlapped among seed lots (Figure 4). The saturation value change displayed significant relationships with blindness, but the more relevant relationship was observed with the NIR variation parameter (PBI-970 nm-SD) (Supplemental Figure S3), which was also somewhat more efficient to separate lots (Figure 4). This feature is derived from the variation in the NIR 970 nm reflectance (970 nm-SD) among dry seeds and exhibited a high correlation with that parameter. Both parameters separated out the three lots with higher blindness+death scores within varieties (lots 1, 3 and 5; Figure 3), suggesting that seeds with the larger initial variation could be linked to higher susceptibility to blindness, adding another signal option to aid sorting. The NIR 970 nm wavelength has been used as an indicator of water status in different substances [29,30], and it could be quantifying moisture distribution in seeds in this study, but further research is required to confirm this. Cracking or splitting of the testa precedes radicle emergence from *Brassica* seeds by a few hours [31], and this could be a factor that would add seed surface variation (e.g., exposure of seed tissues and contrast with the seed coat) as well as some moisture differences that could have been measured between active live and damaged or dead tissues.

Seed respiration during germination after the induction treatment also revealed differences among lots. Seed lots 1 (Variety A) and 3 (Variety B) were ranked with the lowest respiratory potential over several parameters (Figure 7, higher values), which agrees with their higher blindness+death scores (Figure 10). However, seed lots from Variety C (lots 5 and 6) usually had overlapping or inverted results when compared to their blindness+death scores (Figures 7 and 10). This issue can be better visualized with the POD curves (R75-POD Curves, Figure 6), as lot 5 has a fraction of faster respirators while lot 6 has a similar fraction of slower respirators; both lots had approximately 30% of seeds that did not consume 25% of the available oxygen (see also Figure 5). The use of POD curves provides a clear view of all seeds tested and avoids the selective calculation of averages and medians that do not account for seeds that did not reach a certain oxygen level. Additionally, a decreasing number of seeds is used to calculate the parameters when lowering the oxygen level threshold used. The final oxygen level or oxygen level at certain times can address this issue and show a realistic performance comparison at that time for all seeds tested. In this study, the final oxygen level (Final-O_2) was the Q2 parameter with stronger correlation with seedling area and plant performance in the greenhouse (Figure 11). This close connection between oxygen consumption and seedling area was expected and highlights the critical role of respiration in supporting early stages of plant growth [19]. Seed lots 1, 3 and 5 were ranked as the lots with higher final oxygen levels in agreement with their blindness+death scores, but due to the larger overall variance present, seed lot 6 was also included in that group.

The seedling areas determined after the respiration measurements was also efficient in identifying lots 3 and 5 as presenting smaller seedling areas, and the large variation present in lot 1 (Figure 9). As expected, this parameter displayed a significant correlation with viability, with larger seedlings at the time of transplanting to greenhouse trays resulting in more viable plants. The relationship with blindness was also significant, although not as strong as with viability (Figure 11). The MFA analysis also shows how tissue area and Final-O_2 clearly separate dead and normal seedlings on opposite vectors (Figure 12).

To summarize the sorting opportunities at the seed lot level, mean values for chlorophyll, CIELab B, saturation, hue, NIR 870 nm and tissue area were able to distinguish lots with higher susceptibility to blindness in varieties B and C. The mean Q2 parameters (R75 and R50, Final-O_2) were able to separate the more susceptible lots of varieties A and B. Finally, mean values for CIElab L, 375, 435 and 645 nm, and 970 nm-SD were capable of differentiating these lots within all varieties. Most of these parameters displayed significant positive correlations and could be indicative for blindness susceptibility when high values are observed but also negatively correlated with the presence of normal seedlings. The MFA illustrates these contributions and the direction of separation, increasing with higher presence of blind seedlings and dead seeds while lowering towards higher frequency of normal seedlings (Figure 12, Dim1 both panels). On the contrary, tissue area had the opposite relationship with blindness and normal seedling percentages and a clear antagonistic relation with the Final-O_2 parameter (Figure 12 Dim2, both panels). While some of these parameters can be used to rank and sort all seed lots within these varieties, the importance of most parameters to identify blindness susceptibility seem to be variety-dependent. Sorting opportunities at the individual seed level may require larger sample sizes within lots and varieties to increase the pool of reference seeds displaying the phenotype of interest, expand the information available to properly account for the variation in seed physical characteristics and refine traits important for separation to provide higher confidence and accuracy.

5. Conclusions

The methodology and approaches used here demonstrate how a set of relevant parameters correlated to a phenotype of interest can be obtained using analytical instruments to assess individual seeds at different stages, starting from the dry seed through to the manifestation of the phenotype. The collection of the parameters from the different analytical instruments and/or stages combined can give valuable insight on how early or late these relevant parameters can be identified and used for seed lot management or upgrading. High-throughput equipment has been developed to physically separate individual seeds based on CF level (www.seqso.com (accessed on 13 January 2021)). Videometer A/S also recently developed a sorter with less speed and capacity, but capable of sorting seeds individually using multiple seed features or combinations of them. Additionally, new instruments and software are becoming available in which artificial intelligence is used to generate powerful algorithms to analyze seed images based on training sets (e.g., Seed-X, Magshimim, Israel). The procedure described here, of making digital images followed by assessing susceptibility to induction of blindness on a seed-by-seed basis, could be used for such training sets by identifying the greater or less susceptible seeds in the original images. At a minimum, the methods utilized here can efficiently identify lots with potential for injury or blindness in response to cold imbibition, which could then be processed further or pretreated to reduce their susceptibility. While further work is required to more fully confirm these approaches, the data provided here also demonstrate the possibility of sorting *Brassica* seed lots to remove individual seeds most susceptible to blindness following exposure to low temperatures.

Supplementary Materials: The following are available at https://www.mdpi.com/2077-0472/11/3/220/s1, Table S1: Seed parameters database, Table S2: Q2 parameters, Table S3: MFA Eigenvalues, Figure S1: Brassica control seed respiration curves at 20 °C, Figure S2: Brassica control seed respiration curves at 15 °C, Figure S3: Pearson correlation matrix of selected parameters, Figure S4: Full correlation matrix.

Author Contributions: Conceptualization, P.B. and K.J.B.; methodology, P.B. and K.J.B.; software, P.B.; validation, P.B. and K.J.B.; formal analysis, P.B.; investigation, P.B.; resources, P.B. and K.J.B.; data curation, P.B. and K.J.B.; writing—original draft preparation, P.B.; writing—review and editing, P.B. and K.J.B.; supervision, P.B. and K.J.B.; project administration, P.B. and K.J.B.; funding acquisition, K.J.B. All authors have read and agreed to the published version of the manuscript.

Funding: This research was funded by the Western Regional Seed Physiology Research Group.

Institutional Review Board Statement: Not applicable.

Informed Consent Statement: Not applicable.

Data Availability Statement: The work of the second author is supported by the Lomonosov Moscow State University under grant "Modern Problems of the Fundamental Mathematics and Mechanics".

Acknowledgments: The authors thank Corine de Groot for helpful suggestions, technical assistance and providing seed samples. We would like to acknowledge our colleagues Marlen Navarro Boulandier and Vincent Chiu for their initial experimental work and contribution to some of the original data presented here. Thanks to Peter Marks and Aginnovation for providing access to some analytical instruments utilized in these studies. We also thank Allen Van Deynze and Daniel Runcie for critically reviewing the manuscript prior to submission.

Conflicts of Interest: The authors declare no conflict of interest.

References

1. De Jonge, J.; Kodde, J.; Severing, E.I.; Bonnema, G.; Angenent, G.C.; Immink, R.G.H.; Groot, S.P.C. Low temperature affects stem cell maintenance in *Brassica oleracea* seedlings. *Front. Plant Sci.* **2016**, *7*, 800. [CrossRef] [PubMed]
2. Wurr, D.C.E.; Hambidge, A.J.; Smith, G.P. Studies of the cause of blindness in brassicas. *J. Hortic. Sci.* **1996**, *71*, 415–426. [CrossRef]
3. De Jonge, J.; Goffman, F.D.; Kodde, J.; Angenent, G.C.; Groot, S.P.C. A seed treatment to prevent shoot apical meristem arrest in *Brassica oleracea*. *Sci. Hortic.* **2018**, *228*, 76–80. [CrossRef]
4. Still, D.W.; Bradford, K.J. Using hydrotime and ABA-time models to quantify seed quality of Brassicas during development. *J. Am. Soc. Hortic. Sci.* **1998**, *123*, 692–699. [CrossRef]
5. Jalink, H.; Van der Schoor, R.; Frandas, A.; Van Pijlen, J.G.; Bino, R.J. Chlorophyll fluorescence of *Brassica oleracea* seeds as a non-destructive marker for seeds maturity and seed performance. *Seed Scie. Res.* **1998**, *8*, 437–443. [CrossRef]
6. Price, J.S.; Hobson, R.N.; Neale, M.A.; Bruce, D.M. Seed losses in commercial harvesting of oilseed rape. *J. Agric. Eng. Res.* **1996**, *65*, 183–191. [CrossRef]
7. Olesen, M.H.; Nikneshan, P.; Shrestha, S.; Tadayyon, A.; Deleuran, L.C.; Boelt, B.; Gislum, R. Viability prediction of *Ricinus cummunis L.* seeds using multispectral imaging. *Sensors* **2015**, *15*, 4592–4604. [CrossRef]
8. Boelt, B.; Shrestha, S.; Salimi, Z.; Jørgensen, J.R.; Nicolaisen, M.; Carstensen, J.M. Multispectral imaging—A new tool in seed quality assessment? *Seed Sci. Res.* **2018**, *28*, 222–228. [CrossRef]
9. Salimi, Z.; Boelt, B. Classification of processing damage in sugar beet (*Beta vulgaris*) seeds by multispectral image analysis. *Sensors* **2019**, *19*, 2360. [CrossRef] [PubMed]
10. ElMasry, G.; Mandour, N.; Wagner, M.-H.; Demilly, D.; Verdier, J.; Belin, E.; Rousseau, D. Utilization of computer vision and multispectral imaging techniques for classification of cowpea (*Vigna unguiculata*) seeds. *Plant Methods* **2019**, *15*, 1–16. [CrossRef] [PubMed]
11. Shiade, S.R.G.; Boelt, B. Seed germination and seedling growth parameters in nine tall fescue varieties under salinity stress. *Acta Agric. Scand. Sec. B Soil Plant Sci.* **2020**, *70*, 485–494.
12. Cicero, S.M.; Schoor, R.V.D.; Jalink, H. Use of chlorophyll fluorescence sorting to improve soybean seed quality. *Rev. Bras. Sementes* **2020**, *31*, 145–151. [CrossRef]
13. Li, C.; Wang, X.; Meng, Z. Tomato seeds maturity detection system based on chlorophyll fluorescence. In Proceeding of the Optical Design and Testing VII; International Society for Optics and Photonics: Beijing, China, 2016; p. 1002125.
14. Kenanoglu, B.B.; Demir, I.; Jalink, H. Chlorophyll fluorescence sorting method to improve quality of Capsicum pepper seed lots produced from different maturity fruits. *HortScience* **2013**, *48*, 965–968. [CrossRef]
15. Yadav, S.K.; Jalink, H.; Groot, S.P.C.; Van Der Schoor, R.; Yadav, S.; Dadlani, M.; Kodde, J.C. Quality improvement of aged cabbage (*Brassica oleracea* var. *capitata*) seeds using chlorophyll fluorescence sensor. *Sci. Hortic.* **2015**, *189*, 81–85.

16. Wilson, H.T.; Khan, O.; Welbaum, G.E. Chlorophyll fluorescence in developing 'Top Mark' cantaloupe (*Cucumis melo*) seeds as an indicator of quality. *Seed Technol.* **2014**, *36*, 103–113.
17. Van Asbrouck, J.; Taridno, P. Using the single seed oxygen consumption measurement as a method of determination of different seed quality parameters for commercial tomato seed samples. *Asian J. Food Agro-Ind.* **2009**, *2*, S88–S95.
18. Bradford, K.J.; Bello, P.; Fu, J.C.; Barros, M. Single-seed respiration: A new method to assess seed quality. *Seed Sci. Technol.* 2013, *41*, 420–438. [CrossRef]
19. Bello, P.; Bradford, K.J. Single-seed oxygen consumption measurements and population-based threshold models link respiration and germination rates under diverse conditions. *Seed Sci. Res.* **2016**, *26*, 199–221. [CrossRef]
20. Olesen, M.H.; Carstensen, J.M.; Boelt, B. Multispectral imaging as a potential tool for seed health testing of spinach (*Spinacia oleracea L.*). *Seed Sci. Technol.* **2011**, *39*, 140–150. [CrossRef]
21. Draaijer, A.; Jetten, J.; Douma, A.C. A novel optical method to determine oxygen in beer bottles. *J. Inst. Brew.* **1999**, *105*, 155.
22. De Mendiburu, F. *Agricolae: Statistical Procedures for Agricultural Research*; R Package Version 1.3; 2020. Available online: http://CRAN.R-project.org/package=agricolae (accessed on 13 January 2021).
23. Wickham, H. *ggplot2: Elegant Graphics for Data Analysis*, 3rd ed.; Springer: New York, NY, USA, 2016.
24. Revelle, W. *psych: Procedures for Psychological, Psychometric, and Personality Research*; R Package Version 2.012; 2020. Available online: http://CRAN.R-project.org/package=psych (accessed on 13 January 2021).
25. Wei, T.; Simko, V. *R Package "Corrplot": Visualization of a Correlation Matrix*; R Package Version 0.84; 2017. Available online: http://CRAN.R-project.org/package=corrplot (accessed on 13 January 2021).
26. Holm, S. A Simple Sequentially Rejective Multiple Test Procedure. *Scand. J. Stat.* **1979**, *6* 65–70.
27. Le, S.; Josse, J.; Husson, F. FactoMineR: A Package for Multivariate Analysis *J. Stat. Softw.* **2008**, *25*, 1–18.
28. Kassambara, A.; Mundt, F. *Factoextra: Extract and Visualize the Results of Multivariate Data Analyses*; R Package Version 1.0.7; 2020. Available online: http://CRAN.R-project.org/package=factoextra (accessed on 13 January 2021)
29. Penuelas, J.; Filella, I.; Biel, C.; Serrano, L.; Save, R. The reflectance at the 950–970 nm region as an indicator of plant water status. *Int. J. Remote Sens.* **1993**, *14*, 1887–1905. [CrossRef]
30. Liu, J.; Cao, Y.; Wang, Q.; Pan, W.; Ma, F.; Liu, C.; Chen, W.; Yang, J.; Zheng, L. Rapid and non-destructive identification of water-injected beef samples using multispectral imaging analysis. *Food Chem.* **2016**, *190*, 938–943. [CrossRef]
31. Schopfer, P.; Plachy, C. Control of seed germination by abscisic acid. 2. Effect on embryo water uptake in *Brassica napus L. Plant Physiol.* **1984**, *76*, 155–160. [CrossRef]

MDPI
St. Alban Anlage 66
4052 Basel
Switzerland
Tel. +41 61 683 77 34
Fax +41 61 302 89 18
www.mdpi.com

Agriculture Editorial Office
E-mail: agriculture@mdpi.com
www.mdpi.com/journal/agriculture

www.ingramcontent.com/pod-product-compliance
Lightning Source LLC
LaVergne TN
LVHW070952170726
843515LV00005B/974